Building Automation

Control Devices and Applications

AMERICAN TECHNICAL PUBLISHERS, INC.
HOMEWOOD, ILLINOIS 60430-4600

Building Automation: Control Devices and Applications contains procedures commonly practiced in industry and the trade. Specific procedures vary with each task and must be performed by a qualified person. For maximum safety, always refer to specific manufacturer recommendations, insurance regulations, specific job site and plant procedures, applicable federal, state, and local regulations, and any authority having jurisdiction. The material contained is intended to be an educational resource for the user. Neither American Technical Publishers, Inc. nor the National Joint Apprenticeship and Training Committee for the Electrical Industry is liable for any claims, losses, or damages, including property damage or personal injury, incurred by reliance on this information.

American Technical Publishers, Inc., Editorial Staff

Editor in Chief:
Jonathan F. Gosse
Vice President—Production:
Peter A. Zurlis
Art Manager:
James M. Clarke
Technical Editor:
Julie M. Welch
Copy Editors:
Diane J. Weidner
Valerie A. Deisinger
Catherine A. Mini
Cover Design:
Samuel T. Tucker

Illustration/Layout:
Jennifer M. Hines
William J. Sinclair
Eric T. Comiza
Thomas E. Zabinski
Samuel T. Tucker
Gregory A. Gasior
Multimedia Coordinator:
Carl R. Hansen
CD-ROM Development:
Robert E. Stickley
Gretje Dahl
Peter J. Jurek

1 2 3 4 5 6 7 8 9 – 08 – 15

Printed in the United States of America

ISBN 978-0-8269-2000-3

 This book is printed on recycled paper.

Acknowledgments

In loving memory of

Stan Klein

Technical information and assistance was provided by the following companies and organizations:

Acuity Brands Lighting, Inc.
Dwyer Instruments, Inc.
Fire-Lite Alarms
Fireye, Inc.
Fluke Corporation
The Foxboro Company
Furnas Electric Co.
Gamewell-FCI
GE Security
Greenheck Fan Corp.
Honeywell International, Inc.
Kone, Inc.
LEDtronics, Inc.

Leviton Manufacturing Co., Inc.
Lutron Electronics, Inc.
Potter Electric Signal Co.
Siemens
Sioux Chief Manufacturing
The Trane Company
Trerice, H.O., Co.
U.S. Energy Information Administration
U.S. Green Building Council
Watts Regulator Company
WattStopper/Legrand
Waupaca Elevator Company, Inc.

Contents

CD-ROM Contents

- Using the CD-ROM
- Quick Quizzes®
- Illustrated Glossary
- Flash Cards
- Media Clips
- ATPeResources.com

Features

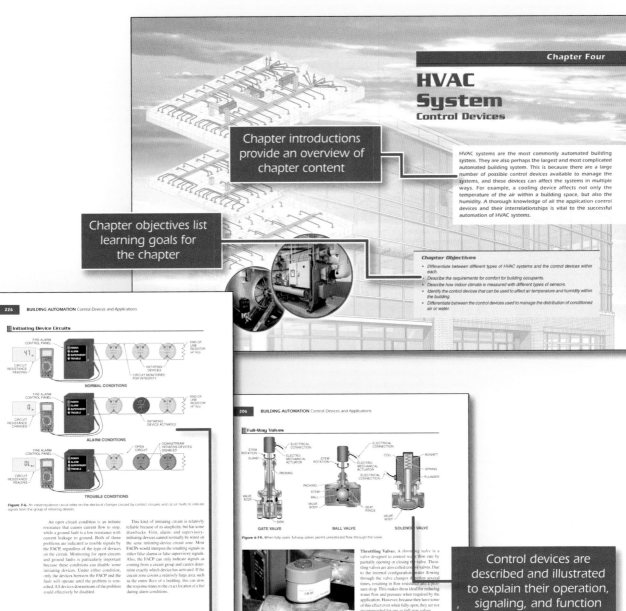

Chapter introductions provide an overview of chapter content

Chapter Four

HVAC System Control Devices

HVAC systems are the most commonly automated building system. They are also perhaps the largest and most complicated automated building system. This is because there are a large number of possible control devices available to manage the systems, and these devices can affect the systems in multiple ways. For example, a cooling device affects not only the temperature of the air within a building space, but also the humidity. A thorough knowledge of all the application control devices and their interrelationships is vital to the successful automation of HVAC systems.

Chapter objectives list learning goals for the chapter

Chapter Objectives

- Differentiate between different types of HVAC systems and the control devices within each.
- Describe the requirements for comfort for building occupants.
- Describe how indoor climate is measured with different types of sensors.
- Identify the control devices that can be used to affect air temperature and humidity within the building.
- Differentiate between the control devices used to manage the distribution of conditioned air or water.

---

226 BUILDING AUTOMATION Control Devices and Applications

Initiating Device Circuits

Figure 7-6. An initiating device circuit relies on the electrical changes caused by contact closures and circuit faults to indicate signals from the group of initiating devices.

An open circuit condition is an infinite resistance that causes current flow to stop, while a ground fault is a low resistance with current leakage to ground. Both of these problems are indicated as trouble signals by the FACP, regardless of the type of devices on the circuit. Monitoring for open circuits and ground faults is particularly important because these conditions can disable some initiating devices. Under either condition, only the devices between the FACP and the fault will operate until the problem is remedied. All devices downstream of the problem could effectively be disabled.

This kind of initiating circuit is relatively reliable because of its simplicity, but has some drawbacks. First, alarm- and supervisory-initiating devices cannot normally be wired on the same initiating-device circuit zone. Most FACPs would interpret the resulting signals as either false alarms or false supervisory signals. Also, the FACP can only indicate signals as coming from a circuit group and cannot determine exactly which device has activated. If the circuit zone covers a relatively large area, such as the entire floor of a building, this can slow the response times to the exact location of a fire during alarm conditions.

The basic organization and operation of each building system is covered in an overview

---

206 BUILDING AUTOMATION Control Devices and Applications

Full-Way Valves

Figure 6-19. When fully open, full-way valves permit unrestricted flow through the valve.

Solenoid valves are compact electrically actuated valves, but are only capable of full-way operation.

Manual valves, including those converted to electronic actuation, often have some way to visually indicate the valve's approximate position at a glance. The stem and/or handwheel may rise when opening, or a long handle may move to one side.

Throttling Valves. A *throttling valve* is a valve designed to control water flow rate by partially opening or closing the valve. Throttling valves are also called control valves. Due to the internal configuration, water flowing through the valve changes direction several times, resulting in flow resistance and a pressure drop. This makes them ideal for reducing water flow and pressure when required by the application. However, because they have some of this effect even when fully open, they are not recommended for use as full-way valves.

Throttling valves are installed on fixture supply pipes for individual fixtures. Examples of throttling valve designs include globe valves and butterfly valves. **See Figure 6-20.** Many throttling valves are required to be installed with the flow direction arrow pointing in the downstream direction.

Throttling valves require analog valves to determine intermediate positions. Many valve actuators accept standard analog signal types. Some are available for use with protocol messaging systems.

Three-Way Valves. A *three-way valve* is a valve with three ports that can control water flow between them. A *port* is an opening in a

Control devices are described and illustrated to explain their operation, signaling, and function

Building Automation: Control Devices and Applications

Control sequences list and illustrate the basic steps to common automation applications

Summaries highlight the key concepts of the chapter

Key terms are listed at the end of the chapter

Review questions test for chapter comprehension

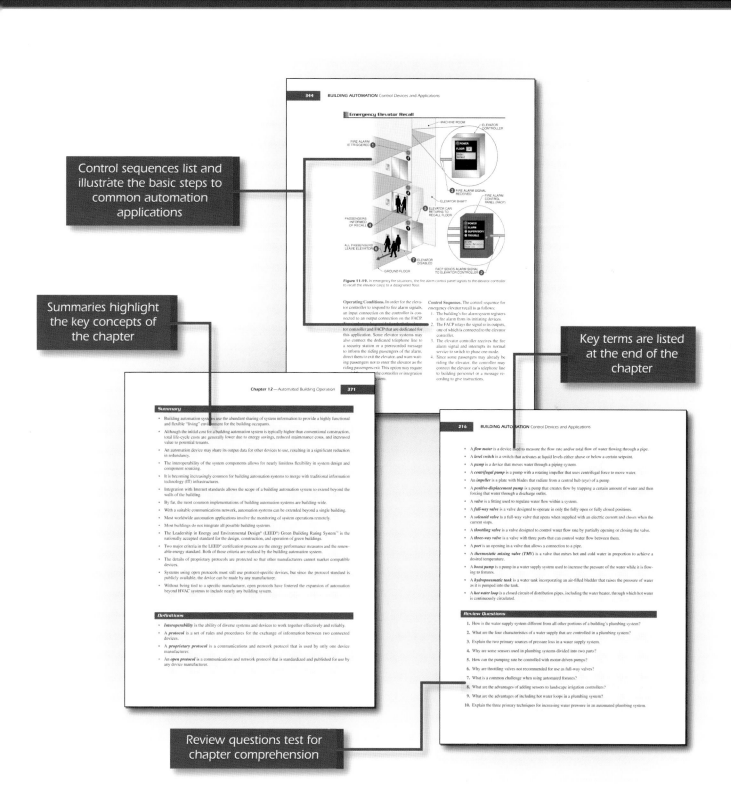

Introduction

Building automation improves productivity, comfort, convenience, safety, and security within living and working spaces while maximizing energy efficiency and minimizing manual control. *Building Automation: Control Devices and Applications* focuses on the devices that monitor and control building systems, such as the HVAC, lighting, and security systems. The textbook covers the operation, signaling, and functions of the common sensors, actuators, and other control devices used in building automation systems for commercial buildings. These modern electronic control devices have considerable capabilities and flexibility, but they require knowledge and skills from the installer, programmer, and operator to attain the most out of a building automation system. This textbook is a solid foundation for a comprehensive training program in building automation.

The textbook is organized by building system, and each system is explained to provide clarity on the function and application of each device. At the end of each system chapter, this foundation is used to introduce the common applications of these control devices, including the integration of multiple building systems together into a sophisticated building automation system.

The *Building Automation: Control Devices and Applications* CD-ROM included at the back of the book features interactive resources for independent study and to enhance learning, including Quick Quizzes®, an Illustrated Glossary, Flash Cards, Media Clips, and ATPeResources.com. The Quick Quizzes® provide an interactive review of key topics covered in each chapter. The Illustrated Glossary is a helpful reference to textbook definitions, with select terms linked to illustrations and media clips that augment the provided definition. The Flash Cards provide a review of terms and definitions related to building automation. The Media Clips provide a convenient link to selected video clips that depict topics covered in the book. ATPeResources.com provides access to a comprehensive array of instructional resources, including Internet links to manufacturers, associations, and ATP resources.

The Publisher

Contributing Writers

Mr. Chuck Sloup began his career as an application engineer for an independent controls contractor. He holds a Bachelor of Science from the University of Nebraska and is a licensed mechanical engineer. For the majority of the last 20 years, he has been a Senior Mechanical Engineer for a large engineering and architecture firm in Omaha, Nebraska. His primary responsibilities include the design and construction administration of HVAC, plumbing, and fire protection systems in hospitals, data centers, labs, and pharmaceutical manufacturing facilities. Specialty assignments include detailed design, commissioning, and troubleshooting fieldwork on these control systems. He has been involved with the American Society of Heating, Refrigerating, and Air-Conditioning Engineers (ASHRAE) as past president of the Nebraska Chapter and as a participant in the code-writing committee for ASHRAE 90.1, *Energy Standard for Buildings Except Low-Rise Residential Buildings*. He is currently starting a new company with faculty from the University of Nebraska to commercially develop jointly held patents dealing with controls optimization.

Mr. Merton Bunker has 23 years of experience as an electrical engineer. He holds a Bachelor of Science in electrical engineering from the University of Maine and a Master of Science from Western New England College. During his career as electrical engineer, he has worked for the U.S. Navy, served as the National Fire Protection Association (NFPA) staff liaison for the National Fire Alarm Code and the National Electrical Code® (NEC®), and taught as an NJATC Instructor. He is a licensed professional engineer, an IAEA (International Association of Electrical Inspectors) Certified Master Electrical Inspector, and an NAFI (National Association of Fire Investigators) Certified Fire and Explosion Investigator. He currently serves on several NFPA committees, including the NEC® Technical Correlating Committee; the NFPA 72 Technical Correlating Committee; and the Technical Committee on Inspection, Testing, and Maintenance of Fire Alarm Systems.

NJATC staff contributors:

Marty Riesberg
Director of Electrical Technologies and Automation
Contributing Writer and Technical Editor

Terry Coleman
Director of Telecommunications Curriculum Development and Training
Contributing Writer

Jim Simpson
Curriculum Specialist
Contributing Writer

Chapter One

Introduction to Building Automation

Automated buildings are often called "smart" buildings or "intelligent" buildings. This is because the building includes the controls to operate its systems optimally with little or no intervention from building personnel. Each individual controller includes the logic to make decisions within its own system. Together, all of these controllers comprise a system that operates smoothly and seamlessly. Building automation also adapts dynamically to changing conditions and needs of the occupants. The benefits of automating a building include significant energy savings, improved control, comfort, security, and convenience.

Chapter Objectives

- Identify the ways in which building automation can improve building efficiency.
- Describe the process in which a building automation system is implemented.
- Differentiate between the types of control devices that are integral to a building automation system.
- Compare the types of information that are shared between control devices.
- Contrast the different types of control logic used for automation programming.
- Describe common building systems that can be integrated in a building automation system.

BUILDING AUTOMATION

Building automation is the control of the energy- and resource-using devices in a building for optimization of building system operations. Building automation uses sensors to measure variables in the building systems and provide the measured values as inputs to a controller. A controller makes decisions on how to optimize the system based on the input information. Output signals are then sent to change some aspect of the system, bringing it closer to the desired conditions. **See Figure 1-1.**

▥ Building Automation

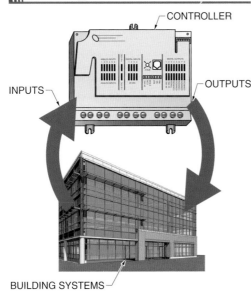

CONTROLLER

INPUTS

OUTPUTS

BUILDING SYSTEMS

Figure 1-1. Building automation control involves controllers processing inputs and producing outputs that control a building system.

Building automation can be implemented in many different ways, with any combination of building systems, and for any type of building. The most common modern systems, though, are electronic control systems in commercial buildings, such as offices, schools, hospitals, warehouses, and retail establishments. These building automation systems make the biggest impact on improving energy efficiency, system control, and convenience.

Building Automation Benefits

Building automation creates an intelligent building that improves occupant comfort and reduces energy use and maintenance. This can result in significant cost savings and improvements in safety and productivity. While building automation can be a significant initial investment, the return usually justifies the expense.

Energy Efficiency. Energy efficiency is probably the most significant benefit of a building automation system. The energy efficiency of a building is improved by controlling the electrical loads in such a way as to reduce their use without adversely affecting their purpose. For example, lighting is only used when rooms or areas are occupied, cutting down on electrical costs and increasing overall efficiency. However, with the automated controls, this reduction in lighting does not negatively impact the use of the space. As the largest consumers of energy in a building, the lighting and HVAC systems are the primary targets for improving energy efficiency through automation.

Building automation also helps conserve other resources, such as water and fuel. Plumbing controls manage water use and HVAC controls reduce fuel consumption by regulating the need for boiler or furnace use.

Improved Control. Most building system functions can be controlled manually, such as turning off lights and adjusting HVAC setpoints, but manual control is not as efficient as automated control. Automated controls can make control decisions much faster than a person and implement much smaller corrections to a system to optimize performance and efficiency. Automated controls work continuously and, if properly implemented, are not subject to common human errors such as inaccurate calculations, forgetting a control step, or missing an important input. This results in much more consistent and predictable control over the systems. For example, indoor air temperature remains consistent throughout the day because the automated controller is constantly monitoring all of the variables that affect this parameter, making small adjustments to the system to maintain the desired temperature. **See Figure 1-2.** Manual control of the same variables would not likely be as effective at providing such consistent results.

▋▋ Improved Control

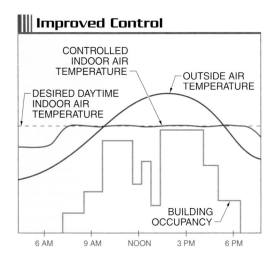

Figure 1-2. *Building automation provides consistent and reliable control of building systems even while other parameters continuously change.*

Automated controls also allow greater flexibility to implement control applications. For example, an output from one system can become an input to another, effectively integrating systems together in new and unique ways. The building systems may also be more secure from accidental or intentional tampering, since the automated controls can limit manual intervention.

Convenience. Building automation provides convenience for building owners, personnel, and occupants. After a system is commissioned and tested, it can operate with little human intervention. It reduces the workload of maintenance and security personnel by providing predictive maintenance information, automatically activating and deactivating some loads, and monitoring the building and its systems for problems. Building occupants find convenience in the automation of routine tasks, such as turning on lights when entering a room or unlocking doors when entering a secure area.

Building Automation Evolution

Early building system control was done manually. For example, an individual may have controlled the temperature in a living environment by adding fuel to a fire or allowing the fire to die down. As structures grew in size and contained a greater number of rooms, hot air from a central fire was regulated manually by opening or closing a diffuser (damper) with an adjusting pulley.

Control systems were developed in the early 1900s to provide automatic control of the building environment, primarily the HVAC system. First were pneumatic systems, but those were later replaced by electronic control systems, which provide better control and greater functionality. Currently, building automation systems are the most popular way to provide automatic control over several building systems.

Pneumatic Control Systems. Pneumatic control systems were the first automated control system for controlling the indoor environment in commercial buildings. A *pneumatic control system* is a control system in which compressed air is the medium for sharing control information and powering actuators. Pneumatic control systems can be separated into four main groups of components based on their function. These groups are the air compressor station, transmitters and controllers, auxiliary devices, and actuators. **See Figure 1-3.**

The air compressor station consists of the air compressor and other devices that ensure that the air supply is clean, dry, oil-free, and at the correct pressure. The compressed air supply is piped through the building to power control devices, mainly HVAC equipment. Transmitters and controllers are connected to the air supply and change the air pressure according to a measured variable, such as temperature or pressure.

The integration of multiple building systems must still comply with all the codes and standards required for stand-alone systems. With each additional system to be automated, more requirements are added.

▌▌▌ Pneumatic Control Systems

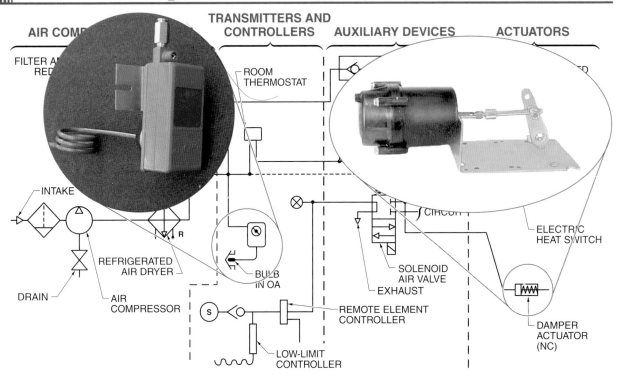

Figure 1-3. Pneumatic control systems include air compressors, transmitters and controllers, auxiliary devices, and actuators for managing an HVAC system.

A pneumatic actuator includes a small cylinder that moves in proportion to the air pressure applied to it.

Auxiliary devices are commonly located between a controller and a controlled device. Auxiliary devices change flow direction, change pressure, and interface between two devices. This modified air supply is routed to the controlled devices, such as dampers, valves, and switches. An actuator accepts an air pressure signal from a controller and causes a mechanical movement, which regulates some aspect of the controlled system, such as water or steam through a valve.

Pneumatic control systems have been primarily used in large commercial buildings such as hospitals and schools. They operate effectively and are very safe, but they may require a great deal of maintenance as well as specialized tools, calibration, and set-up procedures. With newer electronic controls available, pneumatic control systems are becoming increasingly rare.

Direct Digital Control (DDC) Systems. Improvements in electronics allowed for a significant improvement over pneumatic control systems. Electronics technology was adapted for use in building control systems, though still primarily for HVAC systems. A *direct digital control (DDC) system* is a control system in which electrical signals are used to measure and control system parameters.

Similar to pneumatic control systems, DDC systems include sensors, controllers, and actuators, though the electronic versions all operate from low-voltage DC power instead of air pressure. **See Figure 1-4.** Also, DDC controllers allow far greater functionality in making optimal control decisions because each includes a microprocessor to quickly and accurately calculate the necessary output signal based on the information from the input signals. The output of the controller is then sent to an electrically controlled actuator, such as a motor or solenoid.

Direct Digital Control (DDC) Systems

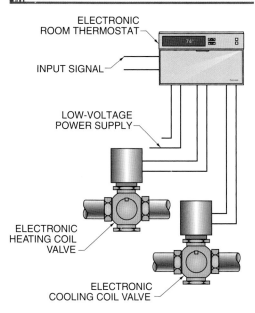

Figure 1-4. Direct digital control (DDC) systems use electronic controllers with their local sensors and actuators to control portions of a system.

DDC control systems are reliable, accurate, and relatively inexpensive. One disadvantage of electronic control systems is that they may require special diagnostic tools and procedures. DDC controllers are also limited to communicating only with the few sensors and actuators it is directly connected to.

> "Building automation" is a broad term covering any automatic control of one or more building systems (such as an HVAC system). A "building automation system," however, specifically refers to a system of microprocessor-based devices that share information on a data network. Building automation systems also make the integration of multiple building systems practical.

Building Automation Systems. In order to share information between building systems, electronic controllers must have a way to communicate between themselves. A *building automation system* is a system that uses a distributed system of microprocessor-based controllers to automate any combination of building systems. Building automation systems can control almost any type of building system, including HVAC equipment, lighting, and security systems.

Building automation system controllers are similar to DDC controllers in that each includes a microprocessor, memory, and a control program. The controller is connected to local sensors and actuators and manages a part of a building system, such as a variable-air-volume (VAV) terminal box. The difference is that the building automation system devices can all communicate with each other on the same, shared network. **See Figure 1-5.** Information is shared in the form of structured network messages that contain details of many control parameters. This information is encoded into a series of digital signals and sent to any other device that requires that information. For example, if one device reads from a sensor measuring outside air temperature, it can share that temperature with any other device, in any building system, that may have use for that information.

Building automation systems are extremely accurate, offer sophisticated features such as data acquisition and remote control, can integrate a variety of building systems, and are very flexible. The disadvantages of building automation systems are that the design and programming can be more involved than with other controls requiring contractors with specialized knowledge of these systems.

Building Automation System Components

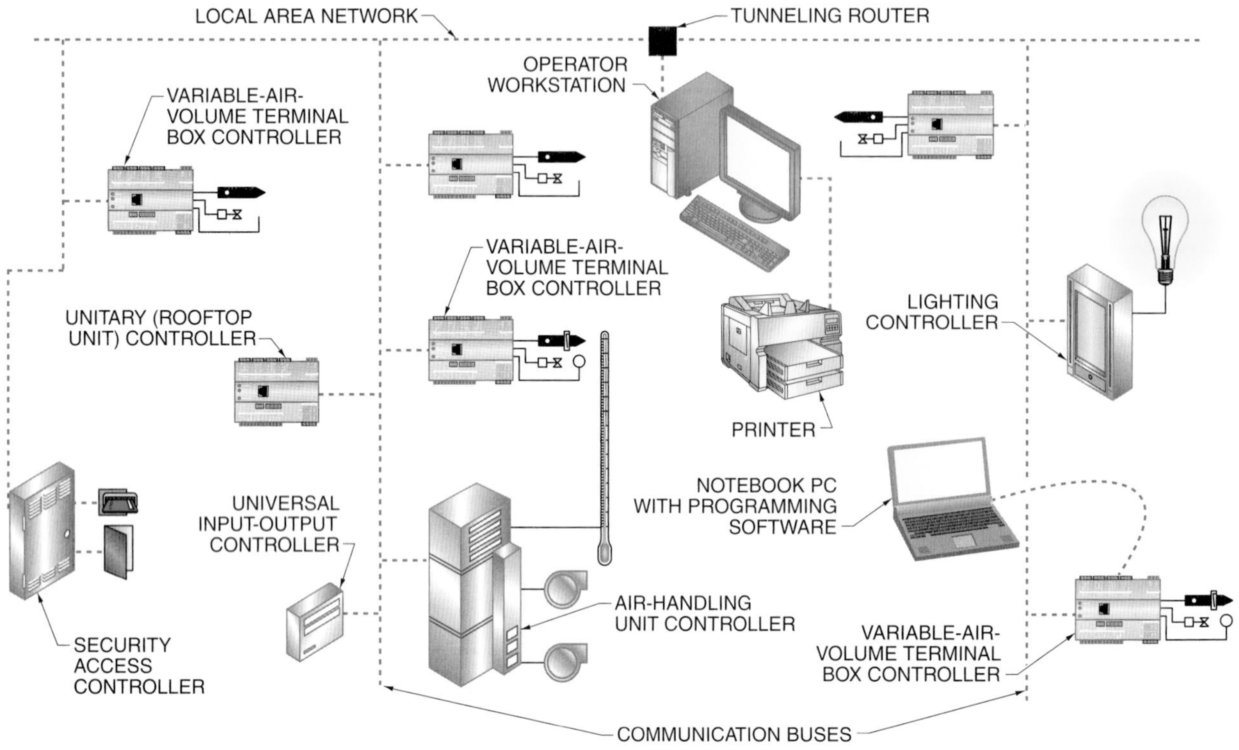

Figure 1-5. Building automation systems network electronic controllers together into a system that can share information between building systems.

Automation Applications

Building automation systems can be implemented in any residential, commercial, or industrial facility. The basic principles of automation are the same, regardless of the building use, though different types of facilities will likely have different requirements for system automation. Device manufacturers have developed different ways to implement automation and often selectively target certain markets based on building use.

Residential Buildings. Residential buildings are the least automated as a group, but automation devices are becoming increasingly affordable for homeowners and landlords, while the installation and programming of the systems are becoming increasingly practical. One common automation technology for residential buildings is X10 technology.

X10 technology is a control protocol that uses powerline signals to communicate between devices. The devices overlay control signals on the 60 Hz AC sine wave, which can be read by any other X10 device connected to the same electrical power conductors. The control signals are used to energize, de-energize, dim, or monitor common electrical loads. This technology is particularly attractive for residential buildings because it uses the existing electrical infrastructure and requires no new wiring. The control devices simply plug into standard wall receptacles. The devices are also very simple to commission and program. **See Figure 1-6.**

For many of the same reasons, wireless devices are also becoming popular for residential control applications. No additional wiring is required and the risk of radio frequency interference in a residential environment is minimal. Also, residential control applications are not likely to be critical systems, so there is little risk to safety or property if there is a communication problem between control devices.

Residential X10 Control Systems

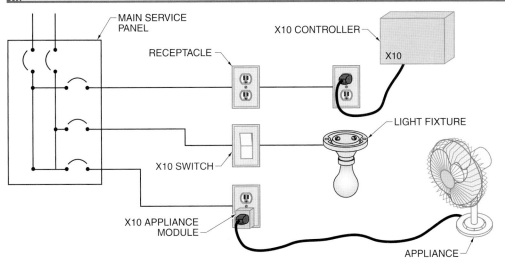

Figure 1-6. X10 is a common residential control system that uses the powerline as a medium for sharing control information between devices.

Commercial Buildings. Commercial buildings include offices, schools, hospitals, warehouses, and retail establishments. With growing emphasis on energy efficiency and "green" buildings, commercial buildings are probably the fastest growing market for implementing new automation systems. **See Figure 1-7.** New construction commercial buildings now commonly include some level of automation and many existing buildings are being retrofitted with automation systems.

Commercial Building Control Systems

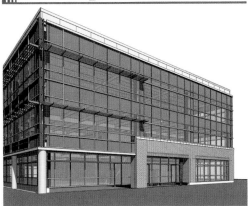

Figure 1-7. Commercial buildings are the most common applications of building automation systems.

Commercial building automation systems can control and integrate any or all of the building's systems, including electrical, lighting, HVAC, plumbing, fire protection, security, access control, VDV, and elevator systems. Building automation systems are inherently flexible, so building owners can choose almost any level of sophistication, from one specific application to all of the building systems. Upgrades or additional applications can be added and integrated into the system at any time in the future.

Industrial/Process Facilities. Automation has a long history in industrial, manufacturing, and process facilities. In addition to the control of lighting, security, and sanitary water, industrial facilities typically automate manufacturing processes, though the building and manufacturing automation systems are typically separate. Manufacturing processes must be so tightly controlled for time, safety, productivity, equipment efficiency, and product quality that automation is essential. **See Figure 1-8.**

Industrial automation systems operate in much the same way as residential or commercial building automation, but for different purposes. The control devices, being in industrial environments, are more rugged. There may be more human interaction with the automation system in order to modify sequences for changing tasks. Industrial automation is commonly applied at the controller level rather than the device level. Primarily, the control sequences are focused on safety and product manufacturing.

Industrial Building Control Systems

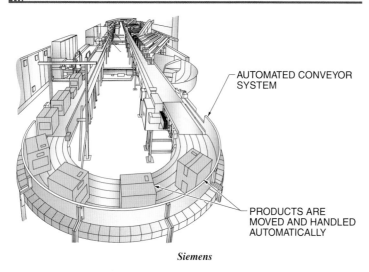

AUTOMATED CONVEYOR SYSTEM

PRODUCTS ARE MOVED AND HANDLED AUTOMATICALLY

Siemens

Figure 1-8. *Industrial buildings include many control systems, but they are typically for manufacturing processes or material handling.*

Building Automation Industry

The building automation industry includes a wide range of individuals and groups that help take requirements, concerns, and ideas for a new or existing building and create a state-of-the-art automated building. The participants in the building automation industry are the building owners, consulting-specifying engineers, controls contractors, and authorities having jurisdiction.

Building Owners. Ultimately, it is the building owner's decision on which building automation strategies to implement based on the desired results. Owners may wish to build a fully automated showcase facility that is highly visible to the public, or they may wish to automate only the few building systems that are most important to the efficient operation of the building. The end use of the building may also affect the automation choices because the building's location, occupancy density, equipment and products, and special on-site processes dictate the automation priorities. For example, lighting and HVAC controls may be especially important

for office buildings, while security and access control may be most important for warehouses full of valuable inventory.

The concerns of building owners also include initial costs, payback periods, maintenance, impact on occupants, system flexibility, system upgradeability, and environmental impacts. For example, automation systems typically return their extra costs within a few years, but still require a greater initial investment. These factors are even more important in retrofitting existing buildings for automation, as the implementation possibilities are typically more limited than in new construction.

Consulting-Specifying Engineers. For designing the most efficient system, the owner meets with consulting-specifying engineers to refine the automation plan. A *consulting-specifying engineer* is a building automation professional that designs the building automation system from the owner's list of desired features. This involves describing the necessary functionality of all the control devices and detailing the interactions between them. The consulting-specifying engineers work with the building owner, architect, and general contractor to make changes as needed to the planned automation system in order to maximize the benefits while working within any aesthetic, financial, or construction constraints.

Consulting-specifying engineers prepare contract specifications that include all of the control information. The *contract specification* is a document describing the desired performance of the purchased components and means and methods of installation. **See Figure 1-9.** Automation applications may be written out as detailed control sequence steps. These contract specifications are then made available to controls contractors to bid the project.

Controls Contractors. Controls engineering has traditionally been considered a subset of mechanical work, especially since HVAC systems are the most commonly automated building systems. However, controls technology has evolved into a highly technical field that has allowed contractors to specialize in automation. This has also given controls contractors the freedom to work with every building system so that they can be seamlessly integrated.

Contract Specifications

1.1   ELECTRONIC DATA INPUTS AND OUTPUTS

A. Input/output sensors and devices matched to requirements of remote panel for accurate, responsive, noise free signal output/input. Control input to be highly sensitive and matched to loop gain requirements for precise and responsive control.
1. In no case shall computer inputs be derived from pneumatic sensors.

B. Temperature sensors:
1. Except as indicated below, all space temperature sensors shall be provided with single sliding setpoint adjustment. Scale on adjustment shall indicate temperature. The following are exceptions to this:
a. The following locations shall have sensor without setpoint adjustment:
1) All electrical and communication rooms.
2) All mechanical rooms.
3) All unit heaters.
4) All public elevator lobbies and entrance vestibules.
5) Elevator equipment rooms.
2. Duct temperature sensors to be averaging type. Averaging sensors shall be of sufficient length (a maximum of 1.8 sqft of cross sectional area per 1 lineal foot of sensing element) to insure that the resistance represents an average over the cross section in which it is installed. The sensor shall have a bendable copper sheath. Water sensors provided with separable copper, monel or stainless steel well. Outside air wall mounted sensors provided with sun shield.

Figure 1-9. Contract specifications are produced by the consulting-specifying engineer and include the device requirements and control sequences for the building automation system.

The controls contractor uses the contract specification to choose the specific components and infrastructure that can accomplish the requirements. For example, the specification may list a temperature sensor that must provide temperature measurements to an air-handling unit controller. The controls contractor chooses the manufacturer and model of the sensor that meets all the requirements for installation type, temperature range, environmental conditions, calibration needs, and signal type. The controls contractor submits a quote for the project based on the costs of the purchased equipment, installation labor, system commissioning, and tuning. If chosen, the controls contractor then performs this work.

Authorities Having Jurisdiction (AHJ). The *authority having jurisdiction (AHJ)* is the organization, office, or individual who is responsible for approving the equipment and materials used for building automation installation. This includes all agencies and organizations that regulate, legislate, or create rules that affect

the building industry including building codes, National Electrical Code® (NEC®) regulations, Occupational Safety and Health Administration (OSHA) regulations, local and municipal codes, fire protection standards, waste disposal regulations, and all levels of government or other industry agencies. These groups may overlap. Building automation is most affected by policies related to energy efficiency and indoor air quality regulations.

Construction Documents

The planning and implementation of a building automation system requires detailed documentation. Controls contractors receive a set of construction documents from the consulting-specifying engineers, giving a comprehensive view of the project. For example, controls contractors must consider a number of trade areas in reviewing the project, including the sheet metal, plumbing, electrical, and piping drawings, as well as project schedules, scope, and other details.

Building automation systems and features are typically determined by the owner and consulting-specifying engineer before and during the building construction.

Contract Documents. A *contract document* is a set of documents produced by the consulting-specifying engineer for use by a contractor to bid a project. Contract documents include all of the various design disciplines: architectural, civil, structural, mechanical, and electrical engineering.

The contract documents consist of construction drawings and a book of contract specifications. The contract specification describes the project requirements in words, rather than graphically. Contract documents are purposely written in a generic manner to accommodate a variety of available products and features. This allows the controls contractor the freedom to specify any manufacturers or models that fulfill the project requirements.

Shop Drawings. The controls contractor's project bid includes his or her shop drawings. A *shop drawing* is a document produced by the controls contractor with the details necessary for installation. The consulting-specifying engineer reviews the shop drawings to ensure that the intent of the contract documents has been met. Shop drawings show much more detail than the contract drawings because they are specific to a particular device manufacturer and model. For example, shop drawings (also known as cut

sheets) include wiring details down to the individual device level. **See Figure 1-10.** In addition to playing a part in the construction process, the shop drawings are used by the owner's building maintenance staff for troubleshooting after the building has been turned over. Shop drawings include two basic categories: components and software/wiring.

The component portion of the shop drawing package includes information on every component and its size, configuration, and location. This allows the other trades workers to coordinate the installation of control components. For example, control valves are installed by pipe fitters long before the controls contractor connects signal wiring to them. Manufacturer data sheets are included to describe the features of the component in detail.

The component shop drawing package is often developed and reviewed by the consulting-specifying engineers first so that purchases with long lead times or decisions that impact other contractors can be made in a timely manner. For example, the controls contractor may choose to have the air terminal unit controllers mounted by the manufacturer at the factory to save installation time.

The software and wiring portion of the shop drawing package addresses all of the control interactions between the components. A building-level overview explains the control system architecture. Wiring interconnection and block diagrams illustrate both signal and power wiring for each panel and device, including information on power sources. Flow charts for each control sequence show interrelationships between inputs, calculations (control loops), and outputs. Sequences of operation describe all flow chart functions in words. A detailed explanation of color conventions clarifies drawing symbology.

Project Closeout Information. The *project closeout information* is a set of documents produced by the controls contractor for the owner's use while operating the building. Project closeout information includes operating and maintenance manuals for all equipment, owner instruction report, and a certified resolution of issues raised during the commissioning process.

Shop Drawings

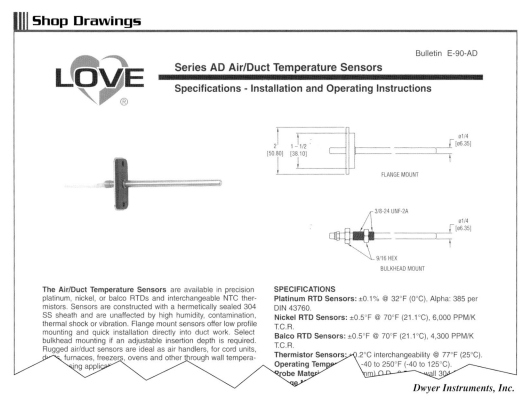

Bulletin E-90-AD

LOVE®

Series AD Air/Duct Temperature Sensors

Specifications - Installation and Operating Instructions

FLANGE MOUNT

3/8-24 UNF-2A
9/16 HEX
BULKHEAD MOUNT

The Air/Duct Temperature Sensors are available in precision platinum, nickel, or balco RTDs and interchangeable NTC thermistors. Sensors are constructed with a hermetically sealed 304 SS sheath and are unaffected by high humidity, contamination, thermal shock or vibration. Flange mount sensors offer low profile mounting and quick installation directly into duct work. Select bulkhead mounting if an adjustable insertion depth is required. Rugged air/duct sensors are ideal as air handlers, for cord units, ducts, furnaces, freezers, ovens and other through wall temperature-sensing applications.

SPECIFICATIONS
Platinum RTD Sensors: ±0.1% @ 32°F (0°C), Alpha: 385 per DIN 43760.
Nickel RTD Sensors: ±0.5°F @ 70°F (21.1°C), 6,000 PPM/K T.C.R.
Balco RTD Sensors: ±0.5°F @ 70°F (21.1°C), 4,300 PPM/K T.C.R.
Thermistor Sensors: ±0.2°C interchangeability @ 77°F (25°C).
Operating Temperature: -40 to 250°F (-40 to 125°C).
Probe Material: mm) O.D. ... wall 304

Dwyer Instruments, Inc.

Figure 1-10. Shop drawings from a controls contractor include specific information on devices that satisfy the automation requirements.

CONTROL DEVICES

A *control device* is a building automation device for monitoring or changing system variables, making control decisions, or interfacing with other types of systems. A *variable* is some changing characteristic in a system. Control devices include sensors, controllers, actuators, and human-machine interfaces.

Sensors

A *sensor* is a device that measures the value of a variable and transmits a signal that conveys this information. The sensor output signal is commonly electrical but may also be conveyed by air pressure, optics, or radio frequency. Even though they output signals, sensors are considered input devices, which is relative to the point of view of the controllers that manage the systems.

As individual devices, sensors allow variables to be measured in any location or type of environment, while other control devices are installed in convenient and centralized places.

For example, a controller may be connected to a number of individual temperature sensors that are installed throughout an HVAC unit. **See Figure 1-11.** Sensors may also be integral with a controller, as in a room thermostat.

Sensors

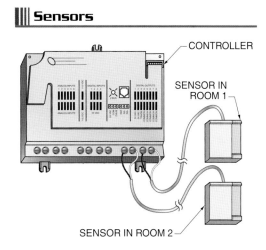

CONTROLLER

SENSOR IN ROOM 1

SENSOR IN ROOM 2

Figure 1-11. Sensors provide the inputs into the building automation controllers.

Sensors must be carefully chosen to provide the appropriate information and in a way that is understandable by the receiving control devices. Factors to consider are sensor type, range, resolution, accuracy, calibration, signal type, power requirements, operating environment, mounting configuration, size, weight, construction materials, and agency listings.

Sensors that are compatible with modern automation systems are typically either electromechanical or electronic sensors. They are distinguished by the way in which the sensors monitor their environment. Electromechanical devices use mechanical means to measure a variable but include electrical components to convert motion into electrical signals and process and transmit the signals. Some variables that may require mechanical capabilities to be measured include pressure, position, and fluid level. Purely electronic devices are typically smaller and more reliable than electromechanical devices, since they have no moving parts. **See Figure 1-12.** The variable is measured by detecting the direct changes the variable has on the electrical properties of a solid-state component. Then other electronics process and transmit the information as a signal. Sensors are connected to terminals on a controller.

▌▌▌ Electronic Sensors

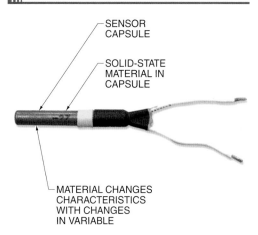

SENSOR CAPSULE

SOLID-STATE MATERIAL IN CAPSULE

MATERIAL CHANGES CHARACTERISTICS WITH CHANGES IN VARIABLE

Figure 1-12. Electronic sensors detect changes in a variable from corresponding changes in the sensor's electrical properties.

Controllers

A *controller* is a device that makes decisions to change some aspect of a system based on sensor information and internal programming. The controller receives a signal from the sensor, compares it to a desired value to be maintained, and sends an appropriate output signal to an actuator.

Modern controllers are all electronic devices with processors, memory, and software that can execute control sequences very quickly and accurately. The electronics also allow relatively simple modification of the programming and settings to optimize building system operation. However, the electronics may be less rugged than some sensors.

Some controllers are designed for specific control applications, such as controlling an air-handling unit. These provide many features and programs developed by the manufacturer to simplify installation and commissioning of the system, but may not allow much flexibility to change inputs, outputs, or decision making. Generic controllers, on the other hand, include only the electronics to read from standard input connections, run software, and write to standard output connections. Any type of device can be connected to the terminals and any type of decision can be made. These devices allow the greatest flexibility in applications, but the program must be custom-written and tested by the controls contractor to ensure correct operation and reliability.

Actuators

An *actuator* is a device that accepts a control signal and causes a mechanical motion. This motion may actuate a switch, rotate a valve stem, change a position, or cause some other change that affects one or more characteristics in a system. From the point of view of the controller, actuators are the output devices. For example, an actuator may open a valve to allow more steam into a heating coil, causing the temperature of the airflow across the coil to rise.

A large number of controlled devices may be used to change system characteristics, such as dampers for regulating airflow, valves for regulating water or steam flow, refrigeration compressors for delivering cooling, and gas valves and electric heating elements for regulating heating. Many of these are actually controlled by just a few different types of actuators. The most common types of actuators are relays, solenoids, and electric motors. **See Figure 1-13.**

▥ Actuators

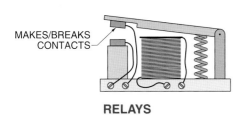

RELAYS

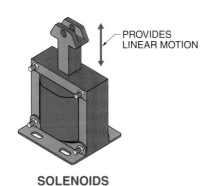

PROVIDES LINEAR MOTION

SOLENOIDS

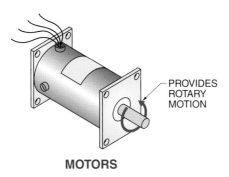

PROVIDES ROTARY MOTION

MOTORS

Figure 1-13. The most common types of actuators found in controlled devices are relays, solenoids, and motors.

A *relay* is an electrical switch that is actuated by a separate electrical circuit. It allows a low-voltage control circuit to open or close contacts in a higher-voltage load circuit. Many relays are electromechanical types, but some are completely solid-state, using no moving parts. Relays can be used with any device that is controlled by turning it ON or OFF. They are common for switching lighting circuits and HVAC package units, as well as sharing ON/OFF information between controllers.

A *solenoid* is a device that converts electrical energy into a linear mechanical force. Solenoids are typically used in quick-acting valves and for locking and unlocking doors. An *electric motor* is a device that converts electrical energy into rotating mechanical energy. Electric actuator motors can be used to turn something completely open or completely closed, but are particularly useful for actuating devices with multiple positions, such as dampers and valves.

Human-Machine Interfaces (HMIs)

A human-machine interface is connected into a building automation system to allow personnel to view and modify the information being shared between control devices. A *human-machine interface (HMI)* is an interface terminal that allows an individual to access and respond to building automation system information. Many HMIs show information graphically and include data over time so that current values and longer-term trends are easily visible. HMIs may also allow a user to interact with the system by manually inputting changes to values or device behavior.

Human-machine interfaces are either hardware-based or software-based. **See Figure 1-14.** Hardware-based HMIs are basically small computers that are specially designed to gather, process, and display system data. They are typically self-contained units with integral monitors, memory, connection terminals, and communication ports. If they are intended to be installed in extreme environments, they may be housed in special dust-resistant, moisture-resistant, rugged enclosures. If the HMI accepts inputs into the system from the user, it typically includes a keypad, touchscreen display, or an I/O port for connecting a separate keyboard or mouse.

Human-Machine Interfaces (HMIs)

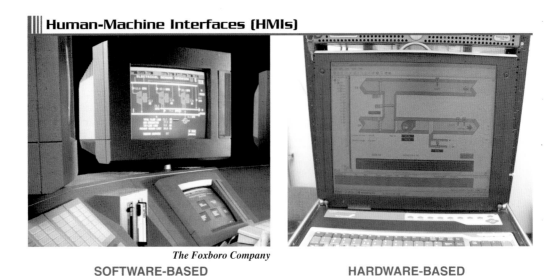

The Foxboro Company

SOFTWARE-BASED   **HARDWARE-BASED**

Figure 1-14. Human-machine interfaces (HMIs) may be software- or hardware-based.

Software-based HMIs may require a special piece of hardware to physically connect with the building automation system and retrieve data, but the interface portion relies on software running on separate personal computers. Some packages require special software (a "thick client" application) on the computer to communicate with the hardware. However, the current trend is to design the interface to be viewed with a standard web browser, which is common software on all computers and requires no additional installation (a "thin client" application). The hardware unit acts as a web server, delivering the building system information in a way that is similar to any other Internet site. These HMIs can be configured to provide this information to only computers within the building or to any computer anywhere in the world. Either type of software-based HMI is capable of being programmed to accept system inputs from the user. Security features ensure that only authorized personnel are able to make these changes.

CONTROL SIGNALS

A *control signal* is a changing characteristic used to communicate building automation information between control devices. Control signals are typically electronic signals but can also be transmitted by other media such as air pressure, light, and radio frequency. There are different types of control signals, depending on the type of information to be shared. The three common types of control signals are digital signals, analog signals, and structured network messages.

Digital Signals

A *digital signal* is a signal that has only two possible states. For this reason, digital signals may also be known as binary signals. Digital signals convey information as either one of two extremes, such as completely ON or completely OFF, or completely open or completely closed.

A digital signal is typically conveyed with a change in voltage. For example, 0 VDC can represent an OFF state and 5 VDC an ON state. **See Figure 1-15.** Alternatively, +12 VDC can represent an ON state and –12 VDC an OFF state. Any pair of two different voltages can be used to send digital signals, though they are commonly DC voltages of 24 VDC or less. Also, different digital signal voltage levels can be used within a system, as long as each device is compatible with the type of signal it is sending or receiving.

The devices also must agree on the voltage levels that define the two states. For example, due to a slight voltage drop in the signaling conductor, a 5 VDC (ON) signal may appear to the receiver as slightly less, such as 4.6 VDC. In order for this voltage to still be registered as an ON state, the two states may be more precisely defined as 0 V to 0.8 VDC for OFF and 2.0 VDC to 5.25 VDC for ON. Voltages between these two levels may be read as an erratically fluctuating ON and OFF, producing unpredictable results.

Digital Signals

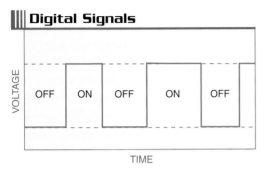

Figure 1-15. Digital signals are produced by a pair of voltage levels that represents either ON or OFF.

Digital signals can be generated by a device with a power supply by applying the necessary voltage to a signaling conductor. **See Figure 1-16.** Alternatively, digital signals can be generated by a set of contacts in an input device, which does not require an external power supply. The contact terminals are wired to a set of controller terminals with a pair of conductors. The controller senses the continuity or discontinuity through the terminals as closure or opening of the contacts in the input device.

Generating Digital Signals

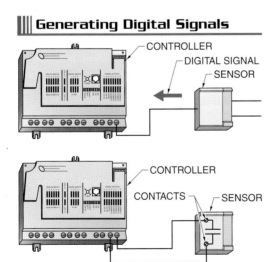

Figure 1-16. Digital signals can be generated by devices with their own power supply or by devices with switch contacts.

Digital signals are commonly used as an output to turn devices ON and OFF, typically through relays that switch the power needed to operate the device. For example, digital signals can turn electric motors, electric heating stages, and valves ON and OFF with digital signals. Digital signals can also be used to initiate different functions in package units. These units are energized by a separate, manual means, but a controller's digital signals may enable/disable the unit's primary function.

Analog Signals

An *analog signal* is a signal that has a continuous range of possible values between two points. **See Figure 1-17.** Analog signals can convey information that has units of measurement, such as degrees Fahrenheit, cubic feet per minute, meters per second, and inches of water column. They can also provide any value between 0% and 100% of some controllable characteristic. For example, analog signals can be used to control fan speed from 0% (stopped) to 100% of its full rated maximum speed.

Analog Signals

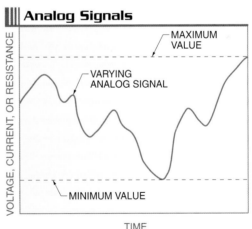

Figure 1-17. Analog signals can vary continuously between two points.

Pulse width modulation (PWM) is a control technique in which a sequence of short pulses (digital signals) is used to communicate analog information. The amount of time the signal is ON versus OFF indicates any value between 0% and 100%.

The most common electrical properties used to convey analog signals are voltage, current, and resistance. The most common analog signal ranges include 0 VDC to 10 VDC, 4 mA to 20 mA, and 0 kΩ to 10 kΩ. Devices may use other analog signal ranges, as long as they are compatible between the sender and receiver and represent the same values. For example, 0 VDC to 5 VDC, 1 mA to 10 mA, and 0 Ω to 1000 Ω ranges are common.

Analog signals differ from digital signals in that small fluctuations in the signal are meaningful. The range of possible signals is mapped to a range of possible values of the measured variable or the range of actuator positions. For example, an analog signal of 4 mA to 20 mA may be used to indicate a temperature between 32°F and 212°F. **See Figure 1-18.** Therefore, the range of 16 mA (20 mA – 4 mA = 16 mA) represents the range of 180°F (212°F – 32°F = 180°F). Each 1°F of change in temperature is indicated by a change of 0.089 mA. Units must be carefully noted in analog signal mapping, as many quantities can be mapped to other scales. For example, the same temperature analog signal also has the characteristic of 0.16 mA/°C (16 mA ÷ [100°C – 0°C] = 0.16 mA/°C).

A disadvantage to analog signals is that electrical noise, which is normal to some degree in any electrical system, can affect the signals by changing their intended value slightly. While digital signals address this problem by providing ranges of values that correspond to the intended information, this cannot be done with analog signals, where every small change in value is potentially important. Depending on the system, noise can cause significant operational problems. Noise can be reduced to some degree by shielded conductors. Proper connections and careful design of the automated system, such as the length of conductor runs and proximity to noise-inducing electrical components, can help mitigate noise problems.

Structured Network Messages

Some control devices can share much richer pieces of information by signaling in the form of structured network messages, which use protocols. A *protocol* is a set of rules and procedures for the exchange of information between two connected devices. The protocol is implemented in both the hardware and software of the devices. It defines the format, timing, and signals for reliable and repeatable data transmission between two devices.

Each structured network message includes information on the sending and receiving devices, the identification of the shared variable, its value, and any other necessary parameters. The information is encoded and transmitted via a series of digital signals, composing one complete message. **See Figure 1-19.** Almost any type of information can be shared in this way, as long as all the devices involved can work with the same protocol.

Analog Values

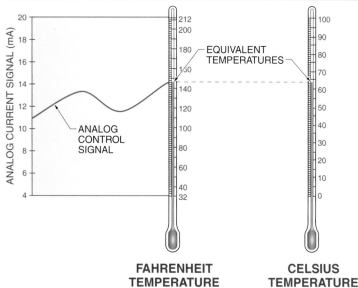

FAHRENHEIT TEMPERATURE **CELSIUS TEMPERATURE**

Figure 1-18. Analog signals are mapped to a certain range for the measured variable in order to determine the current value.

Structured Network Messages

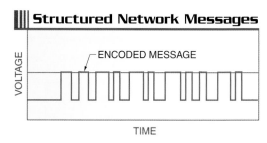

Figure 1-19. Structured network messages are several pieces of control information encoded into a series of digital signals.

Systems using digital and analog signals are typically connected point-to-point. That is, each device is connected with only the devices that need to receive its signal. However, systems using structured network messages are connected together in a similar way to a computer network. In this configuration, any device can communicate with any other device on the network. In fact, some systems can communicate over the same wiring and routing infrastructure as the building's computer local area network (LAN). There are also advantages to keeping the building automation system on a completely separate network, though this increases installation and maintenance costs.

CONTROL INFORMATION

Control devices are used in every type of control application to provide the required information for proper system operation. Hardware and software used to provide control information include control points, virtual points, setpoints, offsets, and deadband.

Control Points

The input information from sensors and output information to actuators creates control points in the building automation system. A *control point* is a variable in a control system. Control points change over time. **See Figure 1-20.** Controllers use input control points to make decisions about changes needed in the system and then determine the best value for the output control points.

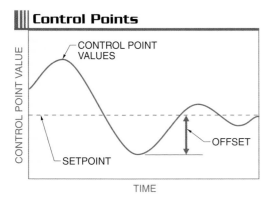

Control Points

Figure 1-20. Controllers modify the building system until the control point variables equal the setpoint.

For example, a temperature sensor in a building space measures a temperature of 74°F, which is an input control point for the HVAC controllers. The information that the controllers generate and use to cause changes in the system is the output control point. For example, a damper actuator accepts a control point signal that corresponds to the position it must maintain.

Virtual Points

A *virtual point* is a control point that exists only in software. It is not a hard-wired point corresponding to a physical sensor or actuator. For example, programmed schedules produce virtual points that correspond to the status of the planned occupancy of a room. The resulting virtual point may have three states: OCCUPIED, UNOCCUPIED, and PRE-OC-CUPANCY (meaning that the room systems are getting ready for OCCUPIED mode). This virtual point may be used by controllers to change the behavior of the HVAC and lighting systems, depending on whether people are expected to be using this room. Since they exist only in software, virtual points can only be monitored with some type of HMI.

Virtual points can also be used to hold snapshots of the values of other control points for different circumstances, such as a timestamp of an event, an extreme (high/low) value of a control point, or the extreme of multiple control points. For example, a series of three control points represent the current temperatures of three different rooms. A separate virtual point may represent the highest temperature any of the rooms reached during the last 24 hr period. Another virtual point may hold the timestamp for that event.

Lists of system inputs and outputs in the contract documents contain both hard-wired control points and virtual points, though they will be distinguished from each other in some way.

Building automation systems can be designed to communicate over a variety of network media, including twisted pair, fiber optic, existing power conductors, or even radio frequency. Multiple media can even be mixed within the same installation.

Setpoints

The purpose of the control system is to achieve and maintain certain conditions, which are quantified as setpoints. A *setpoint* is the desired value to be maintained by a system. Setpoints are matched to a corresponding control point variable, sharing the same variable types and units of measure, such as temperature in degrees Fahrenheit or flow rate in cubic feet per minute. For example, if it is desired to maintain a temperature of 72°F in a building space, the setpoint is 72°F.

There may be more than one setpoint associated with a control point. The currently active setpoint, as in the setpoint that the system is currently trying to achieve, can be changed according to schedules, occupancy, or any other parameter. For example, a cooling setpoint is raised from 74°F during the day to 85°F at night or when a building is unoccupied. This saves energy by reducing the system loads when a building is unoccupied, but still prevents the building conditions from getting excessively far from the occupied setpoint. Alternatively, setpoints can be reset based on other conditions. For example, a hydronic water heating setpoint is raised as the outside air temperature falls.

Offsets

Ideally, a control point is equal to its corresponding setpoint. However, there is often some difference, though it should be minimized as much as possible. The *offset* is the difference between the value of a control point and its corresponding setpoint. Offset values measure the accuracy of the control system. **See Figure 1-21.**

Expectations of control system accuracy have increased as the available technology has become more sophisticated. In the past, an accuracy of ±2°F was considered acceptable. With the widespread use of building automation systems, much tighter accuracies (such as ±0.5°F) are possible. Control system accuracy is a function of the accuracy of the controllers, the precision of the sensors and actuators, and the quality of the system design and tuning.

Offsets

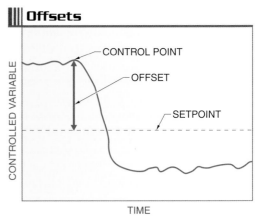

HIGH OFFSET, LOW ACCURACY

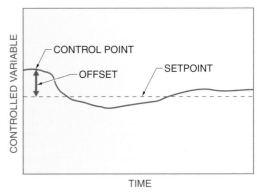

LOW OFFSET, HIGH ACCURACY

Figure 1-21. A control system with large offsets has low accuracy and a system with small offsets has high accuracy.

Deadband

Sometimes it is not necessary to maintain a condition at exactly one setpoint. Rather, the variable must only remain within a certain range. The range is defined by a pair of setpoints. **See Figure 1-22.** For example, the temperature within a space must be maintained between 70°F and 74°F. At 70°F, the heating system operates, and at 74°F, the cooling system operates, with a deadband in between. A *deadband* is the range between two setpoints in which no control action takes place.

As long as this does not adversely affect occupant comfort, then this deadband arrangement reduces energy use by the control system. It also reduces oscillation and frequent cycling of the system.

Deadband

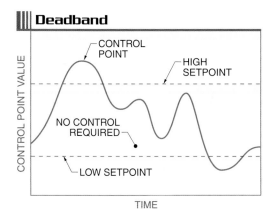

Figure 1-22. Deadband is defined by a pair of setpoints between which no control action is required.

CONTROL LOGIC

The decisions that controllers make to change the operation of a building system involve control logic. *Control logic* is the portion of controller software that produces the necessary outputs based on the inputs. There are many different ways in which these decisions can be made, depending on the inputs used to make the decisions and the algorithms used to produce the results. An *algorithm* is a sequence of instructions for producing the optimal result to a problem. The decision-making process is described with a control loop. A *control loop* is the continuous repetition of the control logic decisions. Control loops are either open loops or closed loops.

Open-Loop Control

Open-loop control is the simpler of the two but has drawbacks. An *open-loop control system* is a control system in which decisions are made based only on the current state of the system and a model of how it should work. An example of an open-loop control system is a controller that turns a chilled water pump ON when the outside air temperature is above 65°F. The controller has no feedback to verify that the pump is actually ON. **See Figure 1-23.** Open-loop control requires perfect knowledge of the system, and assumes there are no disturbances to the system that would change the outcome otherwise. There is no connection between the controller's output and its input.

Open-Loop Control Systems

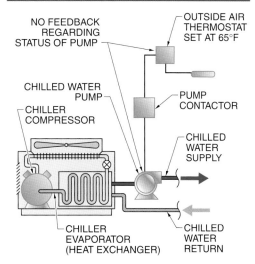

Figure 1-23. Open-loop control systems make changes to a system without verifying its effects.

The most common example of open-loop control in a building automation system is based on time schedules. Time-based control is a control strategy in which the time of day is used to determine the desired operation of a load. Time-based control turns a load ON or OFF at a specific time, without knowledge of any other factors that may affect the need for that load to operate. For example, open-loop time-based landscape irrigation control may activate the sprinklers based on a schedule. However, if it had recently rained, the system is overwatering the landscape and wasting water. Without a moisture sensor to provide an input based on the system output, the system has only open-loop control.

Closed-Loop Control

To address the limitation of open-loop control, most essential control loops include feedback. This makes them closed-loop. A *closed-loop control system* is a control system in which the result of an output is fed back into a controller as an input. For example, a thermostat controls the position of a valve in a hot water terminal device to maintain an air temperature setpoint. The thermostat in the building space provides the feedback of the air temperature that is used to continually adjust the hot water valve. **See Figure 1-24.**

Closed-Loop Control Systems

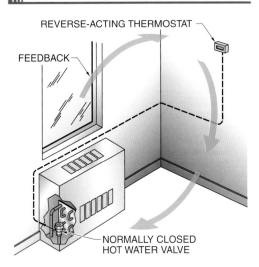

Figure 1-24. Closed-loop control systems provide feedback on system changes to the controller as an input.

Proportional Algorithm

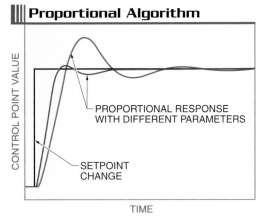

Figure 1-25. In order to bring a control point to a new setpoint, a proportional algorithm adjusts the output in direct response to the current offset.

Algorithms are used to calculate the output value based on the inputs. Different calculation algorithms can be used, each with different characteristics of accuracy, stability, and response time. The correct algorithm must be selected for the system. Common algorithms include proportional, integral, and derivative control algorithms.

Proportional Control Algorithms. Most modern commercial control applications use proportional control. *Proportional control* is a control algorithm in which the output is in direct response to the amount of offset in the system. For example, a 10% increase in room temperature results in a cooling control valve opening by 10%. Proportional controllers output an analog signal, which requires compatible actuators. **See Figure 1-25.**

Proportional control systems have a lower tendency to undershoot or overshoot. While proportional control is used successfully in most applications, it may be inaccurate if not set up properly. When the system reaches the setpoint, the controller outputs a default actuator position, typically a 50% setting. However, a load may require a different position when at the setpoint, resulting in increased offset and energy expenditure.

Integral Control Algorithms. An *integral control algorithm* is a control algorithm in which the output is determined by the sum of the offset over time. *Integration* is a function that calculates the amount of offset over time as the area underneath a time-variable curve. **See Figure 1-26.** This offset area is then used to determine the output needed to eliminate the offset. Integral control algorithms move the system toward the setpoint faster than proportional algorithms.

Integral Algorithm

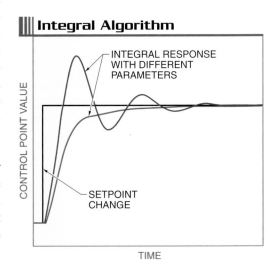

Figure 1-26. In order to bring a control point to a new setpoint, an integral algorithm adjusts the output according to the sum of the offsets in a preceding period.

However, since the integral is responding to accumulated errors from the past, it can cause the present value to overshoot the setpoint, crossing over the setpoint and creating an offset in the other direction. Proportional/integral (PI) control is the combination of proportional and integral control algorithms. These controllers can be more stable and accurate than integral-only controllers.

Derivative Control Algorithms. A *derivative control algorithm* is a control algorithm in which the output is determined by the instantaneous rate of change of a variable. **See Figure 1-27.** The rate of change is then used to determine the output needed to eliminate the offset. However, this calculation amplifies noise in the signal, which can cause the system to become unstable.

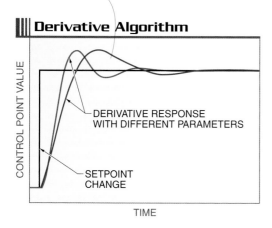

Derivative Algorithm

DERIVATIVE RESPONSE WITH DIFFERENT PARAMETERS

SETPOINT CHANGE

CONTROL POINT VALUE

TIME

Figure 1-27. In order to bring a control point to a new setpoint, a derivative algorithm adjusts the output according to the rate of change of the control point.

Proportional/integral/derivative (PID) control is the combination of proportional, integral, and derivative algorithms. The derivative control algorithm slows the rate of change of the controller output, which is most noticeable close to the controller setpoint. Therefore, derivative control reduces the magnitude of the overshoot produced by the integral component and the combination improves stability. In building automation, only extremely sensitive control applications require PID control. Proportional/integral (PI) control is normally sufficient to achieve a setpoint.

PI and PID control systems must be carefully tuned to ensure accuracy and stability. *Tuning* is the adjustment of control parameters to the optimal values for the desired control response. **See Figure 1-28.** The parameters are the relative contributions of each algorithm to the final output decision and any parameters used in each algorithm calculation. There are several tuning methods by which experienced control professionals can determine the best system values.

Effects of Changing PID Parameters

Parameter	Rise Time	Overshoot	Settling Time	Steady-State Error
Proportional	Decrease	Increases	Small Change	Decreases
Integral	Decrease	Increases	Increases	Eliminates
Derivative	Small Change	Decreases	Decreases	None

Figure 1-28. Tuning involves adjusting the relative contributions of the proportional, integral, and derivative algorithms to quickly stabilize a control point at a setpoint.

BUILDING SYSTEMS

Building automation may involve any or all of the building systems. The level of sophistication of a building automation system is affected by the number of integrated building systems. Common building systems controlled by building automation include electrical, lighting, HVAC, plumbing, fire protection, security, access control, voice-data-video (VDV), and elevator systems.

Electrical Systems

An *electrical system* is a combination of electrical devices and components connected by conductors that distributes and controls the flow of electricity from its source to a point of use. The control of the electrical system involves ensuring a constant and reliable power supply to all building loads by managing building loads, uninterruptible power supplies (UPSs), and back-up power supplies. Sophisticated switching systems are used to connect and disconnect power supplies as needed to maintain building operation and occupant productivity. Electrical systems affect many other systems, such as lighting, security, and voice-data-video circuits.

However, control of those systems is typically handled separately.

Lighting Systems

A *lighting system* is a building system that provides artificial light for indoor areas. Lighting is one of the single largest consumers of electricity in a commercial building. Therefore, improving energy efficiency and reducing electricity costs are the driving factors in lighting system control.

Lighting system control involves switching OFF or dimming lighting circuits as much as possible without adversely impacting the productivity and safety of the building occupants or the security of the building. Lighting systems are controlled based on schedules, occupancy, daylight harvesting, or timers. Specialty lighting control systems can also produce custom lighting scenes for special applications.

HVAC Systems

A *heating, ventilating, and air conditioning (HVAC) system* is a building system that controls a building's indoor climate. Properly conditioned air improves the comfort and health of building occupants. HVAC systems are controlled to operate at optimum energy efficiency while maintaining desired environmental conditions.

© *2008 Lutron Electronics, Inc.*

Lighting systems can be controlled to provide comfortable environments while saving energy.

HVAC systems are the most common building systems to automate but are also the most complicated. They include several parameters that must be controlled simultaneously and that are closely interrelated. Changes in one conditioned air parameter, such as temperature, can affect other parameters, such as humidity. Therefore, the systems use the most sophisticated control logic and must be carefully designed and tuned. A well-automated HVAC system, however, operates efficiently and with little manual input from occupants or maintenance personnel.

Plumbing Systems

A *plumbing system* is a system of pipes, fittings, and fixtures within a building that conveys a water supply and removes wastewater and waterborne waste. Plumbing systems are designed to have few active components. That is, most plumbing fixtures operate on water pressure and manual activation alone. However, there are a few applications in which certain parts of a commercial building's plumbing system can be actively controlled, such as water temperature, water pressure, and the supply of water for certain uses. These systems aim to use water the most effectively and provide a water supply with the optimal characteristics.

Fire Protection Systems

A *fire protection system* is a building system for protecting the safety of building occupants during a fire. Fire protection systems include both fire alarm systems and fire suppression systems. Fire protection systems automatically sense fire hazards, such as smoke and heat, and alert occupants to the dangers via strobes, sirens, and other devices. They also monitor their own devices for any wiring or device problems that may impair the system's proper operation during an emergency.

Fire protection systems are highly regulated due to their role as a life safety system. Fire alarm systems can be integrated with other building systems, but only through the fire alarm control panel (FACP). The fire-sensing devices and output devices may only be connected to the FACP. However, many FACPs include special features that enable integration with other automated systems. For example, some include special output connections that can be used to share fire alarm signals with other systems that have special functions during fire alarms.

Security Systems

A *security system* is a building system that protects against intruders, theft, and vandalism. Security systems are similar to fire protection systems in that they are typically implemented as separately designed package systems and allow connection with other building systems only at the security control panel. The control panel provides special output connections that can be used to initiate special control sequences in other systems, such as unoccupied modes, access, and security monitoring with surveillance systems.

Access Control Systems

An *access control system* is a system used to deny those without proper credentials access to a specific building, area, or room. Authorized personnel use keycards, access codes, or other means to verify their right to enter the restricted area.

Access control systems, through their control panels, can be integrated with other systems in much the same way as fire protection systems and security systems. In fact, due to their similarity, access control and security systems are often highly integrated. When a person's credentials have been verified by the access control system, the control panel can initiate the subsequent response of other systems, such as directing the surveillance system to monitor the person's entry.

Voice-Data-Video (VDV) Systems

A *voice-data-video (VDV) system* is a building system used for the transmission of information. Voice systems include telephone and intercom systems. Data systems include computer and control device networks, but can also carry voice or video information that has been encoded into compatible data streams. Video systems include closed-circuit television (CCTV) systems.

Building automation integration typically involves the VDV systems being controlled by other building systems. For example, a telephone line may be seized by the fire alarm system in order to alert the local authorities to a fire in the building. Alternatively, the CCTV system that is used for surveillance may be controlled by the security or access control systems to monitor certain areas where people have been detected.

Elevator Systems

An *elevator system* is a conveying system for transporting people and/or materials vertically between floors in a building. Elevator systems operate effectively largely on their own and with their own control devices but can accept inputs from other systems to modify their operating modes. A prime example of this is the integration of the fire alarm system with the elevator system. Since elevators can be dangerous to use during a fire, the elevator cars must be removed from service if the fire alarm is activated. An output on the fire alarm control panel (FACP) is connected to the elevator controller. In the event of a fire alarm signal, the elevator controller switches over to a fire service mode and disables the elevator car.

Fire-Lite Alarms
Fire protection systems can be integrated with telephone and public address systems to broadcast safety and evacuation alerts in the event of a fire.

Summary

- The most common modern systems are electronic control systems in commercial buildings.

- Building automation systems improve energy efficiency, system control, and convenience.

- The energy efficiency of a building is improved by controlling the electrical loads in such a way to reduce their use without adversely affecting their purpose.

- Automation results in a much more consistent and predictable control over building systems.

- A pneumatic control system uses compressed air as the medium for sharing control information and powering actuators.

- DDC controllers allow far greater functionality in making optimal control decisions because each includes a microprocessor to quickly and accurately calculate the necessary output signal based on the information from the input signals.

- Building automation system devices can all communicate with each other on the same, shared network.

- One common automation technology for residential buildings is X10 technology.

- New construction commercial buildings now commonly include some level of automation and many existing buildings are being retrofitted with automation systems.

- In addition to the automation of lighting, security, and sanitary water, industrial facilities typically automate manufacturing processes.

- The building owner decides what building automation strategies to implement based on the desired results.

- The consulting-specifying engineers work with the building owners, and also the architect and general contractor, to make changes as needed to the planned automation system to maximize the benefits while working within any aesthetic, financial, or construction constraints.

- The controls contractor submits a quote for the project based on the costs of the purchased equipment, installation labor, system commissioning, and tuning.

- The planning and implementation of a building automation system requires detailed documentation.

- Sensors are input devices, relative to the point of view of the controllers that manage the systems.

- The controller receives a signal from the sensor, compares it to a desired value to be maintained, and sends an appropriate output signal to an actuator.

- The actuator changes some physical characteristic in the system.

- A human-machine interface is connected into a building automation system to allow personnel to view and modify the information being shared between control devices.

- Digital signals convey information as either one of two extremes, such as completely ON or completely OFF.

- Analog signals provide any value between 0% and 100% of some controllable characteristic.

- A structured network message includes information on the sending and receiving devices, the identification of the shared variable, its value, and any other necessary parameters.

- The input information from sensors and output information to actuators creates control points in the building automation system.

- The purpose of the control system is to achieve and maintain certain conditions, which are quantified as setpoints.

- Offset values measure the accuracy of the control system.

- There are many different ways in which control decisions can be made, depending on the inputs and the algorithms used to produce the results.

- In open-loop control, there is no connection between the controller's output and its input.

- In closed-loop control, the result of an output is fed back into a controller as an input.

- Common algorithms include proportional, integral, derivative, and adaptive control algorithms.

- Tuning is the adjustment of relative contributions of each algorithm, and any parameters, to the optimal parameters for the desired control response.

- Building automation may involve any or all of the building systems.

Definitions

- **Building automation** is the control of the energy- and resource-using devices in a building for optimization of building system operations.

- A **pneumatic control system** is a control system in which compressed air is the medium for sharing control information and powering actuators.

- A **direct digital control (DDC) system** is a control system in which electrical signals are used to measure and control system parameters.

- A **building automation system** is a system that uses a distributed system of microprocessor-based controllers to automate any combination of building systems.

- A **consulting-specifying engineer** is a building automation professional that designs the building automation system from the owner's list of desired features.

- The **authority having jurisdiction (AHJ)** is the organization, office, or individual who is responsible for approving the equipment and materials used for building automation installation.

- The **contract specification** is a document describing the desired performance of the purchased components and means and methods of installation.

- A **contract document** is a set of documents produced by the consulting-specifying engineer for use by a contractor to bid a project.

- A **shop drawing** is a document produced by the controls contractor with the details necessary for installation.

- The **project closeout information** is a set of documents produced by the controls contractor for the owner's use while operating the building.

- A **control device** is a building automation device for monitoring or changing system variables, making control decisions, or interfacing with other types of systems.

- A **variable** is some changing characteristic in a system.

- A **sensor** is a device that measures the value of a variable and transmits a signal that conveys this information.

- A *controller* is a device that makes decisions to change some aspect of a system based on sensor information and internal programming.
- An *actuator* is a device that accepts a control signal and causes a mechanical motion.
- A *relay* is an electrical switch that is actuated by a separate electrical circuit.
- A *solenoid* is a device that converts electrical energy into a linear mechanical force.
- An *electric motor* is a device that converts electrical energy into rotating mechanical energy.
- A *human-machine interface (HMI)* is an interface terminal that allows an individual to access and respond to building automation system information.
- A *control signal* is a changing characteristic used to communicate building automation information between control devices.
- A *digital signal* is a signal that has only two possible states.
- An *analog signal* is a signal that has a continuous range of possible values between two points.
- A *protocol* is a set of rules and procedures for the exchange of information between two connected devices.
- A *control point* is a variable in a control system.
- A *virtual point* is a control point that exists only in software.
- A *setpoint* is the desired value to be maintained by a system.
- The *offset* is the difference between the value of a control point and its corresponding setpoint.
- A *deadband* is the range between two setpoints in which no control action takes place.
- *Control logic* is the portion of controller software that produces the necessary outputs based on the inputs.
- A *control loop* is the continuous repetition of the control logic decisions. Control loops are either open loops or closed loops.
- An *open-loop control system* is a control system in which decisions are made based only on the current state of the system and a model of how it should work.
- A *closed-loop control system* is a control system in which the result of an output is fed back into a controller as an input.
- *Proportional control* is a control algorithm in which the output is in direct response to the amount of offset in the system.
- An *integral control algorithm* is a control algorithm in which the output is determined by the sum of the offset over time.
- *Integration* is a function that calculates the amount of offset over time as the area underneath a time-variable curve.
- A *derivative control algorithm* is a control algorithm in which the output is determined by the instantaneous rate of change of a variable.
- *Tuning* is the adjustment of control parameters to the optimal values for the desired control response.
- An *electrical system* is a combination of electrical devices and components connected by conductors that distributes and controls the flow of electricity from its source to a point of use.
- A *lighting system* is a building system that provides artificial light for indoor areas.

- A *heating, ventilating, and air conditioning (HVAC) system* is a building system that controls a building's indoor climate.

- A *plumbing system* is a system of pipes, fittings, and fixtures within a building that conveys a water supply and removes wastewater and waterborne waste.

- A *fire protection system* is a building system for protecting the safety of building occupants during a fire.

- A *security system* is a building system that protects against intruders, theft, and vandalism.

- An *access control system* is a system used to deny those without proper credentials access to a specific building, area, or room.

- A *voice-data-video (VDV) system* is a building system used for the transmission of information.

- An *elevator system* is a conveying system for transporting people and/or materials vertically between floors in a building.

Review Questions

1. Briefly describe the three primary benefits of a building automation system.

2. What is the primary difference between a direct digital control (DDC) system and a building automation system?

3. How do the roles of the consulting-specifying engineer and the controls contractor differ with respect to designing the building automation system?

4. What is the difference between contract specifications and the component portion of the shop drawings?

5. Briefly describe the four primary categories of control devices.

6. What is the difference in the types of information shared between digital signals and analog signals?

7. Why might multiple setpoints be associated with one control point?

8. What is the difference between open-loop control and closed-loop control?

9. What is involved in tuning control loop algorithms for optimal control response?

10. Why are some building systems integrated only through their respective control panels?

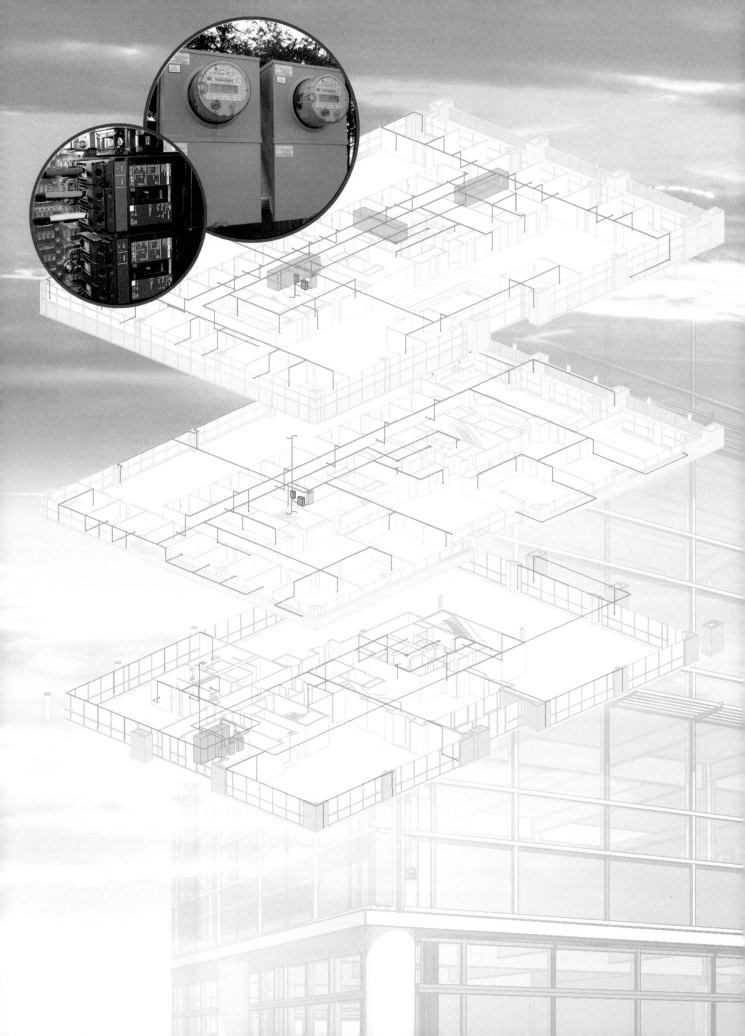

Chapter Two

Electrical System
Control Devices and Applications

Electricity is used for the communication between building automation system devices. Information and instructions are shared through low-voltage analog and digital signals, and control devices are energized or modulated with electrical power. The monitoring and control of the electrical system is required to protect the integrity of automation systems. This ensures the availability of adequate and reliable power, thus making electrical systems central to building automation.

Chapter Objectives

- Identify the typical types of electrical systems in commercial and industrial facilities.
- Differentiate between the various processes of electricity generation and distribution.
- Identify the electrical properties that are typically monitored and/or controlled to optimize the efficiency and reliability of an electrical system.
- Describe common electrical control devices and their functions.
- Compare the methods of switching between available power sources as applications of electrical system controls.

ELECTRICAL SYSTEMS

An *electrical system* is a combination of electrical devices and components, connected by conductors, that distributes and controls the flow of electricity from its source to a point of use. *Electricity* is the energy resulting from the flow of electrons through a conductor. At that point, electricity is converted into some type of useful output, such as motion, light, heat, or sound. **See Figure 2-1.**

Electricity Consumption by Output

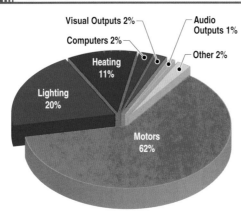

Visual Outputs 2%
Computers 2%
Audio Outputs 1%
Other 2%
Heating 11%
Lighting 20%
Motors 62%

Figure 2-1. The majority of all electricity is consumed by lighting and electric motor loads.

An electrical system includes the distribution equipment and substations that the utility company uses to deliver electricity from power plants to homes and businesses.

Electrical systems encompass the entire electrical infrastructure, from power plants to receptacles. However, it is commonly divided into the sections that are involved with electricity generation, electricity distribution, and the electrical service and related systems within a facility. Building automation systems deal only with the portion of the system within their facility, though the operations are affected by the outside power sources.

Electricity Generation

Most electrical power is generated by converting energy from one form to another until it ultimately becomes useful energy in the form of electricity. Some processes include several steps. For example, coal-fired power plants change chemical energy (energy in the chemical bonds of coal) into heat energy by burning. **See Figure 2-2.** The heat energy is applied to water, which becomes steam, another form of heat energy. The steam is routed to turbines where it moves the blades, which then rotates a shaft, producing mechanical energy. The shaft drives a generator that uses the mechanical energy together with magnetic energy to produce electricity.

Steam is used to convert heat energy into mechanical energy for the generation of electricity, and is used in many types of power plants. Nuclear power plants use nuclear reactions to convert atomic energy into heat energy, which creates electricity by way of steam. Other fossil fuel-powered processes, such as natural gas and oil, also use steam for energy conversion.

Energy sources that begin with mechanical energy, however, do not include steam or heat energy in the conversion cycle. In hydroelectric plants, water pressure and flow drives turbine generators. Turbine generators can also be driven by wind or tidal power, or even the movement of fluids in solar thermal systems.

Direct energy conversion systems produce electrical power without steam or mechanical components. For example, fuel cells use electrochemical processes to convert hydrogen into electrical energy, and photovoltaic (PV) systems utilize semiconductor properties to produce power directly from sunlight.

Primary Power Sources. Most electricity is generated by large power plants that supply millions of consumers. These plants may use a variety of methods to generate electricity, such as coal burning or nuclear reactions, but are all large-scale operations. The power plants are operated by utility companies. A *utility* is a company that generates and/or distributes electricity to consumers in a certain region or state.

Utility-supplied electrical power is very reliable, inexpensive, widely available, and essentially maintenance free. For these reasons, it is the primary, and sometimes only, power source for the vast majority of buildings. The primary power source is the source relied on to provide most or all of the building's power needs. The utility provides a connection to their distribution system, monitors the electricity usage, and bills for the service.

It is possible for buildings to use a nonutility power source as their primary power source, usually when the building's remote location prohibits connection to the utility system. This is extremely rare, however, especially for commercial and industrial facilities.

Secondary Power Sources. In addition to the utility-supplied electrical power, many buildings also have a secondary source for power. This source is intended to supplement the primary electrical service or supply the building with power in the event of an interruption in power from the electric utility. These back-up sources of power come from a variety of owner-supplied sources including engine generators, gas-turbine generators, uninterruptible power supplies (UPSs), and fuel cells.

Secondary power sources are often optional, implemented by the building owner by choice for business or comfort purposes, such as ensuring the continuity of computer systems or production lines. These systems are installed to reduce the potential financial losses caused by a loss of electrical power.

Some secondary power sources are legally required by a governing entity for life safety purposes. These systems provide illumination and power to critical electrical loads, such as egress lighting and fire pumps, in the event of an outage of the primary power source.

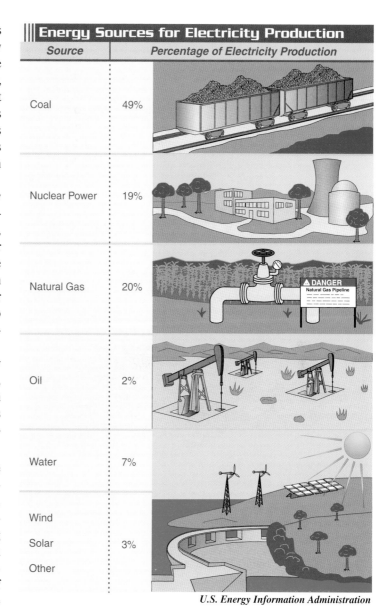

| Energy Sources for Electricity Production ||
Source	Percentage of Electricity Production
Coal	49%
Nuclear Power	19%
Natural Gas	20%
Oil	2%
Water	7%
Wind	
Solar	3%
Other	

U.S. Energy Information Administration

Figure 2-2. Electricity can be produced from a variety of energy sources, though most processes include several intermediate steps.

Electricity Distribution

Electricity distribution is the transmission and delivery of electricity outside of consumers' facilities. For utility-supplied electricity, primary power generation is centralized to take advantage of economies of scale. Electricity is then distributed over a large area in a large-scale supply network. Secondary power sources, however, are typically located near consumer facilities and require very little distribution infrastructure.

Secondary Power Sources

Many sources of electricity can be integrated with building electrical systems and equipment as supplemental or back-up power.

Engine Generators

An engine generator is a combination of an internal-combustion piston engine and a generator mounted together to produce electricity. The engine burns a compressed fuel-air mixture to provide the mechanical power, and the generator converts mechanical energy into electricity by means of electromagnetic induction.

The engines are typically fueled with gasoline, diesel, propane, or natural gas. Small engine generators typically run on gasoline because of its widespread availability. Larger engine generators can be designed for any of these fuels, though diesel is the most common. The type of fuel used affects the load response of the engine generator. Diesel engines are well suited for constant loads, whereas gasoline engines can respond quickly to changing loads. However, because diesel engines have better partial-load efficiency, run slower, are more robust, last longer, and require less maintenance, they are predominately used for large stationary applications.

Gas-Turbine Generators

A gas-turbine generator compresses and burns a fuel-air mixture, which expands and spins a turbine, converting fluid flow into rotating mechanical energy. The shaft power from the turbine is then used to drive a generator to produce electricity. Gas-turbine generators are essentially aircraft-type jet engines developed for stationary applications.

Gas-turbine generators have a better power-to-weight ratio than internal combustion engines, which means that they can generate more power from smaller, lighter equipment. The main disadvantages of gas-turbine generators are high cost and complexity. They typically run on natural gas, liquefied petroleum gas, diesel, or kerosene fuels. Gas-turbine generators are best suited for constant loads.

Photovoltaic Systems

A photovoltaic (PV) system produces electricity directly from the energy of sunlight. Individual PV cells are made from ultrathin layers of semiconductor materials. When exposed to light, one layer releases electrons and the other absorbs electrons. The flow of electrons from one side of the cell to the other can be harnessed as DC electricity. The electrical output of each cell is very small, so they are grouped together to form larger electricity-producing units called modules. Modules are grouped together to form arrays.

PV system arrays can be designed for practically any desired electricity output. These systems produce no waste products, require little maintenance, and include no moving parts. However, since sunlight alone is not a reliable energy source, PV systems either require storage batteries or can only be used for supplemental power.

Wind Turbines

A wind turbine harnesses wind power to produce electricity. The wind rotates blades, and the resulting mechanical energy drives a generator to produce electricity. Wind turbines are viable in areas with an average wind speed greater than 10 mph. Small wind turbines range from 1 kW to 20 kW peak output. Larger commercial wind turbines can produce several hundred kilowatts or even a few megawatts with rotor diameters up to a few hundred feet.

Microhydroelectric Turbines

A microhydroelectric turbine produces electricity from the flow and pressure of water derived from streams and rivers. Moving water acts against turbine blades, rotating a shaft that operates an electrical generator. The principles are the same as for large-scale hydroelectric plants, such as the Hoover Dam, but on a smaller scale. Microhydroelectric systems output less than 100 kW.

Fuel Cells

A fuel cell is an electrochemical device that uses hydrogen and oxygen to produce DC electricity, with water and heat as byproducts. Although similar to batteries, fuel cells are different in that they require a continual replenishment of the reactants (hydrogen and oxygen).

A typical fuel cell element consists of a cathode and anode separated by an electrolytic membrane material. As hydrogen gas flows across the anode, electrons are stripped from the hydrogen and flow through an external circuit, reentering the fuel cell at the cathode. At the same time, positively charged hydrogen ions migrate across the membrane to the cathode, where they combine with oxygen and the returning electrons to form water and heat.

Centralized Generation. *Centralized generation* is an electrical distribution system in which electricity is distributed through a utility grid from a central generating station to millions of customers. By far, most electricity is generated and distributed in this way. The *grid* is the utility's network of conductors, substations, and equipment that distributes electricity from a central generation point to the consumers. **See Figure 2-3.** The grid fans out from the power plants to thousands of homes and businesses within a region. Electricity may travel hundreds of miles before it reaches the end user. Grids may be connected together at certain points so that consumers may still have power if an outage occurs in part of the distribution system. Outages, though rare, do still occur, often due to storm damage or an overloaded system.

Distributed Generation. *Distributed generation* is an electrical distribution system in which many smaller power-generating systems create electrical power near the point of consumption. The electricity may travel only a few feet to the loads. Distributed generation systems include engine generators, PV systems, wind turbines, fuel cells, or other relatively small-scale power systems. **See Figure 2-4.** A distributed generation system may be either a primary or secondary power source, though secondary is by far more common.

Per ANSI/IEEE Standard 1547, *Standard for Interconnection of Distributed Resources with Electrical Power Systems,* there are specific safety and technical requirements for all types of distributed power sources of less than 10 MW.

‖‖‖ Centralized Generation Systems

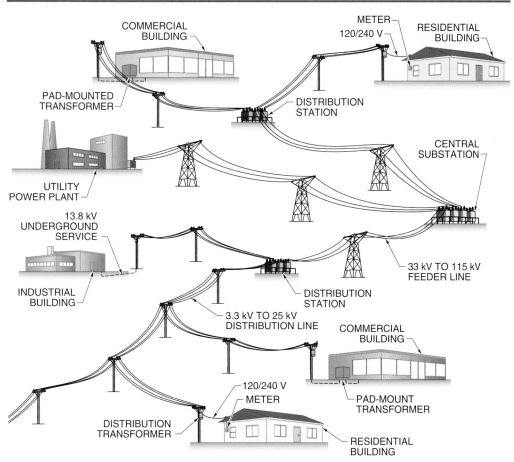

Figure 2-3. Centralized generation systems rely on a large power plant to produce the electricity for many consumers on an interconnected power grid.

Distributed Generation Systems

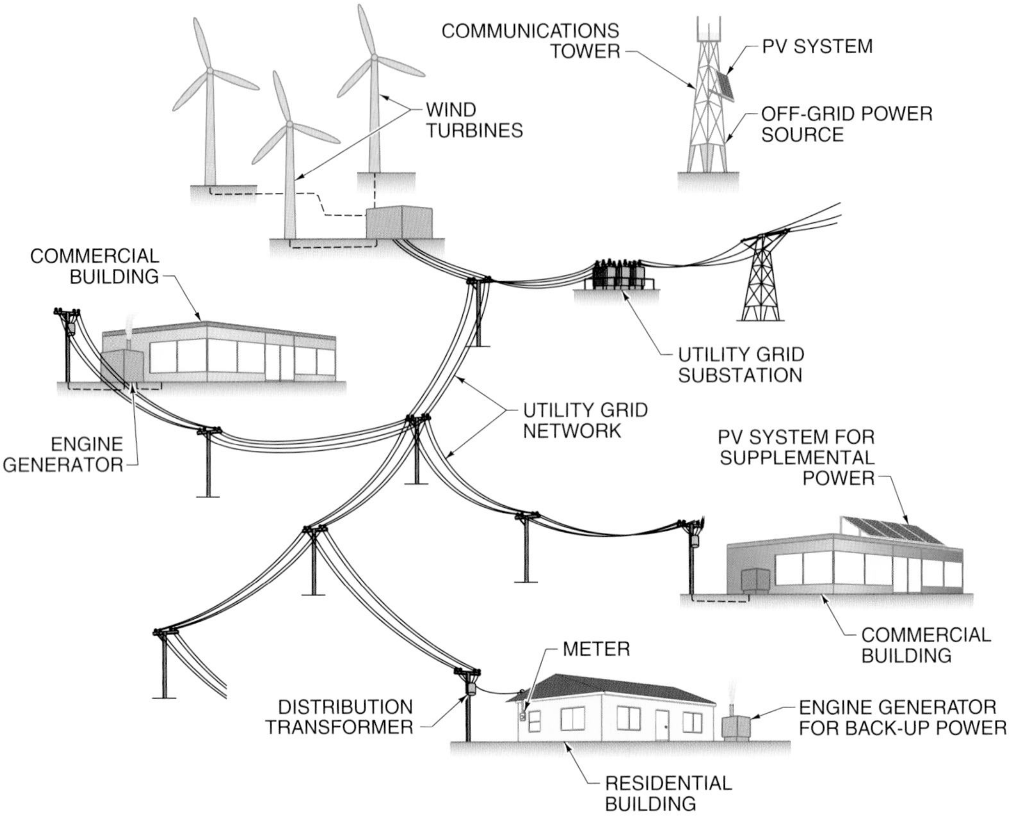

Figure 2-4. Distributed generation systems include many independent power sources to supply electricity close to where it is needed.

If the facility is connected to the utility grid, excess power from the secondary power source can be exported to the grid when it is not needed by the on-site loads. This makes the facility similar to a power plant in a centralized generation system. By metering the amount of exported electricity, in addition to the amount of consumed electricity, the facility receives credit from the utility for the excess power added to the grid. This arrangement can reduce costs for the building owner and increase the utility's capacity to serve customers without building new power plants.

Net metering uses one meter that can measure in either direction, effectively subtracting exported electricity from imported electricity. This determines the net electricity consumption of the utility customer.

Electrical Service

The *electrical service* is the electrical power supply to a building or structure. **See Figure 2-5.** A smaller-scale power distribution system within a building then delivers power from the electrical service to end-use points throughout the building, where it powers individual loads such as motors, lamps, and computers. This system includes switchboards, transformers, panelboards, switches, and receptacles.

Electrical systems must be designed, installed, and maintained in accordance with the National Electrical Code® (NEC®) and/or other applicable codes and regulations adopted by the local authority having jurisdiction (AHJ). All electrical devices must also be rated for both the voltage and current of the application for which they are installed.

Electrical Service

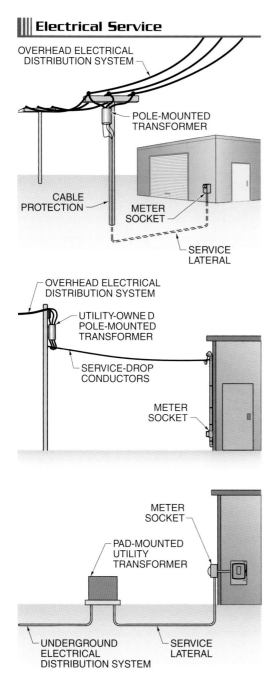

Figure 2-5. The electrical service is the point of connection between the electrical utility company and the electrical system of a building or structure.

Switchboards. A large block of electric power is delivered from a utility substation to a building at a switchboard, where it is broken down into smaller blocks for distribution throughout a building. **See Figure 2-6.** A *switchboard* is the last point on the power distribution system for the power company and the beginning of the power distribution system for the property owner's electrician. The switchboard contains overcurrent protection devices and switches to control the flow of electricity into the building. Then it is up to the building's electrical system to further distribute the power to the end-use points within the facility. **See Figure 2-7.**

Switchboards

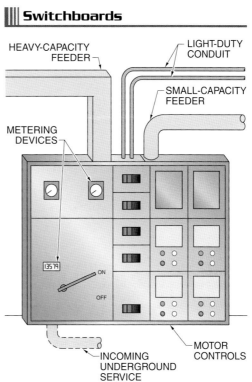

Figure 2-6. As the first component in a building's electrical system, a switchboard receives electricity from the utility and divides it into smaller feeds to other distribution devices.

In addition to distributing the incoming power, a switchboard may contain equipment needed for controlling, monitoring, protecting, and recording the electrical use in a building. For example, the addition of motor starters and controls to the switchboard allows motors to be connected directly to the switchboard. This combination allows these high-current loads to be connected to the source of power without further power distribution.

Building Electricity Distribution System

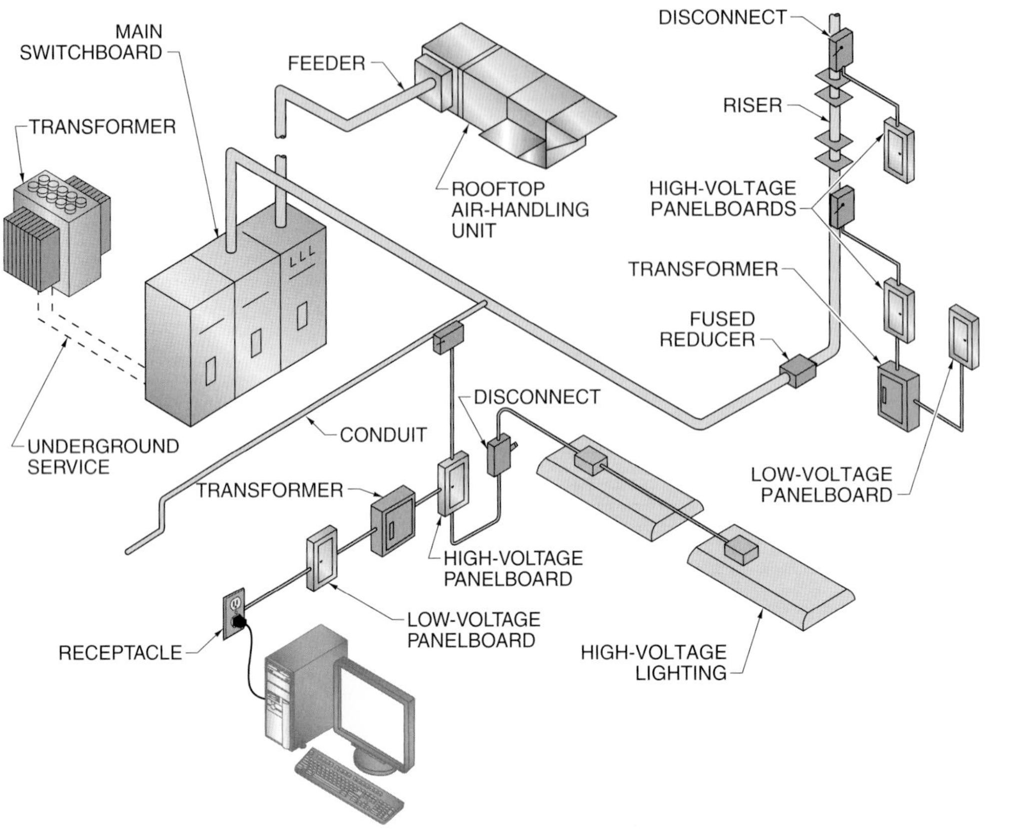

Figure 2-7. The building electricity distribution system includes several parts to control and distribute electrical power among the loads.

Electricity is distributed out from the switchboard in multiple feeders. A *feeder* is the circuit conductors between a building's electrical supply source, such as a switchboard, and the final branch-circuit overcurrent device.

Transformers. A *transformer* is an electric device that uses electromagnetism to change AC voltage or electrically isolate two circuits. A transformer is composed of two windings of conductors around an iron core. The primary winding draws current from the power source, which induces a magnetic field through the iron core. This causes an electric current to flow in the secondary winding, which delivers the power to the load. **See Figure 2-8.**

The ratio of input voltage to output voltage is proportional to the ratio of the number of turns in the two windings. Because the amount of electrical power transferred from one side to the other is nearly constant

(there are only small losses from generated heat), the current is inversely proportional. For example, in a 10:1 transformer, if 120 V at 1 A is applied to the primary side, the secondary side will be 12 V at nearly 10 A. The power on both sides of the transformer is approximately 120 W.

Depending on which side of the transformer is treated as the primary or secondary side, a transformer can either step up or step down the input voltage. Utilities use transformers to distribute large amounts of power over long distances efficiently by raising the voltage. Increasing the voltage reduces the current correspondingly, which reduces losses from voltage drop and allows power distribution through smaller gauge wire. Then, additional transformers throughout the grid lower the voltages down to levels usable by loads.

||||| Transformers

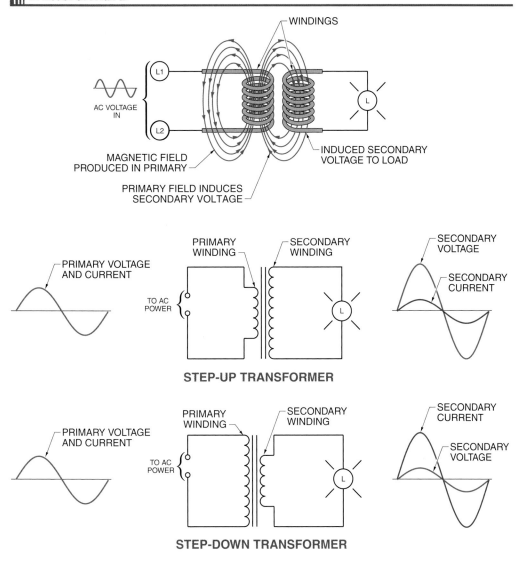

Figure 2-8. Transformers use the magnetic field created by one winding to induce a proportional voltage in another winding.

Transformers are often used similarly within a building's power distribution system. **See Figure 2-9.** High voltage (usually 480 V) is distributed throughout the building, and localized transformers lower the voltage to the 240 V, 208 V, or 120 V levels required by most loads.

Panelboards. A switchboard typically supplies power to multiple panelboards located throughout the building, which divides the power distribution system into smaller units. A *panelboard* is a wall-mounted power distribution cabinet containing overcurrent protective

devices for lighting, appliance, or power distribution branch circuits. **See Figure 2-10.** A *branch circuit* is the circuit in a power distribution system between the final overcurrent protective device and the associated end-use points, such as receptacles or loads.

Although both the voltage and current can be stepped up or down, the terms "step up" and "step down," when used with transformers, always apply to voltage.

Transformer Applications

Fluke Corporation

UTILITY TRANSFORMER   **BUILDING TRANSFORMER**

Figure 2-9. Transformers of different sizes and ratios are used extensively in electricity distribution, both inside and outside of buildings.

Panelboards

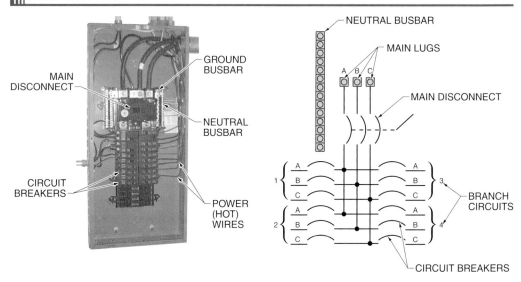

Figure 2-10. Panelboards include overcurrent protection and switching devices for branch circuits.

Switches. The loads on branch circuits are controlled primarily by switches that energize or de-energize the entire circuit. A *switch* is a device that isolates an electrical circuit from a power source. Many switches are operated manually, with a hand-operated lever that opens or closes electrical contacts inside the switch and visually indicates the energized state of the circuit. There are also switches that are operated electrically or operated automatically by changes in their environment, such as temperature or pressure.

Switches are used throughout electrical systems. Common variations of switch types include the method of actuation, voltage and current rating, number and configuration of contacts, and other features. For example, a disconnect is a type of switch commonly used for high-power loads or large circuits. **See Figure 2-11.** Disconnects may also incorporate overcurrent protective devices.

Disconnects

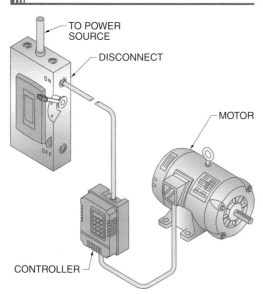

Figure 2-11. A disconnect is used to manually switch electrical power to large loads.

Outlets. Devices carry or control electricity, but do not use it. Devices include outlets, which are installed at convenient access points on the electrical system. An *outlet* is an end-use point in the power distribution system. Outlets allow for the connection of devices such as receptacles, switches, and loads to the circuit. **See Figure 2-12.**

A *receptacle* is an outlet for the temporary connection of corded electrical equipment. Receptacles are available in a variety of current and voltage ratings, connector types, conductor configurations, and colors. Most installed receptacles are of a standard type, though there are many special-use receptacles for certain applications, such as isolated-ground receptacles, hospital-grade receptacles, twist-lock receptacles, and ground-fault circuit interrupter receptacles.

Outlets

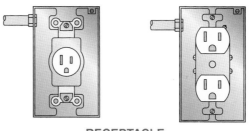

RECEPTACLE

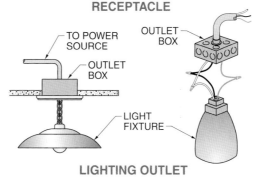

LIGHTING OUTLET

Figure 2-12. Outlets include receptacles and outlet boxes, which allow for the connection of individual loads to the building electrical system.

Grounding System

Grounding is the intentional connection of all exposed noncurrent-carrying metal parts to the earth. Grounding provides a direct path to the earth for unwanted fault current to dissipate without causing harm to persons or equipment. Properly grounded electrical circuits, tools, motors, and enclosures help safeguard equipment and personnel against the hazards of electrical shock.

Most electrical components are bonded to the grounding system with equipment grounding conductors. An *equipment grounding conductor (EGC)* is a conductor that provides a low-impedance path from electrical equipment and enclosures to the grounding system. **See Figure 2-13.** Grounding is then established by connecting the grounding system at the main service equipment to a metal electrode that is in good contact with the ground. A *grounding electrode conductor (GEC)* is a conductor that connects the grounding system to the buried grounding electrode. The total grounding path from any point in the electrical system must be as short as possible, sufficient in conductor size, permanently installed, and uninterrupted from the electrical circuit to the ground. Appropriate grounding methods are specified in the NEC®.

A *grounded conductor* is a current-carrying conductor that has been intentionally grounded. The neutral-to-ground connection is made by connecting the neutral bus to the ground bus at the main service panel with a main bonding jumper. A *main bonding jumper (MBJ)* is a connection at the service equipment that connects the equipment grounding conductor, the grounding electrode conductor, and the grounded conductor (neutral conductor).

▐▐▐ Grounding System

Figure 2-13. *The grounding system consists of several parts that are connected together to provide a low-impedance path to ground for unwanted fault currents.*

Circuit breakers and other overcurrent protective devices open circuits to prevent damage at high temperatures due to an overcurrent condition.

Overcurrent Protection

Electrical equipment must be protected from excessive current flow. An *overcurrent protective device* is a device that prevents conductors or devices from reaching excessively high temperatures from high currents by opening the circuit. High temperatures can damage components and conductor insulation, causing electrical shock and fire hazards. An overcurrent condition can be the result of an overload, ground fault, or short circuit.

Overcurrent protective devices include fuses and circuit breakers that are usually located in panelboards. **See Figure 2-14.** A *fuse* is an overcurrent protective device with a fusible link that melts and opens the circuit when an overcurrent condition occurs. A *circuit breaker* is an overcurrent protective device with a mechanism that automatically opens a switch in the circuit when an overcurrent condition occurs. Two advantages of circuit breakers are that they are resettable after an overcurrent trip and that, in certain circumstances, they can be used as disconnects for installation and maintenance.

CONTROL AND MONITORING OF ELECTRICITY

Electricity is a natural phenomenon that has been developed as a technology to easily store and distribute energy that has potential to do useful work when utilized by electrical appliances. Electricity is monitored and controlled to make the most efficient use of this energy.

Overcurrent Protective Devices

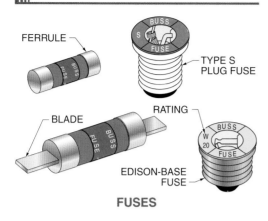

FERRULE

TYPE S PLUG FUSE

BLADE

RATING

EDISON-BASE FUSE

FUSES

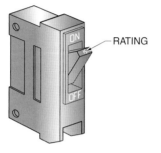

RATING

CIRCUIT BREAKERS

Figure 2-14. Fuses and circuit breakers protect electrical systems from overcurrent conditions that cause equipment damage and electrical shock and fire hazards.

Electrical Switching

The switching of electricity is the alternation between energized and de-energized states by opening or closing ungrounded (hot) conductor(s) in a circuit. *Switching* is the complete interruption or resumption of electrical power to a device. Switching can turn individual electrical appliances or entire circuits ON or OFF. Switching can also be used to alternate between separate power supplies, which can be an important part of managing a reliable electrical system for critical loads.

Switching is accomplished with sets of contacts that make or break the electrical continuity of a circuit. A switch can make or break multiple contacts simultaneously, so there are many possible physical arrangements of contacts. **See Figure 2-15.** The symbols for

contact types are similar to these arrangements. The terms "pole" and "throw" are used to describe these arrangements. **See Appendix.** A *pole* is a set of contacts that belong to a single circuit. A *throw* is a position that a switch can adopt. Contacts may also be either single-break (opening the circuit in one place) or double-break (opening the circuit in two places). The most common types of switches are single-pole, single-throw types.

Switch description abbreviations refer to these terms in a shorthand style. For example, a SPDT-SB switch has a single-pole, double-throw, single-break set of contacts. Some contacts may be normally open (NO) or normally closed (NC) so that when not actuated, they return to an open or closed position, respectively.

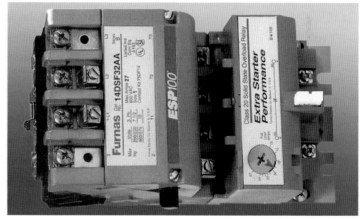

Furnas Electric Co.

Switching devices contain contacts that open and close to break and make electrical connections.

Contacts are two pieces of metal that touch to make a circuit and separate to break the circuit. They are typically made from a material that resists corrosion because most metals corrode into insulating oxides that increase electrical resistance, preventing the switch from working. Other important characteristics of contacts are conductivity, wear resistance, mechanical strength, low cost, and low toxicity.

Electrical Demand

Consumers commonly use more electricity during some periods of the day than others. *Electrical demand* is the amount of electrical power drawn by loads at a specific moment. Electrical demand is measured in kilowatts (kW).

High electrical demands make it more difficult for electric utilities to supply power to all of their customers, so the utilities sometimes change their rates based on demand. Electricity during high-demand periods is more valuable; therefore, the rate is often higher. An electric bill for a commercial or industrial building may reflect changing rates for electricity during the month or include extra charges for high electrical demand. In some cases, the bill for an entire month or an entire year may be based on the rate for the highest demand during that period.

Facilities can reduce demand by selectively shutting down or reducing the use of noncritical loads. This strategy is used to reduce utility costs and to help prevent power outages and brownouts.

Electrical demand is an instantaneous value, analogous to speed. However, for billing purposes, utilities typically average the demand over a time interval of 15 min or 30 min. **See Figure 2-16.** For example, over a 15 min period, the average demand may be 40 kW, though the actual demand ranges from 30 kW to 50 kW. Utilities monitor the demand from commercial and industrial facilities and charge for the demand during a certain time interval each monthly billing period. Methods used to determine the time interval can vary.

The fixed interval method uses a set period to determine electrical demand. For example, electrical demand may always be monitored from 10:00 AM to 10:15 AM. Some utilities send signals on the incoming power service lines that indicate the beginning of each new period.

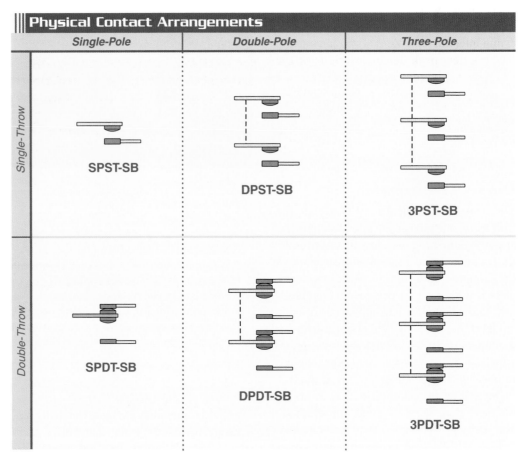

Figure 2-15. Electricity switching is the opening or closing of ungrounded (hot) conductors in one or more circuits with an arrangement of contacts.

Electrical Demand

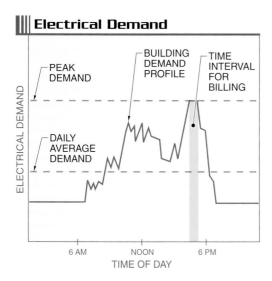

Figure 2-16. Utility billing for electrical demand throughout the billing period is sometimes based on the highest demand interval rather than on daily average demand.

The sliding window method uses the interval with the highest demand. Any 15 min or 30 min interval, regardless of when it occurs, can be a new peak demand period for the month. For example, a sliding window can be from 10:01 AM to 10:16 AM or 10:02 AM to 10:17 AM. The interval may change each month.

Electrical demand over time results in electricity consumption. *Electricity consumption* is the total amount of electricity used during a billing period. Electricity consumption is measured in kilowatt-hours (kWh) and is calculated by multiplying the electrical demand by the amount of time at that rate. For example, if the daily average demand in a building is 40 kW, then over a 24 hr period, the loads consume 960 kWh (40 kW × 24 hr = 960 kWh).

Electricity consumption is metered for the purposes of determining the amount of electricity delivered to (or from) a customer's facility for billing purposes. Meters are installed at the service entrance and establish the transition between utility and customer-owned equipment. Both electrical demand and electrical consumption are important for supplying commercial and industrial facilities.

Power Quality

Power quality is the measure of how closely the power in an electrical system matches the nominal (ideal) characteristics. It is common for actual electrical parameters to vary somewhat, but allowable ranges are typically very small. Good power quality means that the parameters are within acceptable limits for the electrical system. Poor power quality has excessive variations in the parameters that can cause damage to loads and circuit equipment. Power quality is influenced by the performance of the electrical generation and distribution equipment, as well as electrical loads operating on the system.

Power quality can involve many characteristics of the electrical supply, such as voltage, frequency, harmonic distortion, noise, power factor, and unbalanced conditions in 3ϕ power supplies. These can all be monitored and conditioned if improvement is necessary. The most common power quality parameters that are controlled by automated systems, though, are voltage, current, and frequency.

Voltage. *Voltage* is the difference in electrical potential between two points in an electrical circuit. Voltage is the electrical pressure that causes current to flow when the two points are connected by a conductor. Common nominal voltage levels for electrical systems within buildings include 120 V, 208 V, 240 V, 277 V, and 480 V.

Voltages are typically acceptable within the range of +5% to –10% from the nominal voltage. Most electrical devices tolerate some degree of voltage fluctuation, but excessive deviations can cause circuit and load problems. **See Figure 2-17.** Computer equipment is particularly sensitive. Noise or voltage fluctuations can disrupt computers, causing software errors, data loss, and communication failures.

Voltage sags are commonly caused by overloaded transformers, undersized conductors, conductor runs that are too long, too many loads on a circuit, peak power usage periods (brownouts), and high-current loads being turned on. Voltage sags are often followed by voltage swells as voltage regulators overcompensate.

Electrical Demand

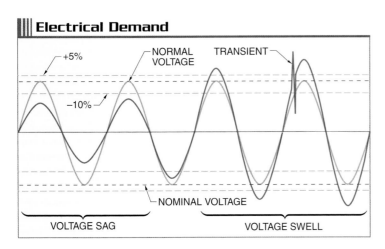

Figure 2-17. Voltage variations outside of the allowable range include voltage sags, voltage swells, and transients.

Voltage swells are caused by loads near the beginning of a power distribution system, incorrectly wired transformer taps, and large loads being turned off. Voltage swells are not as common as voltage sags, but are more damaging to electrical equipment.

A *transient voltage* is a temporary, undesirable voltage in an electrical circuit, ranging from a few volts to several thousand volts and lasting from a few microseconds up to a few milliseconds. Transient voltages are caused by lightning strikes, unfiltered electrical equipment, contact bounce, arcing, and high-current loads being switched ON and OFF. Transient voltages differ from voltage sags and surges by being larger in amplitude, shorter in duration, steeper in rise time, and erratic. High-voltage transients can permanently damage circuits or electrical equipment.

Frequency. *Frequency* is the number of AC waveforms per interval of time. Frequency is measured in hertz (Hz) and is equivalent to cycles per second. The frequency of AC power in the United States is 60 Hz, though it is 50 Hz in many other parts of the world. These frequencies are low enough for efficient electricity transmission, but high enough that the resulting flicker of incandescent lamps is not noticeable.

Frequency of the utility power supply system will vary as load or generating capacity changes. Overloading the system causes the frequency to decrease, while too much generating capacity causes the frequency to increase. Utility generators are constantly adjusted in speed so that the system frequency remains nearly constant. The frequency is so tightly regulated that it typically varies by less than ±1%. This tight tolerance is necessary to be able to synchronize the many generators supplying power to the grid. It also helps maintain accuracy for clocks and motors relying on frequency for their speed.

Electrical systems monitoring for frequency changes may automatically initiate load shedding (demand limiting) when the frequency falls in order to preserve proper functioning for critical systems. Since line frequency determines the speed of AC electric motors, frequency can also be manipulated on individual circuits within a system to control their speeds.

ELECTRICAL SYSTEM CONTROL DEVICES

Electrical system control devices are primarily concerned with monitoring and maintaining a reliable and quality power supply. If necessary, the control devices can either condition the power to improve its quality, or switch the electrical system over to a secondary power source. This switching ability can also be used to respond to high utility rates or requests to shed loads due to high electrical demand.

Electrical Parameter Sensors

Electrical parameter sensors monitor various aspects of electrical demand and power quality. Depending on the measurement, the sensors may use digital (contact closure) signal, analog signals, or structured network messages to share information. For example, if an application requires only verification of voltage, a voltage sensor with a digital output can be used. However, if the exact voltage level is important, an analog voltage sensor can be used so that the voltage level can be monitored at all times.

Demand Meters. Electrical demand and electricity consumption are measured with demand (watt-hour) meters. Electromechanical induction demand meters are conceptually similar to motors. Electricity passing through coils creates a magnetic field, causing a metal disk to rotate. Each revolution accounts for a certain amount of energy transfer, and the rate of disk rotation is proportional to the amount of electrical power passing through the meter. The number of disk revolutions is counted mechanically or electronically, and the corresponding energy use is displayed on a register in digits or dials, or shared with other devices via electronic signals.

Today, many electromechanical demand meters are being replaced by electronic meters using current and voltage transformers and microprocessors to measure, process, and record data. **See Figure 2-18.** Some can record other electrical service information, including peak power demand, power factor, reactive energy, time-of-use consumption, or, in the case of distributed generation, exported energy. Many electronic meters allow the meter data to be read remotely by infrared, radio frequency, telephone, network, or power-line carrier signals.

Electronic Demand Meters

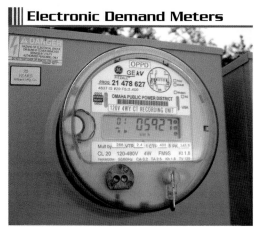

Figure 2-18. Electronic demand meters are capable of recording many different power supply parameters and allow remote access to the data.

One type of electronic demand meter is a pulse meter. A *pulse meter* is a meter that outputs a pulse for every predetermined amount of flow in the circuit or pipe. The pulse can be either dry contact closures or a short application

of voltage (typically 5 VDC). Flow is measured at a register by counting the pulses. A flow unit per pulse is set at the factory. The accumulated number of pulses represents the electricity consumption, and the pulse rate represents the electrical demand. For example, if each pulse corresponds to 1 kWh, 40 pulses means that 40 kWh of electricity were consumed. If the 40 pulses accumulated within an hour, then the average electrical demand during that hour was 40 kW (40 kWh ÷ 1 hr = 40 kW). The use of pulse meters is very common in submetering and demand response applications. Some electric pulse meters can output pulses for both watts and volt-amperes reactive (VAR), enabling the power factor to be calculated and monitored. Meters using a similar pulse-counting method are also used for counting units of fluid flow, such as in water meters.

Demand meters are used by the utility to record the electrical demand and electricity consumption for an entire building or facility. However, some facilities may use similar meters to monitor electricity usage for electrical subsystems, such as certain building areas or production lines, for maintenance or accounting purposes. Additional demand meters can be installed temporarily or permanently throughout the building to provide this data to a building automation system.

Power Quality Sensors. A variety of sensors can be used to monitor the power quality of an electrical power source. These include sensors to measure voltage, current, frequency, harmonics, noise, and unbalance between 3ϕ power supplies. **See Figure 2-19.** Many of these sensors are similar to hand-held and bench-top test instruments, except that instead of displaying the measured value, they output this information via electrical signals to other devices. They also may be permanently installed in the circuit to measure the value continuously.

Most equipment will only operate properly at the frequency for which it was designed. Therefore, most frequency sensors have digital (ON/OFF) outputs because it is not necessary to know the exact frequency value, only if the supplied power is or is not at the correct frequency.

Current Transformers

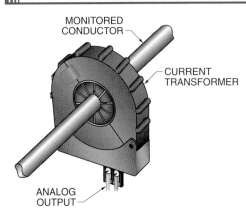

Figure 2-19. Current transformers are electrical sensors for measuring the current flow through conductors.

Relays are packaged in standard sizes and shapes and with standard connection pins so that they can be easily replaced.

Electromechanical relays use the magnetic force generated by a current in a coil to physically move a set of contacts. Solid-state relays can provide the same types of switching but with no moving parts. This means they typically last far longer.

Relays

Electrical circuits can be automatically switched with relays. A *relay* is an electrical switch that is actuated by a separate electrical circuit. The load in the control circuit is the relay coil, which is a winding of a conductor. When energized, the coil becomes an electromagnet, causing an armature to move, switching the state of a set of contacts. **See Figure 2-20.**

The control circuit commonly operates at voltages of 24 V or less, with a current draw of less than 1 A. The load circuit can be hundreds of volts and many tens of amperes or more. The two circuits are electrically separate, so they can also be any combination of AC and DC circuits. This flexibility allows a relay to control a variety of load circuits, such as motors, lights, or heating elements, with a low-voltage control circuit.

Relays can include any arrangement of switch contacts. A common type is a single-pole, double-throw arrangement that allows for both a normally open (NO) and normally closed (NC) set of contacts, sharing a common (COM) third terminal. When de-energized, the NC and COM set is closed, and the NO and COM set is open. When energized, the NO and COM set is closed, and the NC and COM set is open.

Once energized, relays may hold the contacts in position either electrically or mechanically. **See Figure 2-21.** Electrically held relays require a constant application of the control voltage in order to remain in the energized position. This consumes an appreciable amount of power and creates heat, requiring large heat sinks. Mechanically held relays include a latching mechanism that holds the contacts in either position, without constant voltage. Momentary applications of voltage can either open or close the contacts, making mechanically held relays more energy efficient.

General-purpose relays are available for many combinations of voltage types, voltage levels, and contact arrangements. Many are designed for the circuit requirements of certain applications, such as controlling motors or lighting circuits. These may be known by different names, but are really just specialized types of relays. A *contactor* is a heavy-duty relay for switching circuits with high-power loads. Contactors use contacts made from pure silver, which remains a good electrical conductor even after oxidizing from the arcing of switching high currents.

Relays

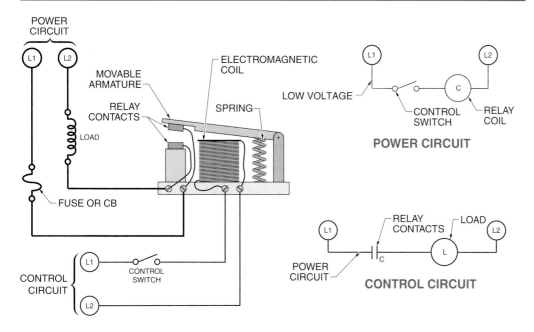

Figure 2-20. A relay allows a low-voltage control circuit to energize or de-energize loads on a higher-voltage power circuit.

Relay Circuits

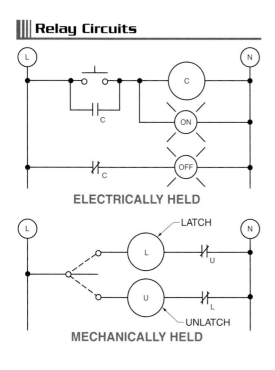

Figure 2-21. Relays may be held in the energized position either electrically or mechanically.

A *magnetic motor starter* is a specialized contactor for switching electrical power to a motor and includes overload protection. Magnetic motor starters commonly control 3ϕ power to electric motors. **See Figure 2-22.** Each output terminal contains an overload contact operated by a heater unit that responds to the heat generated by current flow. Excessive current in a conductor for more than a few minutes causes the heater to trip (open) the NC overload contact in the control circuit, de-energizing the coil and opening the load circuit contacts.

Variable-Frequency Drives

The frequency of a power supply is changed in order to increase or decrease the speed of AC electric motors. A *variable-frequency drive* is a motor controller that is used to change the speed of AC motors by changing the frequency of the supply voltage. In addition to controlling motor speed, variable-frequency drives can control motor acceleration time, deceleration time, motor torque, and motor braking.

Magnetic Motor Starters

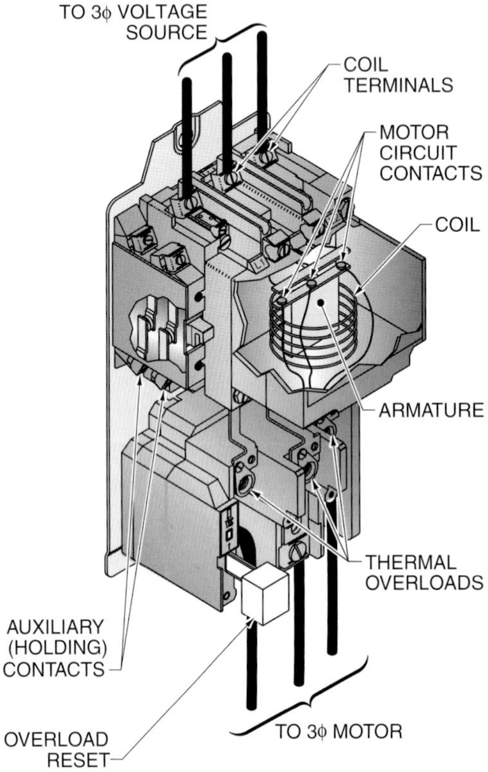

TO 3φ VOLTAGE
SOURCE

COIL
TERMINALS

MOTOR
CIRCUIT
CONTACTS

COIL

ARMATURE

THERMAL
OVERLOADS

AUXILIARY
(HOLDING)
CONTACTS

OVERLOAD
RESET

TO 3φ MOTOR

Figure 2-22. A magnetic motor starter is a specialized contactor that directly energizes or de-energizes an electric motor.

Variable-Frequency Drive Output

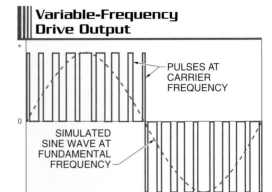

PULSES AT
CARRIER
FREQUENCY

SIMULATED
SINE WAVE AT
FUNDAMENTAL
FREQUENCY

Figure 2-23. Variable-frequency drives simulate fundamental frequencies with pulses at a carrier frequency.

A motor drive changes the frequency of the voltage applied to a motor by converting the incoming AC voltage to a DC voltage, and inverting it back to an AC voltage that simulates the desired fundamental frequency. **See Figure 2-23.** The *fundamental frequency* is the voltage frequency simulated by the changing pulse widths of the carrier frequency. Fundamental frequencies typically range from nearly 0 Hz to over 60 Hz and directly determine the motor speed.

Variable-frequency drives can accept analog signals, commonly 0 VDC to 10 VDC, that are proportional to the desired motor speed. The *carrier frequency* is the frequency of the ON/OFF voltage pulses that simulate the fundamental frequency. The carrier frequencies of most variable-frequency drives range from 1 kHz to about 20 kHz.

Carrier frequency of a variable-frequency drive can be changed to meet particular load requirements. Higher carrier frequencies reduce heat in the motor because the voltage more closely simulates a pure sine wave. This is because higher carrier frequencies have more individual pulses to reproduce each cycle of the fundamental frequency. Higher carrier frequencies also reduce audible noise because the 1 kHz to 2 kHz range is within the range of human hearing. However, the resulting fast switching of the inverter section of the variable-frequency drive produces large voltage spikes that can damage motor insulation over time.

Transfer Switches

A *transfer switch* is a switch that allows an electrical system to be switched between two power sources. This is most commonly implemented in order to keep critical loads operating during utility outages with a secondary power source, such as an emergency engine generator.

Transfer switches are manual or automatic. With a manual transfer switch, the transfer to secondary power is done by manually actuating a switch. **See Figure 2-24.** Automatic transfer switches (ATSs) do not require manual intervention. These switches continually monitor the primary power source and, when the voltage falls below a certain level for a predetermined amount of time, the switch automatically initiates a power transfer operation.

▥ Transfer Switches

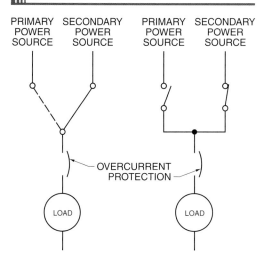

Figure 2-24. Transfer switches are configured to transition from one power source to another.

The transfer operation may include more than just switching power sources. Some secondary power sources, such as engine generators, require time to complete a start-up sequence before they can be loaded. When a power transfer is needed, the automatic transfer switch sends an electronic start-up command to the secondary power source and monitors the resulting voltage and frequency. Only when the secondary power is stable does it complete the transfer by switching the connections. The transfer switch also manages the transfer back to primary power when appropriate.

Automatic transfer switches deal with the resulting transition period between power sources in different ways. In an open-transition transfer (OTT), one power source must be completely disconnected from the electrical system before the other power source is connected. This is known as a "break-before-make" transfer, which prevents the power from back-feeding into the primary system but results in a short power interruption. Depending on the system, the interruption may last from a fraction of a second to several seconds. Sometimes the system's critical loads are supported by power from a UPS battery during the transfer.

In a closed-transition transfer (CTT), however, the transition is smoothed by slightly overlapping the connection of two power sources. This is a "make-before-break" transition. The transfer switch monitors the frequency and phase of each power source. When they are synchronized, the second power source is connected and then the first is disconnected. There is a brief period, typically less than a second, when they are connected in parallel and both are supplying power to the loads. Closed-transition transfer switches used with engine generators require isochronous governors, which keep the generator output frequency constant under any load conditions.

A paralleling load transfer is very similar to a closed-transition transfer, but it parallels the power sources for a longer period, allowing a smoother transfer of loads from one source to the other. These paralleling load transfer systems require load sharing and continuous synchronization equipment and are most often used in large applications paralleling multiple engine generators for "peak shaving" or cogeneration with the utility.

Distributed Generator Interconnection Relay

Distributed generation involves exporting electricity generated with a secondary power source to the utility grid. This can be useful to both the utility and the customer: the utility adds electrical demand capacity to its network and the customer sells excess electricity. However, there are concerns with paralleling two power sources. First, the power sources must be carefully synchronized. Second, if the utility fails while connected, the secondary power source may back-feed into the utility's system, potentially causing equipment damage and safety hazards. This is of particular concern to the utility. Distributed generators must immediately disconnect from the grid if there is a utility outage. Otherwise, utility workers can be seriously injured by unexpectedly energized equipment.

To avoid these problems, distributed generators must interface with the grid through a distributed generator interconnection relay. A *distributed generator interconnection relay* is a specialized relay that monitors both a primary power source and a secondary power source for the purpose of paralleling the systems.

The interconnection relay includes sensors for voltage, current, frequency, direction of power flow, and other parameters. The interconnection relay only allows the paralleling of connections if the two power sources are in phase and immediately opens the connection if they become out of phase.

Distributed Generation Concerns

Electric utilities have concerns about connecting distributed generation equipment to their grid system. For one, the utility distribution system was likely not originally intended to handle distributed power sources. For example, many utility service meters are not designed to monitor two-way power flow. Also, because utilities do not operate, control, or maintain the customer-owned distributed generation equipment connected to their system, there are concerns about the impact it may have on safety and the reliability of utility service to other customers. Distributed generators must work with the utility on the technical, procedural, and contractual requirements for safe and reliable interconnections.

Islanding

Islanding is the undesirable condition where an interactive inverter exports power to the utility grid during a utility outage. Islanding is a serious safety hazard for utility lineworkers working to restore power after an outage. Since the workers expect the grid to be de-energized, they may be injured by the power present in the system from an islanding inverter. Also, islanding can damage utility equipment by interfering with the utility's normal procedures for restoring service following an outage, primarily because the islanded electrical system is no longer in phase with the utility system.

Power Quality

Power quality is also a major concern for interconnected distributed generation equipment. Because electrical loads are designed to operate from ideal electrical power conditions, the electric utility system must be maintained within stringent power quality limits. Otherwise, poor performance or even damage to electrical loads and utility system equipment may result.

Utilities routinely test and monitor their generation, transmission, and distribution equipment for performance and power quality. Correspondingly, utilities also require assurances that interconnected distributed generation systems are operating within these limits and do not adversely affect the quality of utility service to other customers.

UPS Systems

An *uninterruptible power supply (UPS)* is an electrical device that provides stable and reliable power, even during fluctuations or failures of the primary power source. UPSs are used to maintain the operation of critical loads such as computer systems, medical equipment, and sensitive electronics during short interruptions in primary power. Utility power is supplied to the UPS, which supplies conditioned, stable, and disturbance-free output power to the critical loads.

There are several different ways in which UPSs can be used. A small UPS can be a stand-alone device installed at the point of use or a larger piece of equipment integrated into a critical power infrastructure system. Medium-size UPSs protect clusters of sensitive equipment and centrally located UPSs protect larger electrical systems.

Batteries are the most commonly used power source for a UPS, although flywheels and fuel cells are becoming more common for certain applications. The overall functionality of the UPS is the same, regardless of the power source. When utility power is interrupted or the quality is poor, current is drawn from the back-up power source to continue supplying adequate power from the output section of the UPS. The loads continue operating without interruption because they are typically always connected to both power sources. The UPS includes an inverter to convert the DC power to AC power. When acceptable utility power returns, the UPS input returns to using utility power to provide power to the load while also charging the back-up power source (batteries).

Many UPSs have an internal bypass circuit that can connect the utility input power directly to the load. The bypass circuit automatically activates if there is an internal failure or the load draws too much power. The bypass typically operates within 10 ms so that the critical load continues to operate without interruption. The bypass will stay on until the fault or overload is cleared. It can also be manually operated as required to isolate the back-up power source for service or tests. An external bypass circuit can also be incorporated to isolate the entire UPS system while maintaining load operation.

There are several different designs of UPSs, each with relative advantages and disadvantages. The most common designs are standby, line-interactive, rotary, double-conversion on-line, and delta conversion on-line UPSs.

Standby UPSs. Standby UPSs are typically used for small applications (less than about 600 VA) and are commonly used to protect personal computers. This design is also known as an "off-line" UPS because under normal conditions, it simply passes utility power through to the output side and the inverter does not operate. In this case, the back-up power system's inverter is in standby or off-line mode. The UPS provides some power conditioning to suppress surges or filter noise, but only if the utility power falls below minimum requirements does the UPS use back-up power. Then a transfer switch changes completely over to the back-up source. **See Figure 2-25.**

High efficiency, small size, and low cost are the main benefits of this design. However, this type of UPS will switch completely over to the back-up mode for nearly any power quality problem, which can quickly drain the battery. For example, a modest voltage sag can trigger back-up operation, even though some line voltage is present. For this reason, a standby UPS is rarely used for critical commercial loads.

Line-Interactive UPSs. The line-interactive UPS design incorporates a transformer and an inverter/rectifier unit that are always connected to the output of the UPS. **See Figure 2-26.** During normal conditions, the utility power flows through both devices, and a small amount is rectified to charge the back-up power source. When the input power fails, a transfer switch opens and DC power flows from the back-up power source, through the inverter to convert it to AC, and to the UPS output. With the inverter always connected to the output, this design provides additional power filtering and yields fewer switching transients than standby UPS types.

A feature of line-interactive UPSs is the variable transformer that automatically adjusts the secondary winding taps as the input voltage varies. This voltage regulation feature allows the UPS to adapt to a wide range of high- or low-voltage conditions without switching completely over to the back-up power source. This reduces the charge/discharge cycling of back-up batteries, thus extending their useful life.

Standby UPS

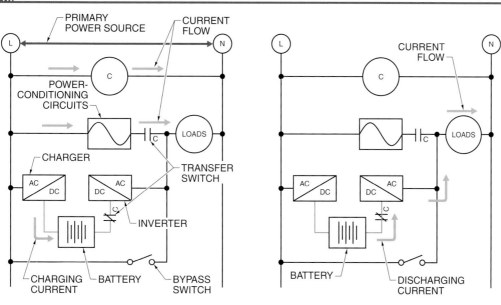

NORMAL CONDITIONS   **LOSS OF PRIMARY POWER SOURCE**

Figure 2-25. The inverter in a standby UPS system does not operate under normal conditions, so this system is known as an off-line UPS.

Line-Interactive UPS

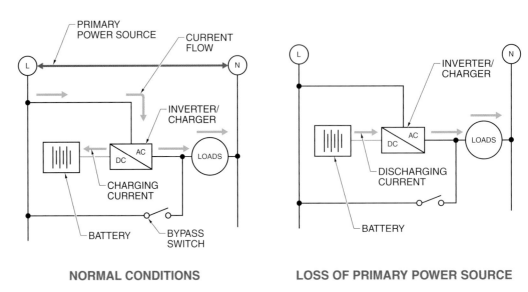

NORMAL CONDITIONS   **LOSS OF PRIMARY POWER SOURCE**

Figure 2-26. The inverter in a line-interactive UPS is always operating, either passing power through during normal conditions or converting DC power to AC while in the back-up mode.

UPS units are used to correct a variety of problems such as power failures, voltage sags, voltage spikes, undervoltage, overvoltage, line noise, frequency variations, transients, and harmonic distortion.

Large UPS systems are used to back up the critical computer server systems in server rooms and data centers.

High efficiency, small size, low cost, and high reliability, coupled with the ability to correct low or high line voltage conditions, make this the dominant type of UPS in the 0.5 kVA to 5 kVA power range. A line-interactive UPS can be either a stand-alone or rack-mounted unit and is common for small business and departmental servers.

Rotary UPSs. The rotary UPS is one of the oldest UPS system designs. There are a few variations of this system, but all include a motor-generator combination. This system uses a rectifier to convert all AC input power to DC. **See Figure 2-27.** This DC power is then used to charge the back-up power source and operate the DC motor. The generator coupled to the motor converts the mechanical rotating energy into AC power. The motor and generator combination operates at all times. If the utility AC input power fails, the DC motor runs on battery power instead.

A rotary UPS provides excellent isolation of the load from utility power and has a high overload and fault-clearing capability. Also, the output from the generator is a perfect sine wave. No other power quality conditioning is needed.

This design, however, includes many moving parts that are subject to wear and mechanical failure, so significant maintenance is required. Therefore, rotary UPSs are most common for very large installations (>200 kVA) with the personnel to accommodate the maintenance requirements.

Double-Conversion On-Line UPSs. A double-conversion on-line UPS is similar to the rotary UPS design in that both perform two conversions of all the power flowing through the UPS: first AC to DC, then DC back to AC. The double-conversion on-line UPS, however, uses an electronic inverter instead of a motor-generator combination.

A rectifier changes all the AC input power to DC power. **See Figure 2-28.** Some DC power is used to charge the back-up power supply, while the majority goes to the inverter to create AC power output. When the input AC power fails, there is no interruption in the output power because the back-up power source is always on-line and ready to supply power to the DC bus. Since all incoming power is rectified to DC, most AC power quality problems become nonissues.

The double-conversion process allows for excellent regulation of the output voltage and frequency. The rectifier may operate as a nonlinear load and, due to the multiple power conversions, each of which can contribute power losses, the efficiency is somewhat lower than other UPS designs. This type of UPS is most common for systems of 5 kVA and larger.

Delta-Conversion On-Line UPSs. The delta-conversion on-line UPS is the most recent UPS design. It includes a circuit that is similar to a double-conversion on-line UPS, but instead of processing all of the incoming power, this circuit only contributes the difference needed to maintain a steady output. For example, if the voltage of the incoming utility power is lower than the minimum level, the inverter delivers extra voltage to the output to make up the difference. **See Figure 2-29.** If the utility power is normal, it passes through the UPS with no further processing, other than basic power conditioning for surges and noise.

Rotary UPS

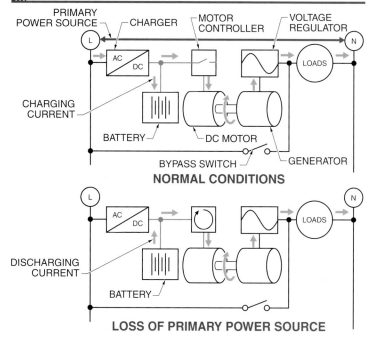

Figure 2-27. A rotary UPS isolates the input and output power by way of a mechanical coupling between a DC motor and an AC generator.

Double-Conversion On-Line UPS

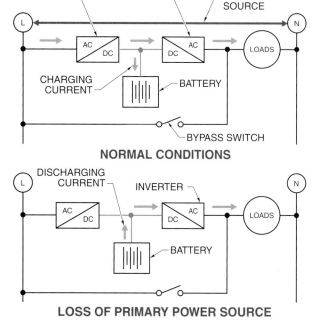

Figure 2-28. A double-conversion on-line UPS converts all input power to DC and then back to AC, providing excellent voltage and frequency regulation.

Delta-Conversion On-Line UPS

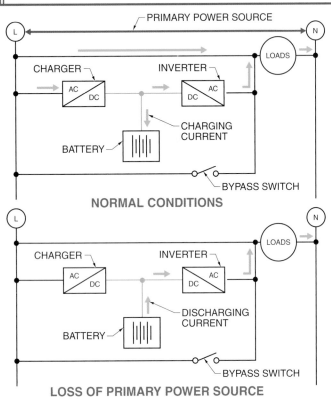

Figure 2-29. During normal operation, a delta-conversion on-line UPS system contributes the "difference" voltage necessary to bring the line voltage within the acceptable range.

The inverter is always on and maintains the load voltage, while the delta converter controls the current to the output. If the utility power completely fails, the back-up power source supplies the inverter with power without interruption.

The delta-conversion system provides excellent voltage regulation and is highly efficient. The UPS operates as a linear device so that there are no harmonic problems related to its operation. It cannot provide frequency conversion. This UPS is also common for systems 5 kVA and larger.

ELECTRICAL SYSTEM CONTROL APPLICATIONS

There are numerous examples of electrical system control applications where electrical loads are switched by relays or electric motors are controlled with variable-frequency drives. These applications are very simple actions, typically embedded, and one control step within a more sophisticated system application, such as the control of an HVAC air-handling unit. For example, relays are used to energize heating or cooling functions, and drives control the speed of blower motors.

Most electrical system control applications deal primarily with the management of the electrical power supply to ensure reliable and adequate electrical power to all the building systems, particularly critical loads. Electrical system controls can also be used to reduce energy costs. These goals can involve reducing the building's overall electrical demand, switching to back-up power sources during utility outages, and operating secondary power sources in parallel with the utility to supplement the energy supply.

Demand Limiting

Electrical demand can have a big impact on utility energy costs, but building automation systems can be used to help avoid high rates or extra charges by limiting the building's overall electrical demand. This is accomplished by temporarily reducing the noncritical electrical loads in the building until overall demand requirements fall. **See Figure 2-30.** For example, if the midday electrical demand for a building during a hot summer day rises too high, the system can be programmed to reduce the power consumption by some of the largest system loads, usually lighting and HVAC systems. For example, the HVAC temperature setpoints may be raised slightly to reduce the cooling loads and the lighting in certain areas may be dimmed.

Operating Conditions. Demand limiting requires the involvement and automation of other building systems in order to reduce the electrical demand. The most common systems to involve are lighting and HVAC, since these are typically the largest consumers of energy and can usually tolerate some operating reductions. Lighting in certain areas of the building may be switched off or dimmed, temperature setpoints may be adjusted by a few degrees to reduce heating or cooling loads, and blowers and pumps may be cycled on and off for limited operation.

It is common for these small changes to make a significant difference in the electrical demand and yet go largely unnoticed by the building occupants for short periods.

Control Sequence. The control sequence for demand limiting with electrical system controls is as follows:

1. Electrical demand is monitored by electricity sensors or meters, such as current transformers and voltage sensors.
2. The electrical demand is continuously compared to the historical demand data. If the demand exceeds a certain threshold (typically 85% to 90%) of the peak demand, the system initiates a demand-limiting program.
3. An electrical system controller signals the appropriate lighting and HVAC controllers to override their normal operating program. Some lighting is switched off or dimmed and HVAC temperature setpoints are adjusted.
4. When the electrical demand falls below a preconfigured percentage of the historical peak demand (typically 70% to 80%), the lighting and HVAC systems are released from the demand-limiting override and allowed to return to their normal operations.

Emergency Back-up Power Source Transfer

Complete power outages or power quality problems (such as very low voltage) of the utility-supplied power are uncommon and typically short in duration, but for buildings with critical loads, even this small risk for problems is intolerable. These facilities use one or more secondary power sources to back up their operations in the event of a utility power problem. These systems must constantly monitor the primary power supply and be ready to switch over to secondary power sources at any time.

▥ Demand Limiting

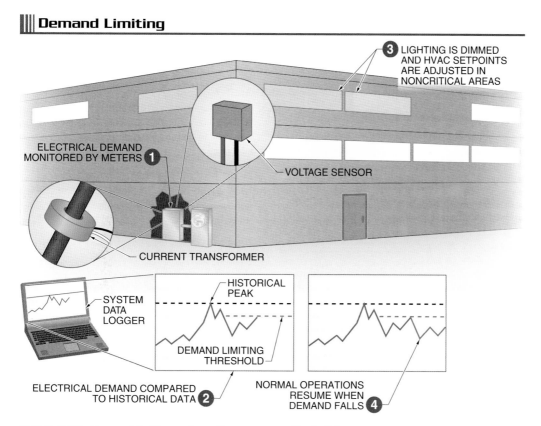

Figure 2-30. Demand limiting reduces the power used by building systems.

Operating Conditions. The alternation of power sources to supply the building's electrical system is managed by an automatic transfer switch (ATS). **See Figure 2-31.** The most common application of a transfer switch uses a diesel engine generator as the secondary power source. An engine requires time to start and come up to speed before being able to sufficiently drive the generator and produce electrical power. Therefore, a UPS system is typically used to power some of the critical building loads for this short transition time. The UPS returns to battery charging when the generator is running.

It is usually not feasible to supply an entire building's electrical system with back-up power because the necessary secondary power source would be too large. Instead, the electrical system is divided into critical and noncritical load circuits. The primary power source would supply electricity to all loads, but the secondary power source would supply electricity to only the critical loads. The critical loads must be grouped together in common switchboards or panelboards. Therefore, this type of arrangement must be planned and executed during the design and construction of the building.

Emergency Back-Up Power Source Transfer

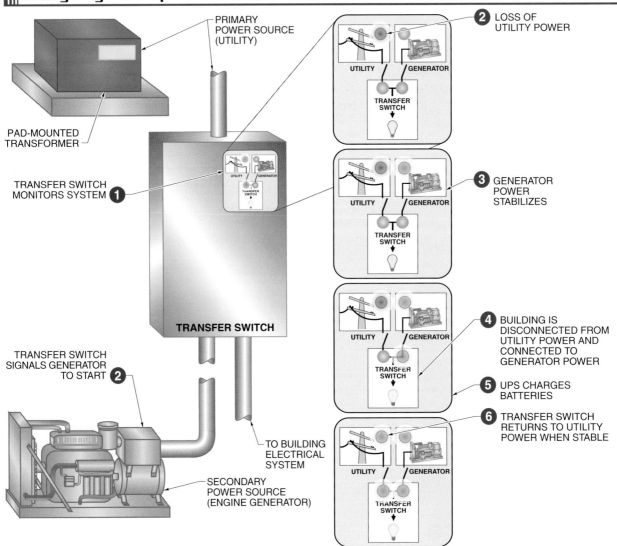

Figure 2-31. Transfer switches transition to secondary power sources during the emergency back-up mode.

To avoid initiating engine startup for very short voltage dips or interruptions, the transfer switch delays the start-up signal for a predetermined period. This time delay is at least 1 sec and may be up to 30 sec. If adequate utility power returns within the delay period, the transfer switch returns to the monitoring mode. Otherwise, it continues with the power source transfer operation.

Control Sequence. The control sequence for emergency back-up power transfer with electrical system controls is as follows:

1. The transfer switch continuously monitors incoming utility power for outages or poor power quality.
2. When the utility power is inadequate, the UPS system provides or conditions power to some of the critical loads while the transfer switch waits for the set delay period. If the power remains inadequate at the end of the delay period, it sends a start-up signal to the engine generator.
3. The transfer switch monitors the power output from the generator. When the engine comes up to speed, the transfer switch senses that the voltage and frequency stabilize and initiates the transfer operation.
4. The transfer switch changes the building electrical system over to the secondary power source, the engine generator. First, the connection between the building electrical system and the utility is opened to avoid back-feeding power onto the grid. Then the generator output is connected to the electrical system.
5. The UPS synchronizes with the input frequency and phase of the generator, maintaining constant power to the load, and then deactivates. The UPS then begins charging its batteries.
6. Upon the restoration of adequate utility power, the transfer switch initiates the process of retransferring the electrical system back to the utility. Another short time delay ensures that the utility power source is stable before the retransfer. Then the generator is disconnected and the utility is reconnected.

On-Demand Distributed Generation

Besides the transfer of power sources for emergency reasons (utility power outage or poor power quality), an automated electrical system can utilize secondary power sources for on-demand distributed generation.

Facilities with a significant secondary power source capability may be able to contract with the utility to be an on-demand distributed generator. This scenario involves the utility signaling to the energy management system at the facility that the overall electrical demand on the grid is high and extra electricity is requested to avoid brownouts. **See Figure 2-32.** Even if the utility-supplied power to the facility is still adequate, the generator starts up, synchronizes with the grid and the building electrical system, and begins contributing electricity when the switch closes. This immediately reduces the building's electrical demand, which helps relieve some of the high-demand problem. If the generator produces enough electricity to cover the building's usage and still export excess electricity, this further helps relieve the high-demand problem.

Operating Conditions. Like emergency back-up applications, this scenario requires an automatic transfer switch and typically involves diesel engine generators, though some other secondary power sources are also compatible. Since this application also tends to activate a secondary power source more often than is required for only emergency needs, it helps keep the systems exercised, which improves reliability.

Control Sequence. The control sequence for on-demand distributed generation with electrical system controls is as follows:

1. After receiving the signal to start distributed generation, a controller signals the engine generator to begin its start-up sequence.
2. The engine starts up and the power output from the generator stabilizes.
3. The distributed generator interconnection relay verifies the voltage, frequency, and phase synchronization of the generator to the utility source.

On-Demand Distributed Generation

PAD-MOUNTED
TRANSFORMER

PRIMARY POWER
SOURCE (UTILITY)

1 UTILITY SIGNALS FOR
ELECTRICITY GENERATION;
TRANSFER SWITCH SIGNALS
ENGINE GENERATOR TO
START

3 DISTRIBUTED GENERATOR
INTERCONNECTION RELAY
INDICATES SYNCHRONIZATION

UTILITY | GENERATOR

TRANSFER
SWITCH

ENGINE GENERATOR
COMES UP TO SPEED **2**

SYNCHRONIZED

TRANSFER SWITCH
PARALLELS TWO
POWER SOURCES **4**

UTILITY | GENERATOR

TRANSFER
SWITCH

POWER SOURCES
ARE CONTINUOUSLY
MONITORED **5**

TRANSFER SWITCH

6 TRANSFER SWITCH
DISCONNECTS GENERATOR
WHEN NO LONGER NEEDED

TO BUILDING
ELECTRICAL SYSTEM

SECONDARY
POWER SOURCE
(ENGINE GENERATOR)

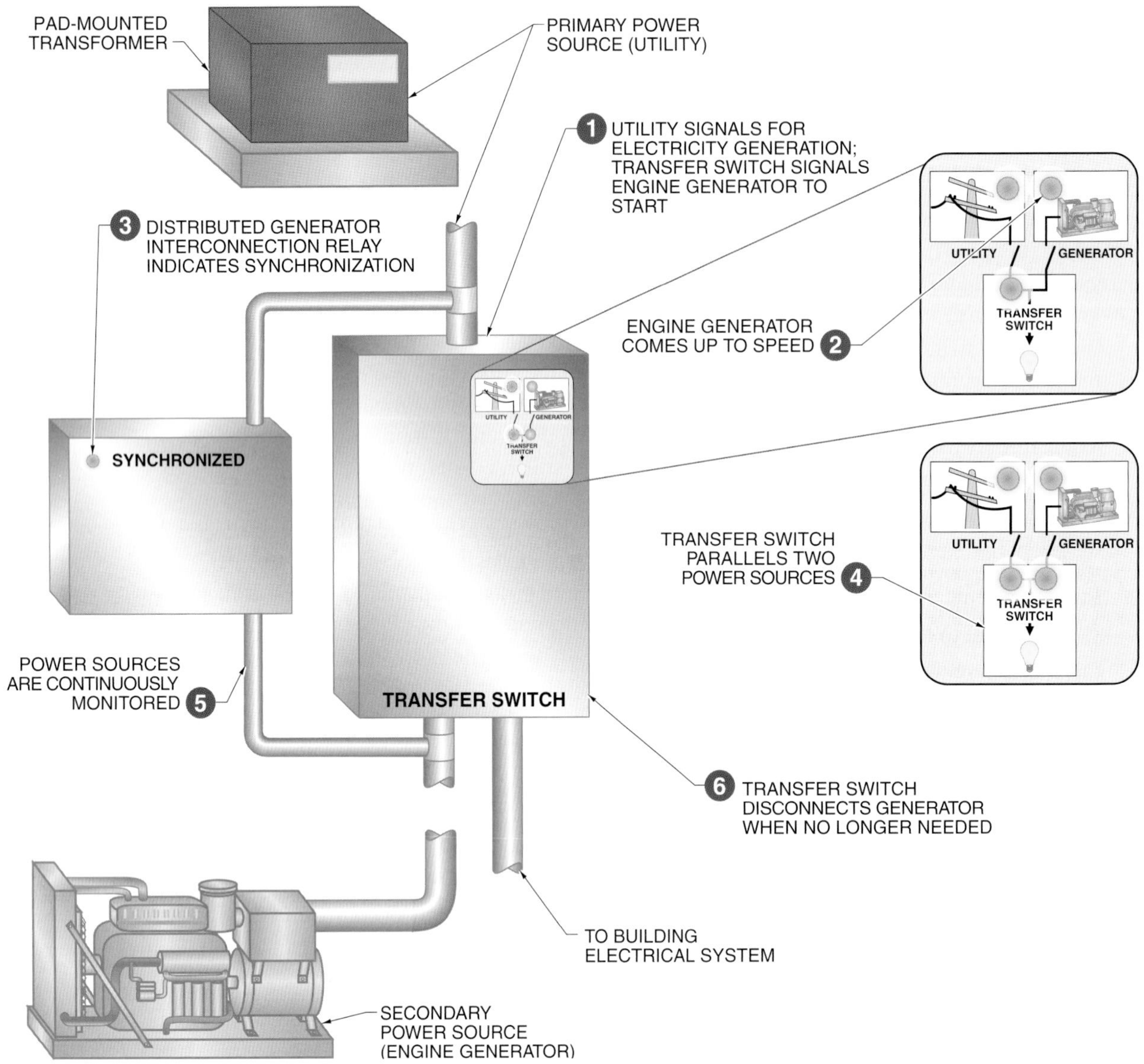

Figure 2-32. On-demand distributed generation exports power to the utility grid through transfer switches.

4. The transfer switch connects both the generator and the utility to the building electrical system.

5. While the two power sources are paralleled, the utility feed is continuously monitored. If there is a problem with the utility feed, the transfer switch opens the utility circuit and the building operates from generator power alone.

6. When the utility no longer needs the distributed generation, it signals the facility's electrical system controllers. The transfer switch closes the connection to the utility power if it is not already closed and the generator connection is opened. This returns the system to normal operation. The engine generator begins its shutdown sequence.

Refer to
Quick Quiz®
on CD-ROM

Summary

- Most electricity is generated by large power plants that supply power to millions of consumers.

- Utility-supplied electrical power is very reliable, inexpensive, widely available, and essentially maintenance free.

- A secondary power source is intended to supplement the primary electrical service or supply the building with power in the event of an interruption in power from the electric utility.

- Electricity distribution deals with the transmission and delivery of electricity outside of consumers' facilities.

- A smaller-scale power distribution system within a building delivers power from the electrical service to end-use points throughout the building, where it powers individual loads such as industrial equipment, motors, lamps, and computers.

- A large block of electric power is delivered from a utility substation to a building at a switchboard, where it is broken down into smaller blocks for distribution throughout a building.

- Transformers are used within buildings to step down high-voltage distribution power to the lower voltage levels required by most loads.

- Receptacles do not use electricity but merely provide a convenient access point on the electrical system from which electrical power can be obtained.

- Properly grounded electrical circuits, tools, motors, and enclosures help safeguard equipment and personnel against the hazards of electrical shock.

- Overcurrent protective devices include fuses and circuit breakers that are located in panelboards.

- Switching can turn electrical appliances or circuits ON or OFF, or alternate between separate power supplies.

- High electrical demands make it more difficult for electric utilities to supply power to all their customers, so they sometimes change their rates based on demand.

- Electrical demand over time results in electricity consumption.

- Power quality is influenced by the performance of the electrical generation and distribution equipment, as well as electrical loads operating on the system.

- Voltages are typically acceptable within the range of +5% to –10% from the nominal voltage.

- The frequency of utility-supplied electricity typically varies less than ±1%.

- Electrical system control devices are primarily concerned with monitoring and maintaining a reliable and quality power supply.

- Electrical demand and electricity consumption are measured with demand meters.

- Additional demand meters can be installed temporarily or permanently throughout the building to provide electricity usage data to a building automation system.

- Power quality sensors are similar to hand-held and bench-top test instruments, except that instead of displaying the measured value, they output this information via electrical signals to other devices.

- Electrical circuits are automatically switched with relays.

- In addition to controlling motor speed, variable-frequency drives can control motor acceleration time, deceleration time, motor torque, and motor braking.

- Transfer switches are used to keep critical loads operating during utility outages and to lower total energy spending by taking advantage of the best utility rates.

- Uninterruptible power supplies (UPSs) are used to maintain the operation of critical loads during short interruptions in primary power.

- Batteries are the most commonly used power source for a UPS, although flywheels and fuel cells are becoming more common for certain applications.

- Most electrical system control applications deal primarily with the management of the electrical power supply to ensure reliable and adequate electrical power to all of the building systems, particularly critical loads.

- Building automation systems can be used to help avoid high rates or extra charges by limiting the building's overall electrical demand.

- Facilities can automatically switch to one or more secondary power sources to back up their operations in the event of a utility power problem.

- Automated electrical systems can utilize secondary power sources for on-demand distributed generation.

Definitions

- An *electrical system* is a combination of electrical devices and components, connected by conductors, that distributes and controls the flow of electricity from its source to a point of use.

- *Electricity* is the energy resulting from the flow of electrons through a conductor.

- A *utility* is a company that generates and/or distributes electricity to consumers in a certain region or state.

- *Centralized generation* is an electrical distribution system in which electricity is distributed through a utility grid from a central generating station to millions of customers.

- The *grid* is the utility's network of conductors, substations, and equipment that distributes electricity from a central generation point to the consumers.

- *Distributed generation* is an electrical distribution system in which many smaller power-generating systems create electrical power near the point of consumption.

- The *electrical service* is the electrical power supply to a building or structure.

- A *switchboard* is the last point on the power distribution system for the power company and the beginning of the power distribution system for the property owner's electrician.

- A *feeder* is the circuit conductors between a building's electrical supply source, such as a switchboard, and the final branch-circuit overcurrent device.

- A *transformer* is an electric device that uses electromagnetism to change AC voltage or electrically isolate two circuits.

- A *panelboard* is a wall-mounted power distribution cabinet containing overcurrent protective devices for lighting, appliance, or power distribution branch circuits.

- A *branch circuit* is the circuit in a power distribution system between the final overcurrent protective device and the associated end-use points, such as receptacles or loads.

- A *switch* is a device that isolates an electrical circuit from a power source.

- An *outlet* is an end-use point in the power distribution system.

- A *receptacle* is an outlet for the temporary connection of corded electrical equipment.

- *Grounding* is the intentional connection of all exposed noncurrent-carrying metal parts to the earth.

- An *equipment grounding conductor (EGC)* is a conductor that provides a low-impedance path from electrical equipment and enclosures to the grounding system.

- A *grounding electrode conductor (GEC)* is a conductor that connects the grounding system to the buried grounding electrode.

- A *grounded conductor* is a current-carrying conductor that has been intentionally grounded.

- A *main bonding jumper (MBJ)* is a connection at the service equipment that connects the equipment grounding conductor, the grounding electrode conductor, and the grounded conductor (neutral conductor).

- An *overcurrent protective device* is a device that prevents conductors or devices from reaching excessively high temperatures from high currents by opening the circuit.

- A *fuse* is an overcurrent protective device with a fusible link that melts and opens the circuit when an overcurrent condition occurs.

- A *circuit breaker* is an overcurrent protective device with a mechanism that automatically opens a switch in the circuit when an overcurrent condition occurs.

- *Switching* is the complete interruption or resumption of electrical power to a device.

- A *pole* is a set of contacts that belong to a single circuit.

- A *throw* is a position that a switch can adopt.

- *Electrical demand* is the amount of electrical power drawn by loads at a specific moment.

- *Electricity consumption* is the total amount of electricity used during a billing period.

- *Power quality* is the measure of how closely the power in an electrical system matches the nominal (ideal) characteristics.

- *Voltage* is the difference in electrical potential between two points in an electrical circuit.

- A *transient voltage* is a temporary, undesirable voltage in an electrical circuit, ranging from a few volts to several thousand volts and lasting from a few microseconds up to a few milliseconds.

- *Frequency* is the number of AC waveforms per interval of time.

- A *pulse meter* is a meter that outputs a pulse for every predetermined amount of flow in the circuit or pipe.

- A *relay* is an electrical switch that is actuated by a separate electrical circuit.

- A *contactor* is a heavy-duty relay for switching circuits with high-power loads.

- A *magnetic motor starter* is a specialized contactor for switching electrical power to a motor and includes overload protection.

- A *variable-frequency drive* is a motor controller that is used to change the speed of AC motors by changing the frequency of the supply voltage.

- The *fundamental frequency* is the voltage frequency simulated by the changing pulse widths of the carrier frequency.

- The *carrier frequency* is the frequency of the ON/OFF voltage pulses that simulate the fundamental frequency.

- A *transfer switch* is a switch that allows an electrical system to be switched between two power sources.

- A *distributed generator interconnection relay* is a specialized relay that monitors both a primary power source and a secondary power source for the purpose of paralleling the systems.

- An *uninterruptible power supply (UPS)* is an electrical device that provides stable and reliable power, even during fluctuations or failures of the primary power source.

Review Questions

1. What are the advantages of having a secondary power source available?

2. What is the primary difference between centralized generation and distributed generation?

3. Briefly explain how electricity is distributed within a building from the electrical service to a receptacle.

4. How is electrical demand used by utilities to affect the billing for electricity usage to some customers?

5. With respect to voltage and frequency, how does power quality affect electrical systems and their loads?

6. Compare two examples of specialized relays with the general-purpose relays.

7. How do variable-frequency drives change the frequency of power supplied to electric AC motors?

8. Compare transfer switches utilizing open-transition transfer (OTT) and closed-transition transfer (CTT).

9. Explain the overall functionality of an uninterruptible power supply (UPS).

10. How can building automation systems be used to limit electrical demand?

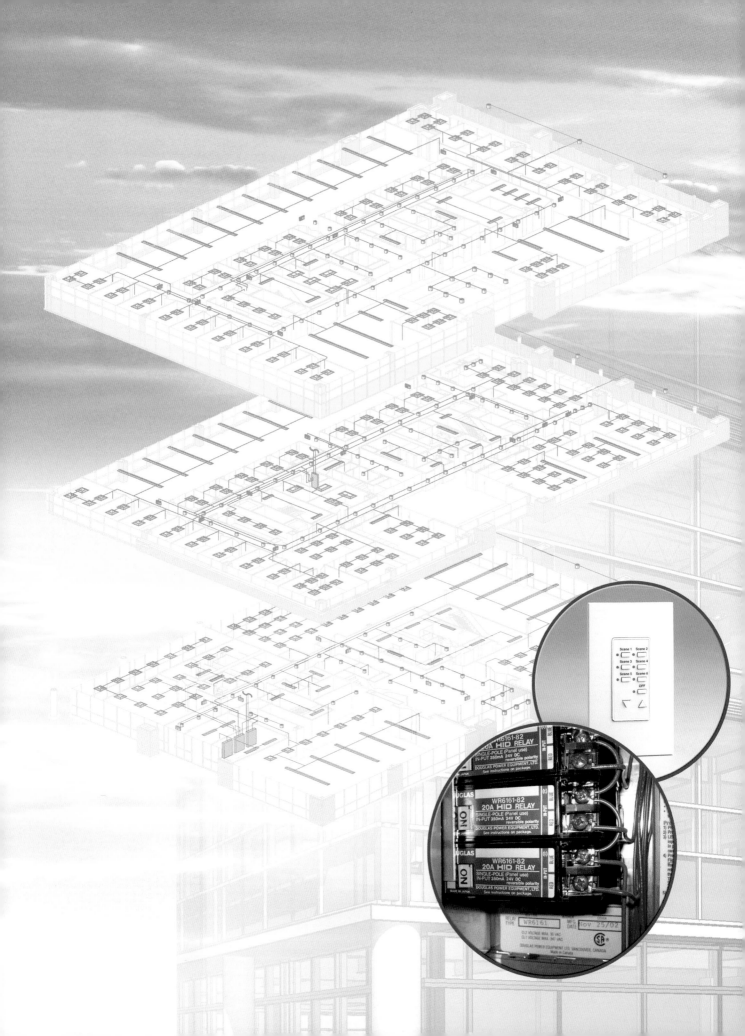

Chapter Three

Lighting System
Control Devices and Applications

A significant portion of a building's energy consumption is used for lighting. A building may also contain a large number of lighting circuits. Fortunately, lighting is also one of the simplest building systems to control. Even the most sophisticated lighting control system must control only switching and dimming devices, and base the control decisions on occupancy, light levels, manual inputs, and any programmed instructions, such as time schedules.

Chapter Objectives

- Differentiate between various methods of lighting system control.
- Compare the different types of electric light sources and the characteristics that affect lighting control.
- Describe the role of occupancy sensors within a lighting control system.
- Compare the functions and types of different switches and dimmers.
- Identify the control sequences in common lighting control applications.

LIGHTING SYSTEMS

Lighting is a major component of energy consumption in buildings, accounting for a significant part of all energy consumed worldwide. **See Figure 3-1.** Lighting should be neither too dark, which can cause eyestrain and safety hazards, nor too bright, which can bother occupants and waste energy. Proper indoor lighting improves the productivity and safety of building occupants.

Commercial Building Electricity Consumption by End Use

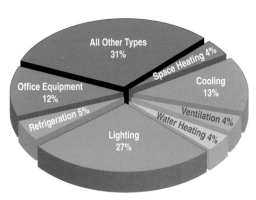

Energy Information Administration

Figure 3-1. Lighting is the largest single use of electricity within buildings, particularly commercial buildings.

Lighting is particularly important for indoor spaces, but can also be implemented outside the building. Outdoor lighting at night improves security, illuminates signage, and enhances landscaping features. Outdoor lighting can be controlled with devices similar to those for indoor lighting, though it may require different control strategies and goals.

Lighting includes both natural daylight and artificial light from lamps. Lighting system control uses as much natural daylight as possible and supplements this with artificial lighting as needed to maintain optimal lighting levels for each building space. Lighting control brings about several benefits. Foremost, lighting control saves energy and can substantially reduce a building's operating costs. Natural daylight is cost-free and can be exploited to

minimize the need for artificial lights. Lights are turned off when not needed, or dimmed to provide adequate additional light without unnecessary excess. Lighting controls also ensure that spaces are lit adequately, which improves occupant productivity and the safety and security of occupants' surroundings.

Automated lighting control allows the system to operate routinely in the background without operator input, adding convenience and saving time. The flexibility of automated lighting systems also accommodates changes to lighting requirements that affect lighting controls. For example, changing schedules, tasks, or workplace configurations can be easily implemented while maintaining lighting control goals.

Centralized Lighting Control

Centralized lighting control is a lighting system that controls lights throughout a building via one main control panel. The lighting branch circuits are not directly connected to any type of operator interface control devices, such as switches or dimmers. Instead, all lighting branch circuits are wired to a central control panel. **See Figure 3-2.** The control panel supplies switched or dimmed power to the lighting circuits individually via the control devices mounted in the panel, such as relays. A control panel may accommodate a large number of lighting circuits.

This control design may still use manually operated switches and dimmers located in convenient locations throughout the building. However, instead of directly controlling the local lighting circuit, these devices send signals to a central control panel. The signals are low voltage (typically 24 VAC) control signals or structured network messages. These signals are then used to operate the control devices.

Microprocessor-based control panels add more flexibility for lighting control. These systems can store control programs based on input signals (such as from occupancy sensors) and/or time schedules. Typically, inputs from manually operated controls override the program so that lighting conditions can also be changed as needed.

||| Centralized Lighting Control

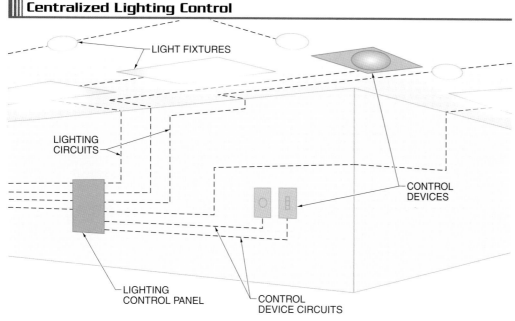

LIGHT FIXTURES

LIGHTING
CIRCUITS

CONTROL
DEVICES

LIGHTING
CONTROL PANEL

CONTROL
DEVICE CIRCUITS

Figure 3-2. Centralized lighting control runs the circuits for all the lighting control devices and lamps to a central panel, where all the control occurs.

A disadvantage of this system is that it requires long runs of conductors to be wired from all the light fixtures and operator interfaces back to a centralized location where the component dimmers and relays are located. Also, the system is susceptible to widespread failure if there is a problem with the central control panel.

Distributed Lighting Control

A distributed lighting control system provides the advantages of centralized control without having a central controller or control panel. *Distributed lighting control* is a lighting system that controls lights directly from local control devices. The switches and dimmers are located throughout the building, allowing shorter lighting control circuits that run directly between the light fixtures and control devices. In a distributed system, each device works independently of the others, providing a more reliable lighting control system.

An automated distributed lighting control system requires special electronic lighting control devices, each with a unique address, that are connected together in a network.

See Figure 3-3. These devices contain microprocessors and programs that manage the control signal communications within the network. The memory in these "smart" devices retains the programs and settings, even during power outages. For example, an operator switch on a distributed lighting control network may contain a program that sends out signals to switch lighting circuits 1, 2, and 3 whenever it is manually actuated.

Many existing lighting devices, such as ballasts, switches, and sensors, can be retrofitted with modules to work with an automated distributed lighting control system. Instead of replacing an entire conventional dimming ballast with a smart one, for example, a smart module can be added to control the ballast through its existing control wiring.

The idea for incandescent electric lights was conceived of as early as 1802, but the filaments used were too dim and too short lived to be practical. After years of experimenting with materials, Thomas A. Edison invented the first practical light bulb in 1879 by using a filament of carbonized cotton thread.

Distributed Lighting Control

CONTROL DEVICE

LIGHT FIXTURES

DIMMERS

CONTROL DEVICE

RELAY PACK

Figure 3-3. *In distributed lighting control, the lighting circuits are directly controlled by the nearby control devices, such as wall switches and dimmers.*

An advantage of distributed lighting controls is that there are many options for designing and building the system. Also, many of the sensing devices used to control the lighting system can be used as inputs in other automated building systems. For example, an occupancy sensor used to control the lights when a person enters the room can also be used to control temperature and airflow, alert security to unauthorized access, trigger closed-circuit television (CCTV) cameras to look at the room, or even help rescue crews to locate occupants in an emergency.

Centralized lighting control runs all lighting circuits to a common lighting control panel.

LIGHT

Light is the portion of the electromagnetic spectrum that the human eye can perceive. Actually, this is more accurately called "visible light," since there are many other types of electromagnetic energy that are invisible to the human eye. **See Figure 3-4.**

Natural light is primarily light from the sun. Artificial light is produced by electric lamps. A *lamp* is an electrical output device that converts electrical energy into visible light and other forms of energy. Electricity passes through the lamp's terminals and electrical and/or chemical reactions inside the lamp produce light.

Only a relatively small percentage of the total energy produced by a lamp is visible light. Approximately 65% to 90% of the total energy produced by lamps is ultraviolet energy, infrared energy, or heat, which are not visible. For example, only about 21% of the total energy of a typical fluorescent lamp yields visible light.

There are two modes of lighting system control used to vary light output. Switching controls when a lamp is lit and dimming controls the amount of light output from a lamp. These modes can be used separately or together in lighting control systems.

Light

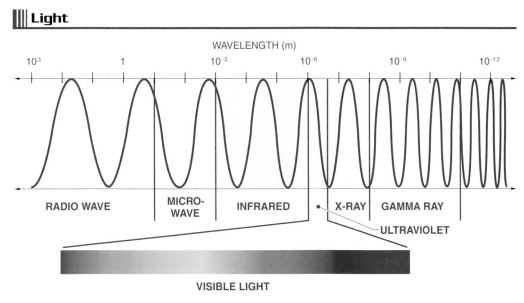

Figure 3-4. *Visible light is only a narrow section of the entire electromagnetic spectrum.*

Lighting Types

Lighting systems can be implemented in many different ways and for many different reasons. Most include a variety of individual lighting applications, which can be classified according to function, such as general, task, or accent. Lighting applications can also be grouped by the way they are installed or implemented.

Lighting Functions

General lighting provides the minimum illumination required for the use of a building space. General lighting is achieved with lighting circuits of relatively high-output lights that produce a diffuse, even light over a large area.

Task lighting is focused on a relatively small area that is used to perform specific tasks. Task lighting is typically used in addition to general lighting for performing work such as parts assembly, reading documents, and other detail-oriented tasks. Task lighting is primarily functional and is the most concentrated type of lighting.

Accent lighting aesthetically enhances the volume of a space or structure. Accent lighting is mainly decorative and is intended to highlight certain pictures, plants, or other elements of interior design or landscaping.

Lighting Installations

Downlighting is the most common lighting installation, with fixtures on or recessed in the ceiling and casting light downward. This is the most common lighting method used in both offices and homes. Although it is easy to design, downlighting causes problems with glare and unnecessary energy consumption due to its large number of fixtures. Recessed lighting is a popular type of downlighting, with fixtures mounted into the ceiling to appear flush with it. These downlights can use wide-angle (for general lighting) or narrow-beam (for task lighting) lamps. Track lighting fixtures can be easily aimed for either task or accent lighting purposes.

Uplighting lights a ceiling, which reflects light back downward, providing diffuse indirect lighting. Uplighting is less common than downlighting, but provides uniform general illuminance levels and minimizes glare on computer displays and other dark, glossy surfaces.

Lighting applications also directly light a surface from various angles. Front lighting is highly effective, but tends to make the subject look flat because it casts almost no visible shadows. Lighting from the side is less common, as it tends to produce glare near eye level. Backlighting either around or through an object is mainly for accent.

Light Switching

Light switching alternates lighting loads between full-rated brightness or completely off, with no intermediate light levels. *Switching* is the complete interruption or resumption of electrical power to a device. Switching electrical power to a light fixture turns the lamp on or off. This is done by opening or closing the ungrounded (hot) conductor(s) in the lighting circuit. **See Figure 3-5.** Wiring many lights together in a single circuit allows them all to be switched simultaneously.

© 2008 Lutron Electronics, Inc.

Lighting control helps save energy while maintaining a pleasant working environment.

Light switching turns lamps on when they are needed and off when they are not. This is an important part of energy management, as many lamps are unnecessarily left on when there are no occupants or situations that require lighting.

Some lighting controls involve frequently switching lamps ON and OFF, which generally shortens a lamp's overall life. However, the overall lifespan of the lamps in lighting control systems is extended due to the fewer hours per day that the lamp is operating, which may be enough to compensate for the lifespan reduction due to switching. Even so, more frequent lamp replacements due to shorter lamp life may be offset by energy savings gained from the lighting system's overall lower energy consumption.

Light Intensity

Light switching alone does not provide precise control over the lighting. Controlling the intensity of the light output from lamps through dimming, however, offers far greater flexibility. *Dimming* is the intentional reduction of electrical power to a lamp in order to reduce its light output.

Light Switching

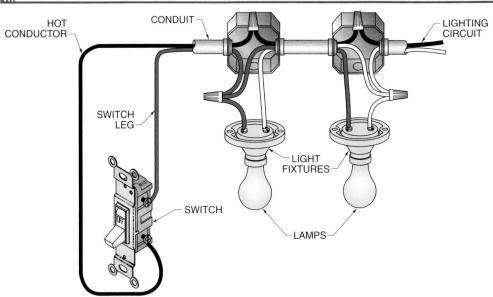

Figure 3-5. Basic light switching involves making or breaking the continuity contacts in the ungrounded (hot) conductor of a lighting circuit.

Dimming adds to the comfort and convenience of the occupants in a lighted space. Reducing light levels reduces glare on computer screens and helps people view audiovisual presentations. Dimming also reduces the amount of energy used by a lamp. This is an important energy conservation technique when full lamp brightness is not required. For example, lamps can be dimmed when natural daylight is available. As the daylight brightens in the morning, the lamp light level is decreased to compensate.

Light output is measured in lumens. A *lumen* (lm) is the measure of the intensity of light radiating from a light source. Since the light output from a lamp can change over time, lamps are rated in initial lumens and in mean lumens. *Initial lumens* is the rated intensity of light produced by a lamp when it is new. *Mean lumens* is the rated average intensity of light produced by a lamp after it has operated for approximately 40% of its rated life.

Light illuminates surfaces. *Illuminance* is the quantity of light per unit of surface area. **See Figure 3-6.** Illuminance is measured in foot-candles (for English measurements) or lux (for metric). A *foot-candle* (fc) is the illuminance from 1 lumen per square foot (lm/ft^2) of surface. A *lux* is the illuminance from 1 lumen per square meter (lm/m^2) of surface. For example, a lamp in the center of a room illuminates the walls, floor, and ceiling. The average illuminance of the surfaces is the total lumen output of the lamp divided by the total area of all the room surfaces in line-of-sight with the lamp.

LIGHT SOURCES

Most artificial light is produced by lamps. There are many different types of lamps, which are differentiated by the method used to produce light. Common types used in commercial buildings for most lighting applications include incandescent, halogen, fluorescent, and high-intensity discharge lamps. A less common lamp type for general lighting is the light-emitting diode (LED), but this application is on the rise.

Lamps are installed within light fixtures. A *light fixture* is an electrical appliance that holds one or more lamps securely and includes the electrical components necessary to connect the lamp(s) to the appropriate power supply. The NEC® refers to a complete lighting unit of lamps and light fixtures as a "luminaire."

Lamps are rated in watts and lumens. The relationship between watts and lumens is an expression of a lamp's efficiency, which is quantified as luminous efficacy. The *luminous efficacy* is the ratio of a lamp's light output (lumens) to the electrical power input (watts). (For each lamp type, the luminous efficacy is given as a range because larger wattage lamps within the same type tend to be a little more efficient.) This information is used to compare lamps and determine the power consumption and cost of certain lighting levels. **See Figure 3-7.** Lamps with a higher luminous efficacy are less expensive to operate.

Illuminance

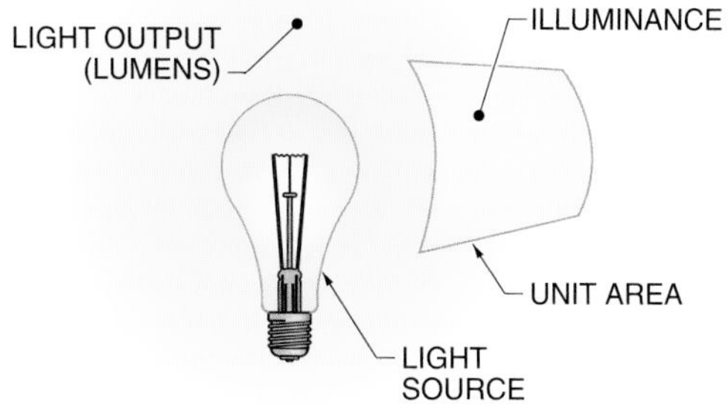

Figure 3-6. Lumen output is the total light output of a lamp, while illuminance is the density of that light on a surface.

Lamp Electrical Efficiency

Lamp Type	Luminous Efficacy*
Incandescent	12 to 18
Halogen	18 to 24
Fluorescent	55 to 100
Low-pressure sodium HID	190 to 200
Mercury vapor HID	50 to 60
Metal-halide HID	65 to 115
High-pressure sodium HID	100 to 150
Light emitting diode	25 to 70

* in lm/W

Figure 3-7. Luminous efficacy is a measure of the efficiency of a lamp type at producing light for each watt of electrical power.

Incandescent Lamps

Incandescent lamps are the most widely used lamps in the world. An *incandescent lamp* is an electric lamp that produces light by the flow of current through a tungsten filament inside a gas-filled, sealed glass bulb. **See Figure 3-8.** A *filament* is a conductor with a high resistance that causes it to glow white-hot from electrical current. The filament produces light, with a substantial amount of heat as a by-product.

Conventional Incandescent Lamp

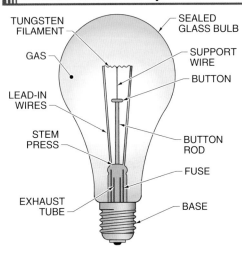

Figure 3-8. Conventional incandescent lamps produce light from the heating of a high-resistance filament by electric current.

Conventional incandescent lamps are relatively inefficient, with luminous efficacies from 12 lm/W to 18 lm/W. Since more-efficient filaments degrade faster, the lifetime of a filament lamp is a trade-off between efficiency and longevity. Lamps for general illumination are designed to provide a lifetime of several hundred hours, even though they are not as efficient.

A *halogen lamp* is an incandescent lamp filled with a halogen gas (iodine or bromine). **See Figure 3-9.** Halogen lamps are also known as tungsten-halogen lamps or quartz-halogen lamps. The gas combines with tungsten vapor from the filament in a reaction that then redeposits the tungsten at the filament hot spots, preventing the early failure of the lamp.

Halogen Lamp

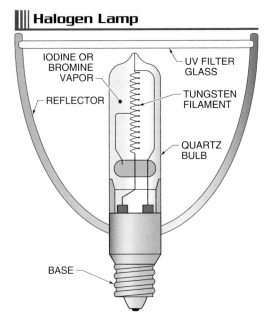

Figure 3-9. Halogen lamps are a variation of conventional incandescent lamps that operate at higher temperatures required for improved luminous efficacy.

Halogen lamps can operate at higher temperatures that would otherwise cause unacceptably short lamp lifetimes in ordinary incandescent lamps. The higher temperature improves luminous efficacy up to about 18 lm/W to 24 lm/W. However, the lamp's envelope must then be made of hard glass or fused quartz instead of ordinary soft glass, which would soften at these temperatures. The lamp lasts about twice as long as a standard incandescent lamp, but can cost up to three times as much.

Incandescent lamps can be operated from AC or DC power, depending on the specific lamp design. Incandescent lamps are connected directly to electrical power sources and their light output is directly determined by the input voltage. Incandescent lamps are rated for a certain light output at a certain voltage. However, they can be operated at other voltages, which will similarly affect the light output. A higher voltage produces more light, but severely limits the lamp's lifespan and is not recommended. However, incandescent lamps are easily dimmable by lowering the voltage, which also increases lifespan, though this lowers the luminous efficacy.

Gas Discharge Lamps

A *gas discharge lamp* is an electric lamp that produces light by establishing an arc through ionized gas. Gas discharge lamps operate from AC power, but cannot be directly connected to electrical power sources because the high inrush currents would destroy the lamps. These lamps must include a ballast in the lighting circuit. **See Figure 3-10.** A *ballast* is a device with a circuit that controls the flow of current to gas discharge lamps while providing sufficient starting voltage. Ballasts include inductors, capacitors, and other components.

‖‖ Ballasts

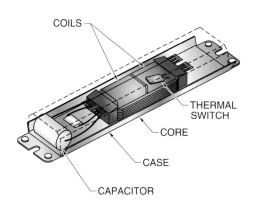

COILS

THERMAL SWITCH

CORE

CASE

CAPACITOR

Figure 3-10. *A ballast is installed in gas discharge lamp fixtures to control the flow of electricity to the lamp.*

Common lamp ballasts are based on either inductors or solid-state electronics. An *inductor* is a coil of wire that creates a magnetic field, which resists changes in the current flowing through the coil. Due to this effect, ballasts based on inductors are often called magnetic ballasts. A transformer may also be necessary to step the voltage up for lamp starting. This design makes large and heavy ballasts, which are built into the light fixture. They can also emit a humming noise due to the changing magnetic field and cause lamps to flicker.

Electronic ballasts use solid-state devices to provide the proper starting and operating electrical power to gas discharge lamps. These ballasts usually change the frequency of the power from the standard 60 Hz to 20,000 Hz or higher, which reduces the flicker and humming effects commonly associated with magnetic ballasts. In addition, the higher frequencies improve lamp light output. Electronic ballasts are generally smaller, lighter, and more efficient (and therefore cooler) than magnetic ballasts. They may even be small enough that some are included with the lamp instead of built into the fixture. Advanced electronic ballasts may also be capable of dimming the lamps with pulse-width modulation and communicate with protocol-based control networks.

The different types of gas discharge lamps have different starting and operating requirements, so ballasts must be designed for the lamp they are used with.

Fluorescent Lamps. A *fluorescent lamp* is an electric lamp that produces light from the ionization of mercury vapor. A fluorescent lamp consists of a cylindrical glass tube sealed at both ends and filled with mercury vapor and an inert gas (normally argon). **See Figure 3-11.** Each end includes a cathode, which makes the electrical connection to the power supply. A *cathode* is a tungsten coil coated with a material that releases electrons when heated. The gas is bombarded by electrons from the cathode, causing the gas to conduct an arc between the two cathodes. This provides ultraviolet light. The ultraviolet light causes a phosphor coating on the inner surface of the bulb to fluoresce, emitting visible light.

Outdoor lighting typically uses gas discharge lamps, particularly low-pressure sodium lamps, because they are long lasting and energy efficient.

Fluorescent Lamps

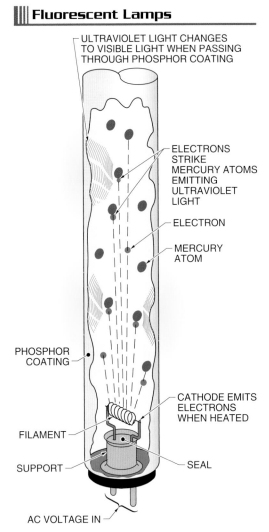

Figure 3-11. Fluorescent lamps emit a stream of electrons that produce ultraviolet light when they strike mercury atoms. The fluorescent light causes coatings on the inside of the lamp to glow in visible light.

Unlike incandescent lamps that produce a large amount of waste heat, fluorescent lamps produce very little heat and are relatively cool to the touch. With 55 lm/W to 100 lm/W, fluorescent lamps are far more efficient than incandescent lamps.

Fluorescent bulbs do not start well in cold climate conditions because the cathode must be hot enough to emit electrons. If the lamp is too cold, the cathode requires more time to heat. Temperature also affects light output. Standard indoor fluorescent lamps are designed to deliver a peak light output

at approximately 75°F ambient temperature conditions. **See Figure 3-12.** Moderate changes in ambient temperature (50°F to 105°F) reduce the light output by 10% or less. However, temperatures lower than 50°F or higher than 105°F have a significant affect on light output.

Fluorescent Lamp Output

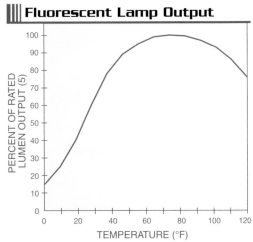

Figure 3-12. Lumen ratings for fluorescent lamps are designed to operate optimally at 75°F. Operating at temperatures above and below 75°F reduces the light output.

Outdoor rated fluorescent lamps include an outer glass jacket. The jacket helps hold in heat, which shifts the peak output to a lower ambient temperature point. Jacketed lamps are available with different peak output points. Jacketed lamps are recommended for many outdoor applications and certain indoor applications such as freezer warehouses, subways, and tunnels where cold and windy conditions exist.

High-Intensity Discharge Lamps. A *high-intensity discharge (HID) lamp* is an electric lamp that produces light by striking an electrical arc across tungsten electrodes housed inside an arc tube. The arc tube is a specially designed, fused quartz or fused alumina tube that contains metal and gas vapors and electrodes. The gas helps the arc form. Then, the metals produce the light once they are heated to a point of evaporation. The arc tube is enclosed within an outer bulb that may include coatings to improve color rendering, increase light output, and reduce surface brightness.

This startup process can require several minutes, so HID lamps are typically used for general lighting that is switched on a schedule (which can allow for startup time) or that is operated for long periods.

HID lamp types include low-pressure sodium, mercury-vapor, metal-halide, and high-pressure sodium lamps. **See Figure 3-13.** Each has different starting, operating, and light output characteristics. HID lamps can be dimmed, though this requires special ballasts.

A *low-pressure sodium lamp* is an HID lamp that produces light by an electrical discharge through low-vapor-pressure sodium. The arc tube includes a small amount of solid sodium metal, which is vaporized by the starting arc. This lamp has the highest efficacy rating of any lamp, up to 200 lm/W. This is 10 times the output of an incandescent lamp. Low-pressure sodium lamps do not require a starting electrode, but the ballast must provide a very high voltage to start the arc.

▐▐▐ High-Intensity Discharge (HID) Lamps

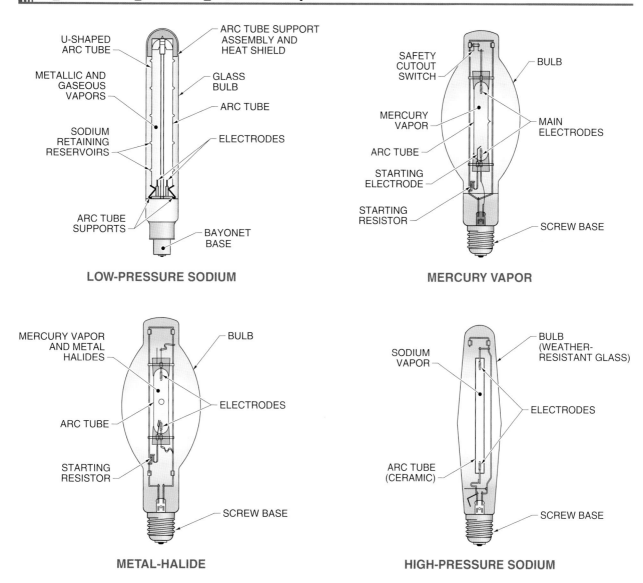

Figure 3-13. *While there are several designs of HID lamps, they share some basic features.*

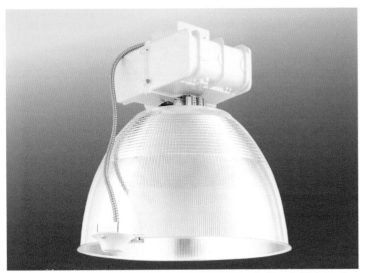

Acuity Brands Lighting, Inc.
HID lighting is typically used for applications requiring high levels of light over large areas.

A *mercury-vapor lamp* is an HID lamp that produces light by an electrical discharge through mercury vapor. A mercury-vapor lamp contains an extra starting electrode. An electrical field is set up between the starting electrode and one main electrode when power is first applied to the lamp. The electrical field causes current to flow and an arc to strike. Current flows between the two main electrodes as the heat vaporizes the mercury. Mercury-vapor lamps have luminous efficacy ratings of about 50 lm/W to 60 lm/W.

A *metal-halide lamp* is an HID lamp that produces light by an electrical discharge through mercury vapor and metal halide. Metal-halide lamps produce about 65 lm/W to 115 lm/W and have shorter life spans than the other HID lamps. Power interruptions, even brief, will extinguish the arc and the high vapor pressure in the hot arc tube will prevent re-striking the arc. The lamp must cool down for 5 min to 10 min before the lamp can be restarted.

A *high-pressure sodium lamp* is an HID lamp that produces light when current flows through sodium-vapor under high pressure and high temperature. High-pressure sodium lamps do not have a starting electrode. The ballast delivers a high-voltage pulse to start and maintain the arc. High-pressure sodium lamps produce about 100 lm/W to 150 lm/W.

Light-Emitting Diodes

A *light-emitting diode (LED)* is a semiconductor device that emits a specific color of light when DC voltage is applied in one direction. LEDs are not yet common for general lighting in commercial or residential applications, but some LED-based lighting products are available and further growth is expected. LED lighting has potential for smaller, colorful, smarter, more flexible lighting applications. LEDs also have extremely long life spans of 50,000 hours or more and are resistant to physical shock and vibration.

Like other types of diodes, LEDs consist of a junction of two semiconductor materials, which electrons can cross in only one direction. **See Figure 3-14.** LEDs use particular materials that cause electrons to emit a small amount of light energy as they cross the junction. The wavelength of the energy emitted, and therefore the color, depends on the junction materials. The most common LED colors are reds, ambers, yellows, and greens. Blue, violet, and white LEDs are less common and significantly more expensive.

Light-Emitting Diodes

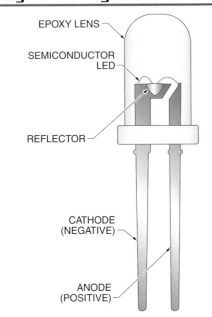

EPOXY LENS

SEMICONDUCTOR LED

REFLECTOR

CATHODE (NEGATIVE)

ANODE (POSITIVE)

Figure 3-14. Light-emitting diodes are semiconductors that produce light from very small DC currents.

Color Rendering

Color rendering is the appearance of a color when illuminated by a light source. Light sources produce light of different colors, so this directly affects the color rendering of objects. For example, a red color may be rendered light, dark, pinkish, or yellowish depending on the type of light source under which it is viewed.

The color of light is determined by its wavelength. Wavelengths of visible light range in color from red, orange, and yellow to green, blue, and violet. When a light source, such as the sun, produces energy over the entire visible light spectrum in approximately equal quantities, the combination produces white light. Light from many other sources, however, is concentrated at certain wavelengths, producing light of a nonwhite color.

Light output color is primarily determined by the type of lamp and the way in which it produces light. However, many lamp types allow minor variations in design that affect the resulting light color. Lamp manufacturers use these variations to adjust light output in an effort to produce the more desirable light characteristics. Lamp choice is often a balance between acceptable light characteristics and operating costs.

Fluorescent Lamps

Typical cool white fluorescent lamps produce a pale, blue green whiteness. These fluorescent lamps are suggested for higher lighting requirements, such as industrial and commercial workspaces, because they are an efficient light source. However, the light color can be harsh and give skin an unpleasing tone.

Warm white fluorescent lamps produce a yellow white color that gives a warm feeling and enhances brighter colors. This lamp type is recommended for lower lighting levels. For example, restaurants may use warm, white fluorescent lamps to help set a cozy mood.

White fluorescent lamps produce a very pale yellow whiteness that is often labeled as a "natural light" or "sunlight" color. These lamps are midway between the blueness of cool white and the yellowness of warm white. This lamp is used in areas where a more neutral effect is desired, such as supermarkets, libraries, classrooms, and where fluorescent lighting is mixed with incandescent lighting.

Low-Pressure Sodium Lamps

Low-pressure sodium lamps produce a yellow to yellow orange light, which significantly distorts the true color of objects viewed under the lamp. However, this distortion is usually acceptable for outdoor lighting such as street, highway, parking lot, and floodlight applications, given the very high luminous efficacy for these lamps, which makes them inexpensive to operate.

Mercury-Vapor Lamps

Mercury-vapor lamps produce a distinctly bluish white light, creating a poor rendering of reds and a purplish rendering of blues. Light from mercury vapor lamps is especially harsh for skin tones, which tend to look unhealthy and bloodless. Special phosphor coated bulbs improve the color rendering by converting some of the invisible ultraviolet light produced by the mercury into red light, somewhat evening the total wavelength distribution to make the light more white.

Metal-Halide Lamps

Metal-halide lamps produce a light yellow to white light, providing good overall color rendering. Metal halide lamps are recommended for many sport, indoor, and outdoor lighting applications because they are a good balance between high luminous efficacy and true color appearance. When dimmed, however, the light output color shifts toward blue, producing a light similar to that of mercury-vapor lamps. Also, the light color changes slightly both during the startup process and over the life of the lamp. This effect can usually be seen in the various colors of lamps in multiple-lamp fixtures.

High-Pressure Sodium Lamps

High-pressure sodium lamps produce a golden white light, providing reasonably good color rendering, though reds, greens, blues, and violets appear somewhat muted. High pressure sodium lamps are very efficient and produce only moderate color distortion, making them well suited for use in parking lots, street lighting, shopping centers, exterior buildings, and storage areas.

Individually, LEDs produce little light output, but their power consumption is so low that the luminous efficacy is about 25 lm/W to 70 lm/W, comparable to halogen and fluorescent lamps. General lighting applications require a large number of LEDs to be grouped together into an array to provide the necessary lumen output. **See Figure 3-15.** While individual LEDs do not generate much heat, many high-output LEDs grouped together can become hot enough to degrade the performance and lifespan of the LEDs. Cooling LEDs requires heat sinks and/or forced ventilation.

⫿⫿⫿ LED Lamps

LEDtronics, Inc.

Figure 3-15. Individually, LEDs output too little light to be useful for most general lighting applications. However, LEDs can be easily grouped into large arrays.

Most LED lighting devices require an LED driver, which has a role similar to ballasts. An *LED driver* is a circuit that provides a constant DC voltage source and protection from line voltage transients, such as surges and sags. LED drivers can be wired into traditional lighting controls for both switching and dimming applications.

LIGHTING SYSTEM CONTROL DEVICES

Control devices for lighting systems include two types of devices corresponding to inputs and outputs within the system. The first group, the inputs, includes sensors that detect a need for a change in the lighting within a space. The second group, the outputs, includes the devices that enact those changes.

Occupancy Sensors

Occupancy sensors are the most common form of automatic lighting control used in buildings today. An *occupancy sensor* is a sensor that detects whether an area is occupied by people. It is used to save energy by deactivating lights when no one is in the room or area. Lighting controllers use occupancy sensors to automatically turn lights ON when a room becomes occupied, keep the lights ON without interruption while the controlled space is occupied, and turn the lights OFF within a preset time period after the space has been vacated.

Occupancy sensors are mounted on the ceiling or at a high level on a wall. Occupancy sensors provide a set of contacts that is used to control a relay or send a digital signal to a controller. This is an ON or OFF device based on the presence of people in the area. Occupancy sensor devices often include a power supply and relay unit that both pulls power from the lighting branch circuit to operate the sensor and switches the lights. **See Figure 3-16.**

⫿⫿⫿ Occupancy Sensor Circuits

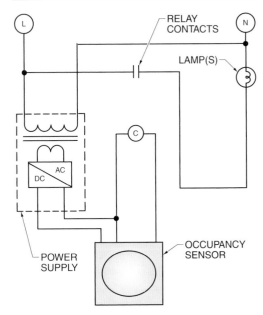

Figure 3-16. Occupancy sensor devices typically include components to provide power to the sensor and a relay to switch the lighting circuits.

Occupancy sensors may use one or more technologies in order to sense people within a space, including passive infrared, ultrasonic, and microwave.

Passive Infrared (PIR) Occupancy Sensors. A *passive infrared (PIR) sensor* is a sensor that activates when it senses the heat energy of people within its field-of-view. **See Figure 3-17.** Infrared occupancy sensors are line-of-sight devices and require the entire detection area to be within view. Any obstructions between the sensor and people in the room may make an infrared occupancy sensor ineffective.

Passive Infrared (PIR) Occupancy Sensors

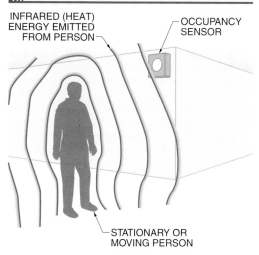

Figure 3-17. Passive infrared (PIR) occupancy sensors detect people by their heat energy.

The PIR sensor detects the rapid change in temperature caused by a person passing through the detection area. A rapid temperature rise of several degrees is required to detect motion. The device that detects the change in temperature within a PIR sensor is a pyroelectric sensor. A *pyroelectric sensor* is a sensor that generates a voltage in proportion to a change in temperature. Electronics within the PIR analyze the signal and if it exceeds a certain threshold, the device switches the contacts.

The infrared energy from the detection area is focused on the pyroelectric sensor with either a Fresnel lens or a parabolic mirror that determines the field-of-view. A *Fresnel lens* is a lens with special grooved patterns that provide strong focusing ability in a compact shape. However, the Fresnel lens is prone to dead zones. **See Figure 3-18.** A *dead zone* is an area within the field-of-view of an occupancy sensor that is not covered by one of the detection zones on the lens. Parabolic mirror technology is a newer and more advanced technology, allowing continuous floor to ceiling coverage.

PIR Occupancy Sensor Field of View

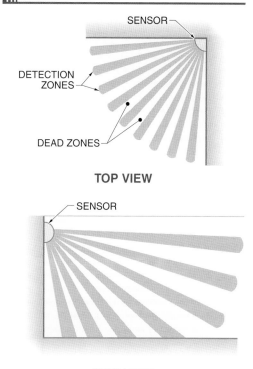

Figure 3-18. Fresnel lenses focus a large overall area onto a PIR occupancy sensor but include dead zones in the field-of-view.

The field-of-view can be modified by masking (with a special tape) or replacing the focusing element. For example, different Fresnel lenses are available to change the size and shape of the sensor's detection area. Lenses can be designed to focus infrared energy from a long, narrow area, such as a corridor.

Manufacturers are improving the reliability of PIR sensors. For example, signal processing circuits evaluate size, speed, shape, and duration of the signal to determine whether the PIR sensor has detected an actual person. Some PIR sensors use more than one pyroelectric sensor. Other technologies look at the pulse generated by the pyroelectric sensor and evaluate whether the signal is the same on both the positive side of the pulse and the negative side of the pulse to determine the validity of the detection. Many of the detectors also have adjustable settings to select how many pulses are required to trigger a detection.

Ultrasonic Occupancy Sensors. An *ultrasonic sensor* is a sensor that activates when it senses changes in the reflected sound waves caused by people moving within its detection area. The sensor emits low-power, high-frequency sound waves and listens for a phase shift in the reflected sound returning to the sensor, which indicates a moving target. **See Figure 3-19.** This is very similar to how radar works. This even works to some degree around corners, so the detection area is larger than the line-of-sight field-of-view. (The hard surfaces of the walls will reflect the sound pattern around the room).

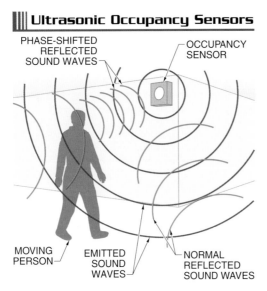

Figure 3-19. *Ultrasonic occupancy sensors detect people by the changes in sound waves reflected by their movement.*

Since they detect people by their motion and not necessarily their presence only, ultrasonic occupancy sensors are also widely known as motion detectors or motion sensors. The reliance on motion for sensor activation can also cause false deactivations when occupants are very still for long periods.

Ultrasonic sensors are more sensitive than most infrared sensors, but this can cause unintentional detection signals in unoccupied rooms from incidental moving objects, such as fan blades or curtains. Sensitivity levels on an ultrasonic occupancy sensor can be adjusted to help avoid these problems. Ultrasonic sensors are well suited for restrooms with stalls or in open office environments with short-walled cubicles.

Microwave Occupancy Sensors. A *microwave sensor* is a sensor that activates when it senses changes in the reflected microwave energy caused by people moving within its field-of-view. **See Figure 3-20.** Like ultrasonic occupancy sensors, microwave occupancy sensors are also commonly known as motion detectors because they detect people by their movement.

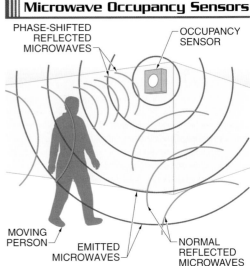

Figure 3-20. *Microwave occupancy sensors detect movement by the changes in the reflected microwave energy.*

Microwave sensors are not affected by temperature changes, sunlight, air turbulence, or humidity. Microwaves will penetrate most building materials, including glass. However, microwave occupancy sensors have some limitations. They have difficulty detecting slow moving objects. They can cause false activations from vibrations of the mounting surface or moving metal objects, such as fan blades, within the field-of-view. Also, since microwave travel distance is theoretically unlimited, the sensors typically include a range control feature. This allows the sensor to respond only to motion within a certain distance from the sensor.

Dual-Technology Occupancy Sensors. Since each occupancy-sensing technology has both advantages and disadvantages, many occupancy-sensing devices include a combination of two sensors together. This takes advantage of the best of each technology to maximize accuracy. By activating the switch only when both sensors within the device sense people within the space, false activations are greatly reduced. The most common combination within dual-technology occupancy sensors is a passive infrared sensor with an ultrasonic sensor, though combinations with microwave-sensing technology are also possible. However, dual-technology devices are typically more expensive than single-technology occupancy sensors.

Light-Level Sensors

Light-level sensors continuously monitor the intensity of light within an area so that a lighting controller can adjust the lighting system accordingly. A *light-level sensor* is a control device with a photocell that measures the intensity of light it is exposed to. A *photocell* is a small semiconductor component that changes its electrical characteristics, such as output current or resistance, in proportion to the light level. Two common types of photocells are photoresistors and phototransistors. **See Figure 3-21.**

Photoresistor. A *photoresistor* is a photocell that changes resistance in proportion to its exposure to light. The resistance increases for lower light levels and decreases with higher light levels. In other words, a photoresistor is a variable resistor that is controlled by the light level. Most photoresistive cells are made of cadmium sulfide (CdS) and are often used in outdoor lighting control. Photoresistors are also known as photoconductors.

Photocells

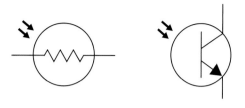

PHOTORESISTOR   PHOTOTRANSISTOR

Figure 3-21. Photocells used in light-level sensors include photoresistors and phototransistors.

Phototransistors. A *phototransistor* is a phototocell that controls a current in proportion to the light level. Otherwise stated, this is simply a transistor that is controlled by light level. A transistor is a junction of two types of silicon that controls current flowing through its two terminals in response to a control input, similar to a valve. Instead of having a third terminal used for electronic control signals like most transistors, phototransistors use light as the control input. Phototransistors are becoming more common than photoresistors because silicon is more resistant to high temperatures and airborne contamination. This makes it ideal for new outdoor lighting control applications or to replace failed photoresistors.

Electrical changes in a photocell are interpreted by the sensor or controller, sometimes a separate unit, and translated into light-level information. The output may be an ON/OFF signal (in relationship to a setpoint) or a variable value in the form of an analog signal or structured network message, depending on the intended use of the sensor. ON/OFF-type light-level sensors use an internal relay or switching transistor to directly control a low-power lighting circuit or operate an external relay or contactor to control higher power lighting circuits. Analog-type light-level sensors may directly control the dimming of one or more light fixtures or share information with a central controller, which then provides more sophisticated controls.

A light-level sensor can be used for either open-loop or closed-loop lighting control. Each of these control schemes can be implemented in either top-lit or side-lit installations.

The installation of multiple light-level sensors within the space improves the accuracy of light-level control.

In the open-loop method, the light-level sensor views daylight directly and does not sense light from the artificial lighting system. **See Figure 3-22.** Therefore, the resulting light-level information does not include the effect of the controlled lighting. In a top-lit installation, daylight enters the room via skylights or light wells. The light-level sensor is placed in the well or in direct view of the skylight. The sensor is placed on a south-facing surface so as not to view the sun directly at any time during the day. In side-lit installations, light enters the room through vertical windows. The sensor is placed on the ceiling 6″ to 36″ from the window and should view as little of the artificial lighting as possible. Covering part of the sensor may be necessary to limit the field-of-view to daylight only.

Open-Loop Lighting Control

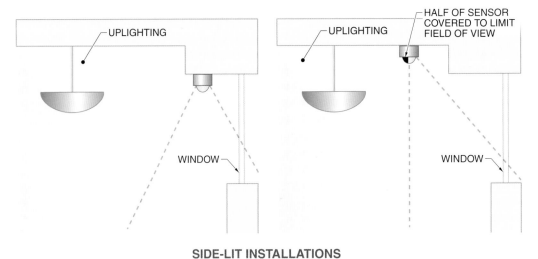

TOP-LIT INSTALLATIONS

SIDE-LIT INSTALLATIONS

Figure 3-22. The open-loop lighting control method uses light-level sensors to determine only the contribution of daylight to indoor lighting.

In the closed-loop method, the light-level sensor views an area of the room that represents the overall lighting level from all sources, including artificial light and daylight. **See Figure 3-23.** Therefore, the resulting light-level information is a feedback of the lighting system control. The top-lit installation is used when daylight falls off significantly in the controlled space because of obstructions such as shelving or partitions. Multiple sensors are distributed throughout the space in order to view different combinations of artificial and daylight contributions. Side-lit installations, however, assume that the space has a fairly even overall contribution of daylight, and sensors are placed 48″ or more from the windows. In some circumstances, covering part of the sensor may be needed to avoid viewing a light source directly.

Closed-Loop Lighting Control

TOP-LIT INSTALLATIONS

SIDE-LIT INSTALLATIONS

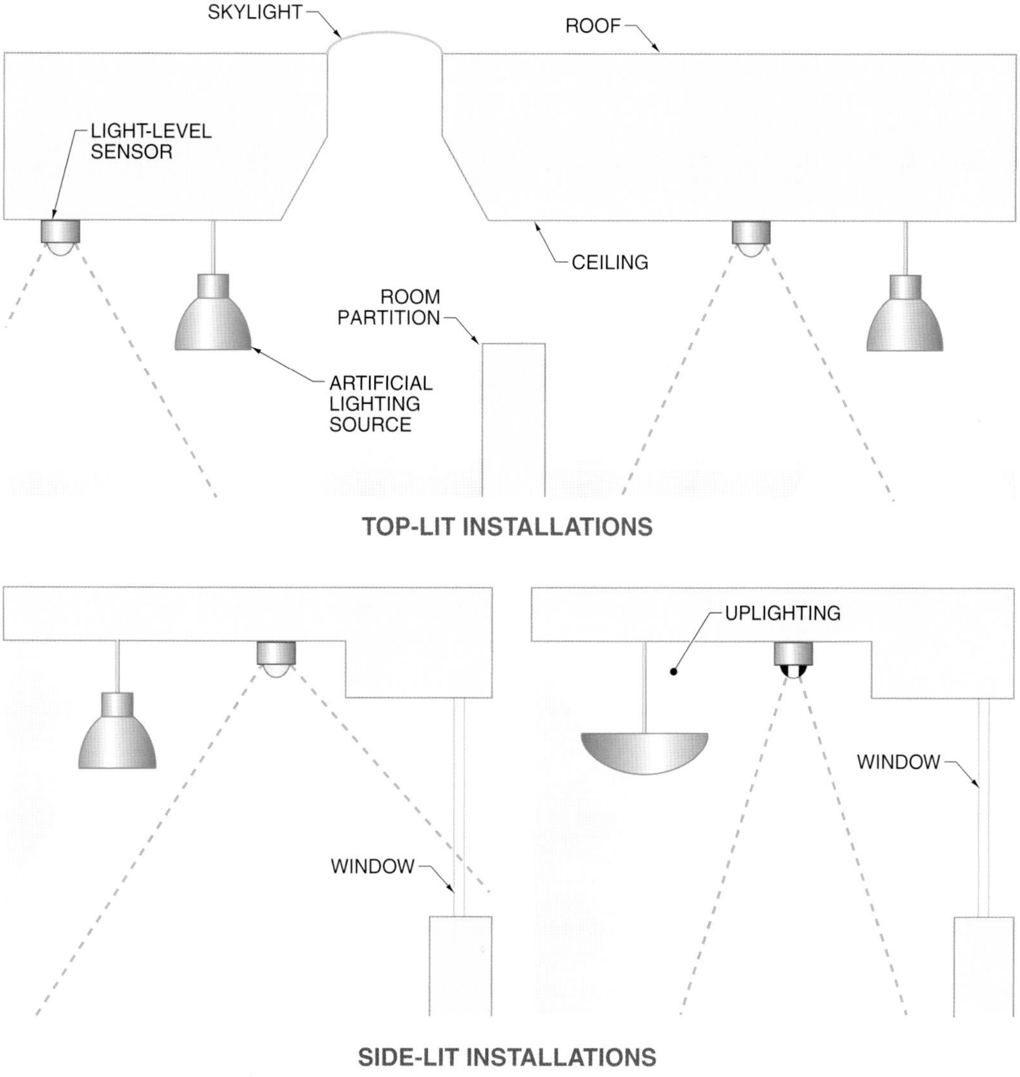

Figure 3-23. *The closed-loop lighting control method uses light-level sensors to determine the overall lighting resulting from all sources of light, including artificial and natural sources.*

Switches

Switching controls the ON/OFF state of a lighting load or any other type of electrical load. This may be done with a variety of manual or automatic devices, but the basic idea of switching is the same. Switches cannot dim lamps; they can only alternate the lamps' state between completely ON and completely OFF.

Manual light switching is familiar to everyone. The light switch on the wall is included in the same circuit as the light in the ceiling. Manually actuating the switch makes or breaks contacts with the ungrounded conductor supplying electrical power to the light. This interrupts the flow of current to the lamp, causing it to de-energize and go dark.

Automated light switching accomplishes the same result but by different means. Electronically controlled relays make or break electrical power contacts when they receive signals from lighting controllers. The controllers switch lighting circuits based on user inputs, programs, or schedules. Manually operated light switches are still common within automated lighting control systems. Signals from these devices are sent to the controller, where they may override the automated program.

Lighting Contactors. Automated switching can be done with lighting contactors. **See Figure 3-24.** A *lighting contactor* is a solenoid-operated, high-power relay used with lighting circuits. A control signal voltage is applied to the lighting contactor, which energizes the coil, creating a magnetic field that pulls a set of normally open contacts closed. These contacts switch a lighting circuit, anywhere between 24 V and 480 V. Some lighting contactors include pilot lights to indicate the switched status.

Lighting Contactors

Products Unlimited

Figure 3-24. Lighting contactors are similar to other types of electromechanical relays but are designed specifically for lighting loads.

A lighting contactor is very similar in operation to a magnetic motor starter but differs in contact configuration and other features. Most motor starters are three- or four-pole contactors (for 3ϕ motors) and include overload protection to prevent overcurrent conditions in the motor.

In comparison, lighting contactors switch 1ϕ power, but can switch multiple lighting circuits simultaneously, so they may include anywhere from 2 to 12 sets of contacts. Lighting contactors also do not include overload protection.

Most lighting contactors use double-break contacts to open and close the lighting circuit. Double-break contacts provide better arc suppression than single-break contacts when opening high-power lighting circuits, such as those with high wattage lamps or inductively ballasted fluorescent or HID lamps on the same branch circuit. Lighting contactors may be either electrically held or mechanically held.

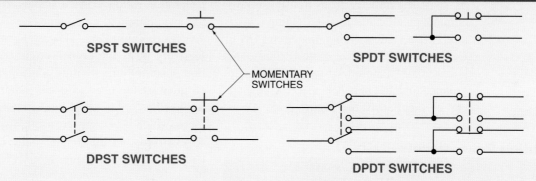

Switches are also described by their behavior after being actuated. A maintained contact switch is a switch that remains in position after actuation. A momentary contact switch is a switch that always returns to its normal position after actuation. The normal position can be either open or closed. Any type of switch can be designed as a maintained or momentary contact switch.

Single-Pole Switches

The most common method of switching is the single-pole switch. A single-pole switch is a switch that makes or breaks the connection of a single conductor.

A single-pole, single-throw (SPST) switch switches a single conductor in a single branch circuit. This switch has two screw terminals and ON/OFF designations.

There are variations of single-pole switches with more than one throw. A single-pole, double-throw (SPDT) switch is a switch that makes or breaks the connection of a single conductor with either of two other single conductors. This switch has three terminal screws, so it is commonly called a "three-way switch."

SPDT switches are often used in pairs to control lighting circuits from more than one location, such as switching lights in rooms with multiple entrances. It has no ON/OFF markings because either toggle position may be ON or OFF, depending on the position of the other switch.

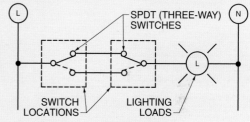

Double-Pole Switches

A double-pole switch is a switch that makes or breaks the connection of two circuit conductors.

A double-pole, single-throw (DPST) switch switches two circuit conductors in a single branch circuit. These switches are used for 240 V load circuits that have two ungrounded (hot) conductors. This switch has four terminal screws and ON/OFF markings. A DPST switch is equivalent to two SPST switches that are actuated by the same mechanism.

A double-pole, double-throw (DPDT) switch is a switch that makes or breaks the connection of two conductors to two separate circuits. This switch has six terminal screws and may have a center OFF position. A DPDT switch is equivalent to two SPDT switches that are actuated by the same mechanism.

A special type of DPDT switch shares the poles between the two circuits by cross-connecting them when actuated. This is known as a "four-way switch" because it has four terminal screws. Four-way switches are easily confused with double-pole, single-throw (DPST) switches because they share the same number of terminals, but they have very different functions. Four-way switches are wired between a pair of SPDT (three-way) switches to control circuits from a third location. In fact, any number of four-way switches can be used in this way to add switching locations. This switch has no ON/OFF markings.

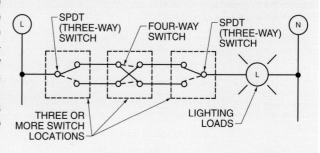

Lighting Relays. Lighting circuits can also be switched directly from lighting relays, eliminating all contactors. A lighting relay is a relay that switches a lighting circuit ON or OFF when it receives a control signal. This is essentially a motorized switch. Relays are grouped together in panels that resemble circuit breaker panels. **See Figure 3-25.**

Lighting Relay Panel

Figure 3-25. A panel of lighting relays serves as a lighting control center for switching lighting circuits.

In some cases, a lighting relay is combined with a circuit breaker to both control the circuits and provide overcurrent protection in the same device. These are known as remote-controlled circuit breakers or power-operated circuit breakers.

A digital controller inside the lighting panel utilizes low-voltage signals to control the switching and communicate with other devices or controllers. This arrangement effectively turns the branch circuit distribution panel into the lighting controller panel, eliminating the need to install other panels with contactors or other light switching devices, which is particularly useful for retrofit applications. In many cases, the existing panel can remain in place and the old circuit breakers and backplanes are simply replaced with new remote controlled circuit breakers and backplanes.

However, not just any circuit breaker can be used for switching in this manner. It must have special ratings. Circuit breakers suitable for switching applications shall be UL listed for HID (high-intensity discharge lighting) and SWD (switch-duty) loads. Such circuit breakers are marked with "SWD" or "HID." The breakers may come in single two-pole and three-pole versions with ratings up to 30 A.

Time Clocks. Scheduling is a common mode of lighting control, particularly for large open office areas. *Scheduling* is the automatic control of devices according to the date and time. Scheduling lighting control systems automatically turn lights ON and OFF at certain times of the day. This saves energy by ensuring that lights are not left on when the building is unoccupied. It is also convenient, since the lights do not require manual switching. This type of control requires time clocks.

A *time clock* is a switch that automatically changes state (switches) at certain times. Time clocks may be either mechanical or digital and may have a variety of time-related features. **See Figure 3-26.** A basic time clock turns the lighting system ON at a certain time and then OFF at a certain time. However, since sunrise and sunset times change throughout the year, lighting controlled by this type of switch may sometimes be ON unnecessarily. For example, a basic time clock controlling an outdoor lamp at fixed times will waste energy in the summer because the sun rises earlier and sets later, providing the necessary light while the lamp is still switched ON. **See Figure 3-27.**

Time Clocks

MECHANICAL

Leviton Manufacturing Co., Inc.

ELECTRONIC

Figure 3-26. Mechanical time clocks are common for simple applications, but electronic clock controls are far more versatile for sophisticated lighting schedules.

Fixed-Schedule Time Clocks

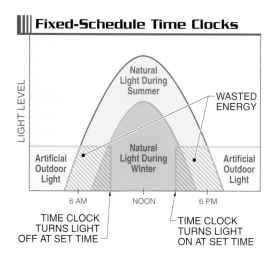

Figure 3-27. Fixed-schedule time clocks can be inefficient for outdoor lighting because they cannot adjust for the seasonal change of sunrise and sunset.

The solution to this limitation is a time clock that automatically adjusts for sunrise and sunset. Some mechanical clocks offer this feature, but digital time clocks provide the greatest flexibility because they may compensate for sunrise and sunset, daylight saving time, and even anticipated changes in building occupancy. Digital time clocks can be programmed for different day schedules depending on the day of the week, special holidays, or time of the year. For example, a lighting circuit may be controlled to be ON from 8 AM to 6 PM, Mondays through Fridays, but not on holidays.

Dimmers

Dimmers are devices that vary the light level (intensity) of lamps by controlling the amount of power that is delivered to the lamp. Like switching, lamp dimming can be controlled by either manual or automatic means.

There are two methods of dimming: reducing the voltage to the lamp to some lower value (voltage modulation) and reducing the amount of time that full voltage is applied to the lamp (phase control). Each of these methods corresponds to a single major component found within these dimmers, though the devices typically include other circuits and electronic components to support this function. The applicable dimming method must be compatible with the lamp type.

Many automated dimmers are also capable of gradually changing between two different light levels. This is especially common with many dimmer switches, which provide soft-ON and fade-OFF switching. The soft-ON feature gradually raises light levels to the desired brightness. This soft-ON feature is recommended to avoid overwhelming occupants with sudden, bright lighting. It also limits the inrush current, prolonging lamp life. The fade-OFF feature gradually dims lamps until they are completely off. The rate at which the light levels change may be linear or non-linear, depending on the dimmer design. **See Figure 3-28.** Some dimmers may include multiple dimming curves to choose from. The inverse-square dimming curve is most comfortable to people's eyes.

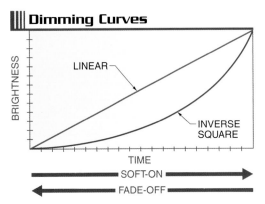

Figure 3-28. Some automated dimmers are capable of ramping light levels up or down according to mathematical dimming curves.

Dimmer circuits may be included in individual light fixtures, allowing fixtures to be dimmed separately, or in dimming controls for entire lighting circuits, allowing multiple fixtures to be dimmed simultaneously.

Autotransformer Dimmers. An autotransformer is used in dimmers to step down line voltage to a lower voltage. **See Figure 3-29.** The AC voltage sine wave frequency remains constant, but with a lower magnitude. Since the lamp is a fixed resistance, the current is also reduced. Therefore, the power consumed by the lamp is reduced, which dims the lamp.

An *autotransformer* is a transformer with only one winding, which is tapped by both the primary and secondary sides at various points to produce varying voltages. A variable autotransformer makes the secondary connections with a brush sliding along the winding, allowing for an almost continuously variable turns ratio. **See Figure 3-30.** This provides a smooth control of the output voltage from 0 V to the full line voltage. Variable AC autotransformers are often known as "variacs."

Variable Resistors

A simple and effective way to dim lamps is to use a resistance to change the voltage available to the lamp. A variable resistance is used to vary the lamp's voltage from 0 V to the full-line voltage, and any point in between, which provides dimming control. There are two types of variable resistors used in this way: potentiometers and rheostats.

A potentiometer is a three-terminal variable resistor that can be used to divide a supply voltage across its two resistances. A wiper, connected to the middle terminal, moves across a resistive material to divide it into two resistances that change in relation to one another. The voltages across these two resistances are similarly proportioned. Therefore, potentiometers are commonly known as voltage dividers. Lighting loads are connected to the resulting voltages in parallel.

A rheostat is a two-terminal variable resistor that limits the current and voltage available to loads connected in series. Both potentiometers and rheostats operate in similar ways, though rheostats are commonly designed for larger power applications, such as lighting circuits. By not connecting one of the three terminals, a potentiometer can be wired as a rheostat, provided it can handle the power requirements.

While variable resistors are effective dimmers, they must absorb power from the lighting circuit in order to reduce power to the lamps. This means that they are very inefficient. This power is also converted to heat, which must then be dissipated by large heat sinks. Variable resistance dimmers were common until solid-state electronics made dimming possible without producing significant heat. Now they are rarely used for lighting control.

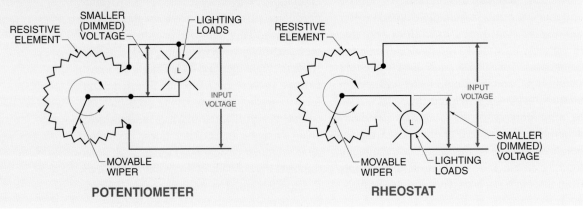

Autotransformers are relatively efficient dimmers and their dimming effect is independent of the applied load. They deliver a true sine wave output with no radio frequency noise. However, automated control of autotransformers is impractical, usually involving coupling a motor to a shaft for adjusting the secondary brush contacts. Precise motor controls must slowly and steadily adjust the output voltage, reducing or increasing the brightness of the attached lamps. Therefore, with solid-state controls available, autotransformer dimmers are becoming less common.

Phase Control Dimmers. The other method of dimming is phase control. *Phase control* is the frequent switching of AC voltage to limit the power in a circuit. This switching occurs at twice the line frequency (60 Hz), so the human eye cannot see the ON/OFF periods, only a dimming of the lamp in proportion to the amount of time the power is ON. For example, for a dimming of approximately 50%, each half cycle of the AC sine wave is switched OFF and ON so that only about 50% of the power is available to the lamp. Phase control distorts the shape of an AC sine wave, clipping it at the edge of each half cycle.

The extremely fast switching required for phase control dimming is accomplished with solid-state switches, such as triacs and transistors. These semiconductor switches allow the flow of a large current through two terminals only when a small trigger current is applied to a third. Since each semiconductor switch allows current flow in only one direction, a pair of switches in inverse parallel is needed to control and conduct current for AC power. **See Figure 3-31.**

Phase control dimming must take into account type of transformer used to supply AC power to the lighting circuit. A magnetic transformer is an inductive load and tends to cause the current to lag behind the voltage. Electronic transformers, however, are usually capacitive, which causes a leading current. These lagging or leading currents influence when the dimmer should turn the voltage ON and OFF.

Voltage Modulation Dimming

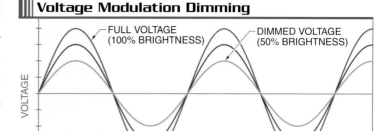

Figure 3-29. One method of dimming involves using autotransformers to reduce the amplitude of the voltage sine wave. This does not distort the sine wave shape.

Autotransformers

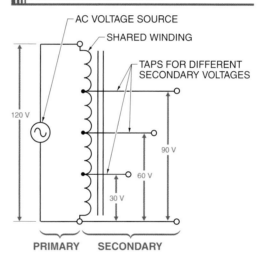

Figure 3-30. Autotransformers use only one winding, which is tapped at different points on the secondary side to provide lower voltages.

Phase Control Circuit

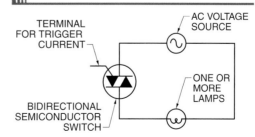

Figure 3-31. Phase control dimmers switch an AC voltage ON and OFF frequently to reduce the amount of power to the lighting circuit, which dims the lamps.

Dimmers used with lagging currents are known as leading-edge (LE) or forward phase control (FPC) dimmers. A leading-edge dimmer switches the voltage ON partway through the half cycle, then switches the voltage OFF at the zero crossing. **See Figure 3-32.** Once triggered, the current flowing through the switching device keeps it conducting until the zero crossing. However, with lagging current, it is possible that the trigger current through the switch will not reach its threshold level before the trigger pulse ends. This can result in unacceptable dimming performance. Therefore, leading-edge dimmers use a triggering technique known as a "hard firing," which maintains the trigger pulse long enough to ensure that the current reaches the switching device's threshold level for conducting. Magnetic, wire-wound, low-voltage ballasts can only be dimmed using leading-edge dimmers.

Phase Control

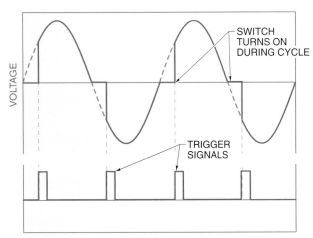

Dimmed to Approx. 85% Brightness

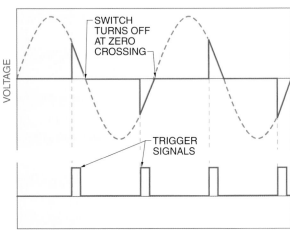

Dimmed to Approx. 15% Brightness

LEADING-EDGE

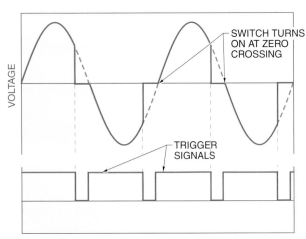

Dimmed to Approx. 85% Brightness

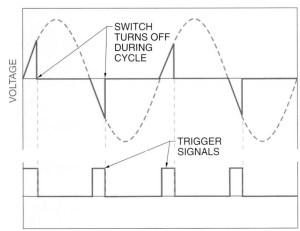

Dimmed to Approx. 15% Brightness

TRAILING-EDGE

Figure 3-32. *Depending on the type of power sources and lighting loads, phase control dimmers use either leading-edge or trailing-edge switching to clip the voltage sine wave.*

Dimmers used with leading currents are known as trailing-edge (TE) or reverse-phase control (RPC) dimmers. Trailing-edge dimmers control the switching at the end (trailing edge) of each cycle. A trailing-edge dimmer applies line voltage to the load beginning at the zero crossing, then switches off at some point in the half cycle. Trailing-edge dimmers are used with electronic ballasts.

Incandescent lamp loads can be dimmed with either type of phase control.

Semiconductor switches are the primary component of a dimmer, but there are other electronic components required to interpret the signal indicating the desired dimming level (such as an analog input signal and produce the appropriate trigger signals.

Dimming Ballasts. Gas discharge lamps, such as the common fluorescent lamps, can be dimmed, but require special considerations because of their starting characteristics. Regular phase control dimming does not work well for these lamps, since they rely on the sustained heating of vapors to maintain an arc at low-power levels.

Therefore, the dimming of gas discharge lamps requires special dimming ballasts, which are phase control devices, but keep the lamp's cathodes fully heated even as the arc current is reduced. This allows the lamp to quickly illuminate when power is switched back on. Still, only rapid start lamps can be dimmed.

The dimming function for these ballasts is built into the ballast fixture; they cannot be used with an external dimmer. However, the control signal that governs the dimmer can be shared among many dimming ballasts, provided it is compatible with the ballast inputs. Analog ballasts use a 0 VDC to 10 VDC signal to control lamp intensity. Digital ballasts can be used with digital systems sharing structured network messages.

LIGHTING SYSTEM CONTROL APPLICATIONS

Lighting system control has two primary functions: provide adequate lighting for the tasks performed within a lighted space, and conserve energy by automatically turning lamps off when they are not needed. To achieve these goals, automated lighting systems use one or more control functions that determine appropriate lighting levels based on the electrical power demand, time of day, available natural lighting, occupancy, and use of the space.

Poorly implemented lighting control systems can cause problems. Over-dimming can irritate occupants by forcing them to constantly override the system to provide adequate lighting. Under-dimming wastes energy, resulting in lower energy savings. Frequent cycling of dimming or switching can also be an annoyance. Poor scheduling leaves lights on when not needed, which wastes energy.

Manual control of lighting circuits is still possible, and commonly available, in automated systems. However, if the system has been properly designed and installed, this type of control will be minimal.

Bi-Level Switching

Dimming can sometimes be accomplished with switching, which is typically simpler to implement. *Bi-level switching* is a technique to control general light levels by switching individual lamps or groups of lamps in a multilamp fixture separately. This technique requires multilamp fixtures that are specifically wired so that lamps are ballasted or powered separately. The lamps can then be switched by any switching control means, such as lighting contactors, relays, or line voltage switching.

Operating Conditions. Most commercial multilamp fluorescent fixtures are capable of bi-level switching. For example, in a typical three-lamp fluorescent fixture, the outer two lamps are switched separately from the middle lamp, allowing the user to switch ON one, two, or all three lamps. This makes three different light levels available without requiring dimming control devices.

In order to implement bi-level switching, however, the middle lamp must be on a separate lighting switch leg than the other two. **See Figure 3-33.** For example, in a room with several three-lamp fixtures, all of the middle lamps are on one lighting switch leg and all of the outside pairs are on another.

||| Bi-Level Switching

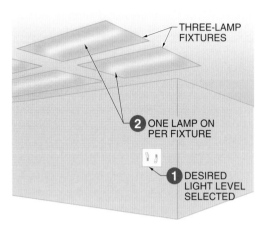

LOW-LIGHT LEVEL

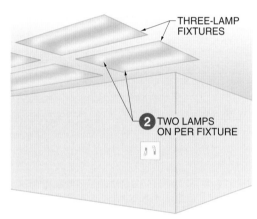

MODERATE-LIGHT LEVEL

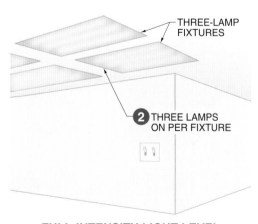

FULL-INTENSITY LIGHT LEVEL

Figure 3-33. Bi-level switching is a technique to provide simple brightness control by selectively switching lamps within multilamp light fixtures.

Control Sequence. The control sequence for bi-level switching with lighting system controls is as follows:

1. The desired light level is selected based on occupant input (wall switch) or lighting control program.

2. If the low-light level, corresponding to approximately 33% of the full intensity, is desired, the lighting circuit connecting the center lamps is switched ON. The other lighting circuit remains or is switched OFF.

   If the moderate-light level, corresponding to approximately 67% of the full intensity, is desired, the lighting circuit connecting the outer lamps is switched ON. The other lighting circuit remains or is switched OFF.

   If the full-intensity light level is desired, both lighting circuits are switched ON.

Demand Limiting

During peak demand periods, utilities often charge higher prices for electricity, which compensates for the correspondingly greater distribution expenses and encourages lower power consumption. Building automation systems can respond by developing demand-limiting programs to reduce utility costs. *Demand limiting* is the automated shedding of loads. *Load shedding* is the deactivation or modulation of noncritical loads in order to decrease electrical power demand. Demand limiting helps building owners respond quickly to rate changes and utility requests to help avoid power outages. Load shedding can save the building owner money and may even earn cash rebates from the utility.

Load shedding is often implemented more in other building systems, but is becoming more common in lighting systems. Dimming and bi-level switching can be used effectively to reduce power consumption temporarily without significantly affecting the building occupants. However, the affected lighting circuits must be chosen carefully to avoid reducing the lighting level for critical applications, such as exit lighting and emergency systems.

Operating Conditions. The signals to load-shed are usually transmitted via data link from the utility. Load shedding requires an energy management system (EMS) and/or compatible lighting controller. **See Figure 3-34.** (EMS devices can interface with other building systems for controlling significant electrical loads, such as HVAC systems.)

These controllers have preprogrammed control scenarios to progressively shed loads based on utility billing rates, power demand, and the power utilization within the building. These programs switch off or dim certain lighting zones as required, which can significantly reduce demand. For example, a moderately high power demand may require that only a few lighting circuits are affected, while a very high demand requires a greater reduction in loads, affecting many lighting circuits.

For some systems, a separate load shed controller is provided by the utility to process the incoming utility signals and forward the load shed instructions to the building's EMS or lighting controller. This signal to the local EMS or lighting controller could be in the form of a basic contact closure or a structured network message.

Control Sequence. The control sequence for demand limiting with lighting system controls is as follows:

1. During periods of peak demand, the utility sends a signal to the EMS mounted on or near the building's service entrance equipment.

2. The interprets the signals and forwards commands to the EMS lighting controller to dim or disconnect some lighting loads.

Demand Limiting

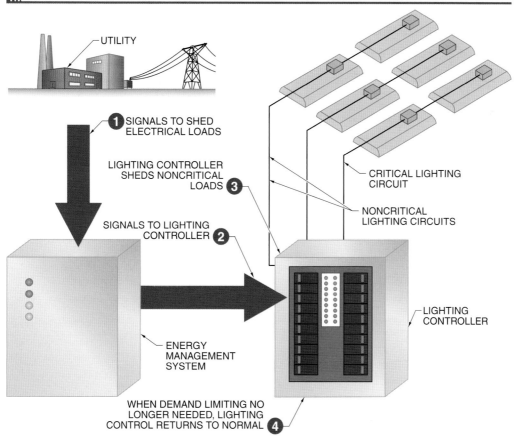

Figure 3-34. Systems that utilize demand limiting reduce power demand at peak periods by selectively shedding noncritical lighting loads.

3. Based on the utility's information, the lighting controller chooses the appropriate load shedding program and begin switching or dimming certain lighting circuits to reduce the power demand.

4. When demand levels or utility rates are lower, the controller restores the lighting loads to normal levels.

Photo Control

The large number of outdoor lighting applications, which provide adequate nighttime illumination, makes manual control impractical. *Photo control* is an automatic lighting control that uses a light-level sensor to turn lamps ON around dusk and turn lamps OFF at dawn. **See Figure 3-35.** This ensures that lighting is adequate at night without wasting energy by having the lamps on during daylight. This simple control scheme is used extensively for roadway and parking lot lighting.

Photo Control Sensor

Acuity Brands Lighting, Inc.

Figure 3-35. Photo control devices include a small photoresistor sensor that changes resistance in response to the ambient light levels.

Operating Conditions. Light-level sensors used with simple photo control employ a photoresistor, which changes resistance based on the intensity of the light it is exposed to. The photoresistor is in series with the coil of a normally closed relay. **See Figure 3-36.** During the day, the sensor's resistance is low, which allows enough current through the circuit to energize the coil and hold the lighting circuit

contacts open. At night, the resistance is high, blocking enough current that the relay returns to its normally closed state, energizing the lighting circuit.

Photo Control Circuit

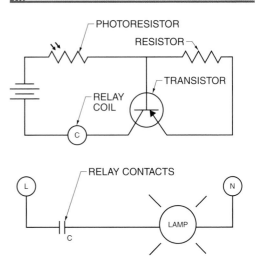

Figure 3-36. Photo controls use a simple relay to switch a lighting circuit ON or OFF according to the ambient light levels. An ON-delay timing relay may be included to avoid false deactivations.

Light fixtures may be controlled individually by their own light-level sensor, or one sensor may control a number of fixtures on a lighting circuit, by using a lighting contactor or computerized controller.

Photo controls usually have a built-in time delay of a few seconds before turning lamps OFF. This is so that they will not be triggered OFF at night by momentary light sources such as car headlights or lightning strikes because the sudden darkness is a safety hazard. False de-energizing of HID lamps is especially problematic since they may take several minutes to return to full light output. Sudden darkness during the daytime is not an issue, however, so there is no time delay to turn the lamps ON.

Control Sequence. The control sequence for photo control with lighting system controls is as follows:

1. During the day, light level is high enough—usually a minimum of 1.5 fc

(foot-candles) for most outdoor applications—to energize the relay and open the lighting circuit. This keeps the lights off.

2. At dusk, the lighting level falls, increasing the resistance of the photoresistor until the relay coil is no longer energized and the held-open contacts return to their normally closed state. This energizes the lighting circuit.

3. If the lighting level increases, the decrease in resistance of the sensor allows current that activates a timing circuit. Lighting control is determined by the state of the sensor at the end of the timing interval, typically a few seconds.

4a. If the light level is still high, then the increase is likely due to the sunrise. The low resistance of the photoresistor then allows current to energize the relay coil, opening the normally closed contact. This de-energizes the lighting circuit.

4b. If the light level falls, then the increase was due to an intermittent light source and it is likely still nighttime. The high resistance of the photoresistor keeps the relay coil de-energized, leaving the lighting circuit energized.

Daylighting

A control scheme similar to outdoor photo control is applicable to indoor spaces, but the lighting must be controlled more carefully than only ON or OFF. Indoor applications call for a smooth transition from natural to artificial light, requiring the lighting circuits to be dimmed at various levels throughout the day. This also minimizes the power consumption of the lighting system by adding only enough artificial light needed to maintain the desired range of illumination.

Daylighting is a lighting control scheme that measures the total amount of illumination in a space from all sources and adjusts the artificial lighting to maintain a minimum level. The necessary light output from the artificial lighting sources changes throughout the day to maintain a constant light level. **See Figure 3-37.** Daylighting is also known as "automatic daylight dimming" or "daylight harvesting."

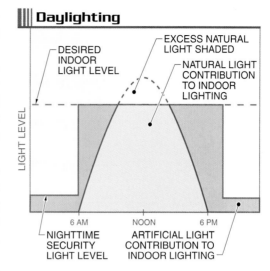

Figure 3-37. *Daylighting controls exploit natural light to conserve energy by adding only enough artificial light to maintain desired light levels.*

Daylighting systems also include time clocks switches and/or occupancy sensors, so that the daylighting function operates only when there are occupants in the building.

Operating Conditions. Daylighting systems require light-level sensors that output a signal indicating the actual light level, either with an analog signal or structured network message. This information is then to be used by a daylighting controller to increase or decrease the artificial lighting level until the reading matches the desired level. Daylighting functions can be combined with other lighting control features, such as occupancy-sensing or scheduling, to operate the lamps only when necessary.

Relying on natural light for indoor lighting may require additional considerations. Large windows designed to allow a lot of sunlight indoors also affect the building's thermal balance. Sunlight can heat a space, which may add additional cooling loads in the summer. Window glass may also allow indoor heat to escape during the winter, requiring more heating. New construction designed for daylighting may include architectural features or special window treatments to manage the sunlight infiltration. Therefore, daylighting systems may also involve controlling motorized window blinds or curtains to balance the impacts of energy consumption by the lighting and HVAC systems.

WattStopper/Legrand

Lighting control circuits are often grouped together into relay panels.

Control Sequence. The control sequence for daylighting with lighting system controls is as follows:

1. The light-level sensor continuously monitors the total light intensity within an area from all light sources, both natural and artificial. The sensor outputs an analog or structured network message signal proportional to the light level.

2. The daylighting controller receives the light level information and compares this with the desired level to determine the amount of additional illumination needed.

3a. If the light level is below the desirable range, the daylighting controller signals the switching and/or dimming devices, increasing the contribution of artificial light to the general lighting of an area.

3b. If the light level is above the desirable range, the daylighting controller signals the switching and/or dimming devices, decreasing the contribution of artificial light to the general lighting of an area.

Scene Lighting

Most rooms contain a number of lamps, which may include a variety of lamp types. If the room is used for a variety of purposes, the desirable lighting settings may change. Occupants can change the lighting manually as needed, but a common lighting control system feature can change between these settings automatically.

Scene lighting is a lighting control scheme that switches and/or dims groups of lamps together to a mixture of predetermined levels. The levels can be ON, OFF, or any light intensity in between. A *scene* is a group of settings for all the lamps in a space that correspond to a certain use for the space. For example, a multi-purpose room may include both fluorescent lamps for bright, general lighting and small recessed incandescent lamps for focused-task lighting. When the room is used as a classroom, all of the general lighting lamps are set to a high level and the task lights are turned off. **See Figure 3-38.** During a multimedia presentation, however, this lighting scene would make it difficult for the audience to see the presentation. A more appropriate scene turns off the general lighting and sets the task lights to a mid-range setting, which makes the presentation highly visible and allows just enough light so the audience can take notes.

Operating Conditions. Scene control in an automated lighting system allows occupants to change between preprogrammed lighting scenes with corresponding buttons. A scene controller is connected to the lighting controls and stores the settings for all the necessary lighting circuits as a scene program. **See Figure 3-39.** For example, a "Presentation" scene button instructs the scene controller to switch or dim all of the room's lamps to the settings defined for that use of the room. Controllers can be programmed to switch between several scenes with very different lighting requirements manually, on a schedule, or from signals from other systems.

Scene controllers can also control nonlamp devices related to scene changes, such as closing motorized window treatments, lowering projector screens, and activating an integrated audio system. Alternatively, a non-lighting system can integrate with scene controllers for special applications. For example, VDV systems can initiate presentation scenes.

Control Sequence. The control sequence for scene lighting with lighting system controls is as follows:

1. A room occupant chooses a lighting scene appropriate for the use of the room from a switch, wall-mounted selector panel, or wireless remote.

2. The scene controller initiates the program of lamp settings based on the selected scene.

3. The scene controller interfaces with the lighting circuit control devices, such as lighting contactors, relays, and dimmers, to set the desired lighting level for each circuit. Signals to these devices may be ON/OFF, analog, or structured network message signals.

4. If there are any nonlamp devices associated with the initialization of a lighting scene, such as window treatments, the controller activates or controls those devices as well.

5. If a new scene is selected, the controller will change all the lighting circuit and nonlamp circuit settings to those in the new program.

Scene Lighting Controller

Leviton Manufacturing Co., Inc.

Figure 3-39. Scene controllers include user-operated keypads for one-touch switching between different lighting scenes.

Lighting Scenes

"SET UP" SCENE

"MEETING" SCENE

"PRESENTATION" SCENE

"UNOCCUPIED" SCENE

© 2008 Lutron Electronics, Inc.

Refer to Quick Quiz® on CD-ROM

Figure 3-38. Lighting scenes are combinations of settings for different lighting circuits in a room to provide optimal lighting for certain tasks.

Summary

- Lighting represents a major component of energy consumption, accounting for a significant part of all energy consumed worldwide.

- Proper indoor lighting improves the productivity and safety of building occupants.

- Lighting includes both natural daylight and artificial light from lamps.

- Lighting system control uses as much natural daylight as possible and supplements this with artificial lighting as needed to maintain optimal lighting levels for each building space.

- In centralized lighting control, all lighting circuits are wired to a central control panel, which controls the switching or dimming of each lighting circuit individually.

- With a distributed lighting control system, dimmers and relays are located throughout the building, allowing shorter lighting control circuits that run directly between the light fixtures and control devices.

- Light switching alternates lighting loads between full-rated brightness or completely off, with no intermediate light levels.

- Dimming controls the intensity of the light output from lamps.

- Common lamps used in commercial buildings for most lighting applications include incandescent, halogen, fluorescent, and high-intensity discharge lamps.

- Luminous efficacy is used to compare lamps, determine their power consumption, and therefore determine the cost required to achieve certain lighting levels.

- Incandescent lamps are connected directly to electrical power sources, and their light output is directly affected by the input voltage.

- Gas discharge lamps operate from AC power but cannot be directly connected to electrical power sources because the high inrush currents would destroy the lamps. These lamps must include a ballast in the lighting circuit.

- Occupancy sensors save energy by deactivating lights when no one is in the room or area.

- Occupancy sensors may use one or more different technologies in order to sense people within a space, including passive infrared, ultrasonic, and microwave.

- Light-level sensors continuously monitor the intensity of light within an area so that a lighting controller can adjust the lighting system accordingly.

- Switches cannot dim lamps—they only alternate a lamp's state between completely ON and completely OFF.

- Electronically controlled relays make or break electrical power contacts when they receive signals from lighting controllers.

- Scheduling lighting control systems automatically turn lights ON and OFF at certain times of the day.

- Dimmers are devices that vary the light level (intensity) of lamps by controlling the amount of power that is delivered to the lamp.

- Gas discharge lamps, such as the common fluorescent lamps, can be dimmed, but they require special considerations because of their starting characteristics.

- Lighting system control has two primary functions: provide adequate lighting for the tasks performed within a lighted space and conserve energy by automatically turning lamps off when they are not needed.

- Automated lighting systems use one or more control functions that determine appropriate lighting levels based on the electrical power demand, time of day, available natural lighting, occupancy, and use of the space.

- Automated load shedding helps building owners respond to rate changes and utility requests to help avoid power outages.

- Photo control is a simple control scheme used extensively for roadway and parking lot lighting.

- Daylighting necessitates a smooth transition from natural to artificial light, requiring the lighting circuits to be dimmed at various levels throughout the day.

- Scene lighting allows occupants to easily change between programs of lighting settings designed for specific room uses.

Definitions

- **Centralized lighting control** is a lighting system that controls lights around a building via one main control panel.

- **Distributed lighting control** is a lighting system that controls lights directly from local control devices.

- **Light** is the portion of the electromagnetic spectrum that the human eye can perceive.

- A **lamp** is an electrical output device that converts electrical energy into visible light and other forms of energy.

- **Switching** is the complete interruption or resumption of electrical power to a device.

- **Dimming** is the intentional reduction of electrical power to a lamp in order to reduce its light output.

- A **lumen** (lm) is the measure of the intensity of light radiating from a light source.

- **Initial lumens** is the rated intensity of light produced by a lamp when it is new.

- **Mean lumens** is the rated average intensity of light produced by a lamp after it has operated for approximately 40% of its rated life.

- **Illuminance** is the quantity of light per unit of surface area.

- A **foot-candle** (fc) is the illuminance from 1 lumen per square foot (lm/ft^2) of surface.

- A **lux** is the illuminance from 1 lumen per square meter (lm/m^2) of surface.

- A **light fixture** is an electrical appliance that holds one or more lamps securely and includes the electrical components necessary to connect the lamp(s) to the appropriate power supply.

- The **luminous efficacy** is the ratio of a lamp's light output (lumens) to the electrical power input (watts).

- An **incandescent lamp** is an electric lamp that produces light by the flow of current through a tungsten filament inside a gas-filled, sealed glass bulb.

- A **filament** is a conductor with a high resistance that causes it to glow white-hot from electrical current.

- A **halogen lamp** is an incandescent lamp filled with a halogen gas (iodine or bromine).

- A **gas discharge lamp** is an electric lamp that produces light by establishing an arc through ionized gas.

- A **ballast** is a device with a circuit that controls the flow of current to gas discharge lamps while providing sufficient starting voltage.

- An **inductor** is a coil of wire that creates a magnetic field, which resists changes in the current flowing through the coil.

- A *fluorescent lamp* is an electric lamp that produces light from the ionization of mercury vapor.

- A *cathode* is a tungsten coil coated with a material that releases electrons when heated.

- A *high-intensity discharge (HID) lamp* is an electric lamp that produces light by striking an electrical arc across tungsten electrodes housed inside an arc tube.

- A *low-pressure sodium lamp* is an HID lamp that produces light by an electrical discharge through low-vapor-pressure sodium.

- A *mercury-vapor lamp* is an HID lamp that produces light by an electrical discharge through mercury vapor.

- A *metal-halide lamp* is an HID lamp that produces light by an electrical discharge through mercury vapor and metal halide.

- A *high-pressure sodium lamp* is an HID lamp that produces light when current flows through sodium-vapor under high pressure and high temperature.

- A *light-emitting diode (LED)* is a semiconductor device that emits a specific color of light when DC voltage is applied in one direction.

- An *LED driver* is a circuit that provides a constant DC voltage source and protection from line voltage transients, such as surges and sags.

- An *occupancy sensor* is a sensor that detects whether an area is occupied by people.

- A *passive infrared (PIR) sensor* is a sensor that activates when it senses the heat energy of people within its field-of-view.

- A *pyroelectric sensor* is a sensor that generates a voltage in proportion to a change in temperature.

- A *Fresnel lens* is a lens with special grooved patterns that provide strong focusing ability in a compact shape.

- A *dead zone* is an area within the field-of-view of an occupancy sensor that is not covered by one of the detection zones on the lens.

- An *ultrasonic sensor* is a sensor that activates when it senses changes in the reflected sound waves caused by people moving within its detection area.

- A *microwave sensor* is a sensor that activates when it senses changes in the reflected microwave energy caused by people moving within its field-of-view.

- A *light-level sensor* is a control device with a photocell that measures the intensity of light it is exposed to.

- A *photocell* is a small semiconductor component that changes its electrical characteristics, such as output current or resistance, in proportion to the light level.

- A *photoresistor* is a photocell that changes resistance in proportion to its exposure to light.

- A *phototransistor* is a photocell that controls a current in proportion to the light level.

- A *lighting contactor* is a solenoid-operated, high-power relay used with lighting circuits.

- A *remote-controlled circuit breaker* is a circuit breaker that switches a circuit ON or OFF when it receives a control signal.

- *Scheduling* is the automatic control of devices according to the date and time.

- A *time clock* is a switch that automatically changes state (switches) at certain times.

- An *autotransformer* is a transformer with only one winding, which is tapped by both the primary and secondary sides at various points to produce varying voltages.

- *Phase control* is the frequent switching of AC voltage to limit the power in a circuit.

- *Bi-level switching* is a technique to control general light levels by switching individual lamps or groups of lamps in a multilamp fixture separately.

- *Demand limiting* is the automated shedding of loads.

- *Load shedding* is the deactivation or modulation of noncritical loads in order to decrease electrical power demand.

- *Photo control* is an automatic lighting control that uses a light-level sensor to turn lamps ON around dusk and turn lamps OFF at dawn.

- *Daylighting* is a lighting control scheme that measures the total amount of illumination in a space from all sources and adjusts the artificial lighting to maintain a minimum level.

- *Scene lighting* is a lighting control scheme that switches and/or dims groups of lamps together to a mixture of predetermined levels.

- A *scene* is a group of settings for all the lamps in a space that correspond to a certain use for the space.

Review Questions

1. Contrast the location of the control devices in centralized and distributed lighting control systems.

2. How does lamp switching affect lamp life?

3. Why do gas discharge lamps require ballasts, while incandescent lamps can be connected directly to an electrical power source?

4. Why do both ultrasonic and microwave occupancy sensors require motion to detect people within their field-of-view?

5. Compare the operations of photoresistors and phototransistors.

6. What is the difference between the light measured by light-level sensors used in open-loop versus closed-loop control installations?

7. Describe the two methods of changing the voltage sine wave to produce dimming.

8. What are some of the problems associated with poorly implemented lighting control systems?

9. How can daylighting systems affect a building's HVAC systems?

10. Explain how lighting scenes are changed with a scene controller.

Chapter Four

HVAC System
Control Devices

HVAC systems are the most commonly automated building system. They are also perhaps the largest and most complicated automated building system. This is because there are a large number of possible control devices available to manage the systems, and these devices can affect the systems in multiple ways. For example, a cooling device affects not only the temperature of the air within a building space, but also the humidity. A thorough knowledge of all the application control devices and their interrelationships is vital to the successful automation of HVAC systems.

Chapter Objectives

- *Differentiate between different types of HVAC systems and the control devices within each.*
- *Describe the requirements for comfort for building occupants.*
- *Describe how indoor climate is measured with different types of sensors.*
- *Identify the control devices that can be used to affect air temperature and humidity within the building.*
- *Differentiate between the control devices used to manage the distribution of conditioned air or water.*

HVAC SYSTEMS

An *HVAC system* is a building system that controls a building's indoor climate. HVAC functions are also often called "climate control." The primary functions of an HVAC system are heating, ventilating, and air conditioning (cooling), which give the system its name. **See Figure 4-1.** Additional functions include air filtration and humidity control. These functions are closely interrelated and use some of the same components.

The HVAC acronym is sometimes written as "HVACR" or "HVAC&R" to include refrigeration, another common function of a climate-control system. Also, sometimes the ventilation portion is dropped, creating the acronym "HACR." All of these terms refer to the same category of building systems.

Precise control of HVAC systems maintains a comfortable indoor climate for the building occupants, optimizes the indoor conditions for operating equipment or stored inventory, and does these functions with minimum energy use during all times of the year. Automated HVAC systems also reduce air infiltration and maintain pressure relationships between spaces.

HVAC systems, particularly the heating and cooling functions, rely on the basic principles of thermodynamics and heat transfer. HVAC systems transform and redistribute heat energy to create the most desirable indoor climates. HVAC systems are classified into two groups based on the method used to distribute heat energy throughout the building: forced-air and hydronic systems.

HVAC Systems

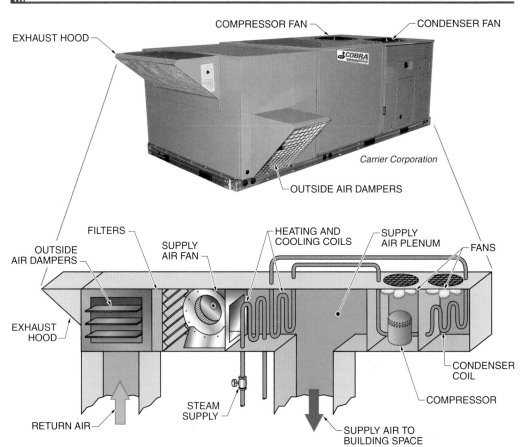

Figure 4-1. Many commercial HVAC systems are package units that provide all of the primary HVAC functions. These units are often installed on the building rooftop.

Thermodynamics

Thermodynamics is the science of thermal energy (heat) and how it transforms to and from other forms of energy. The two laws of thermodynamics apply to the heating and cooling of air in a building space.

First Law of Thermodynamics

The first law of thermodynamics states that energy cannot be created or destroyed but may be changed from one form to another. This is also known as the law of conservation of energy. An example of this law is the combustion process. As fuel burns, the hydrogen and carbon, which are found in all fuels, combine with oxygen in the air. This chemical reaction releases some of the chemical energy in the elements in the form of thermal energy. The hydrogen, carbon, and oxygen recombine to form new compounds that are the products of combustion. Two of the new compounds are carbon dioxide (CO_2) and water vapor (H_2O). During the process, energy has changed from chemical energy to thermal energy, but no energy has been created or destroyed.

Another example is within an air conditioning system. An electric motor uses electrical energy to drive a compressor, converting electrical energy into mechanical energy. The mechanical energy compresses the refrigerant in the system. As the compressed refrigerant expands, it produces a cooling effect. Most of the electrical energy used to drive the compressor results in the cooling effect of the system. Some of the mechanical energy is converted to thermal energy because of friction, but no energy is created or destroyed in the process.

Second Law of Thermodynamics

The second law of thermodynamics states that heat always flows from a material at a higher temperature to a material at a lower temperature. This flow of heat is natural and does not require outside energy to facilitate the process. The second law of thermodynamics applies to all cases of heat transfer. Heat transfer is the movement of heat from one material to another. The rate of heat transfer increases with the temperature difference between two substances. For example, air in a furnace is heated by the products of combustion. Heat flows from the hot burner flame to the cool air. Air conditioning systems use energy to control the movement of heat. Heat flows from warm room air into the cold refrigerant, cooling the air. Then, the now-hot refrigerant gives up this heat to the warm outside air. The refrigerant flow is driven by a compressor, but the flow of heat from the room to outside, via the medium of the refrigerant, is a natural example of the second law of thermodynamics.

The three methods of heat transfer are conduction, convection, and radiation. Conduction is the transfer of heat from molecule to molecule through a material. For example, if one end of a metal rod is heated, heat is transferred by conduction to the other end. Convection is the transfer of heat from warm to cool regions of a fluid from the circulation of currents. For example, as air is warmed by a fire, the warm air rises and is replaced by cool air. The movement of air creates a current that continues as long as heat is applied. Radiation is the transfer of heat between nontouching objects through radiant energy (electromagnetic waves). Radiant energy waves move through space, but they only produce heat when they contact an opaque object. For example, the radiant energy from an electric heating element passes through the air without heating it. This energy heats a person or object that comes into contact with it.

Forced-Air HVAC Systems

A *forced-air HVAC system* is a system that distributes conditioned air throughout a building in order to maintain the desired conditions. Forced-air HVAC systems use a system of ductwork, dampers, and fans to move conditioned air into the building areas where it is needed. *Conditioned air* is indoor air that has been given desirable qualities by the HVAC system. (This should not be confused with "air conditioning," which refers only to cooling the air.) For example, during the winter, the indoor air is conditioned by the heating functions of the HVAC system. Hot air is introduced into a building space where the air is cool. The net result is a warm indoor climate.

The hub of a forced-air HVAC system is an air-handling unit. This device combines many HVAC control devices together to fully condition the air at a central location before it is distributed throughout the building. An *air-handling unit (AHU)* is a forced-air HVAC system device consisting of some combination of fans, ductwork, filters, dampers, heating coils, cooling coils, humidifiers, dehumidifiers, sensors, and controls to condition and distribute supply air. **See Figure 4-2.** Depending on the application, air-handling units may vary in size and configuration, but most are arranged similarly.

Air-Handling Unit

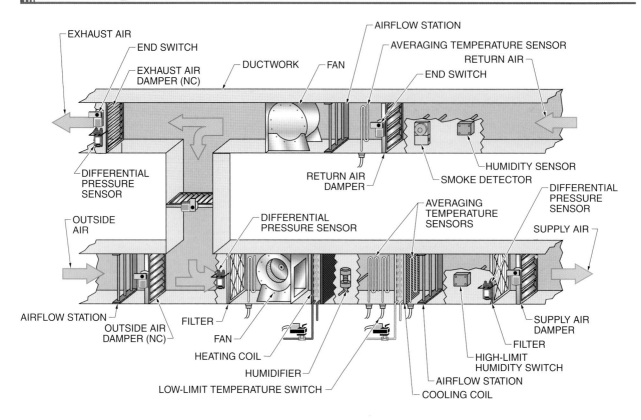

Figure 4-2. An air-handling unit includes all of the devices needed to condition and distribute supply air throughout a building.

Air-handling units manage five types of air. Entering the air-handling unit are the return air and outside air. *Return air* is the air from within the building space that is drawn back into the forced-air HVAC system to be exhausted or reconditioned. *Outside air* is fresh air from outside the building that is incorporated into the forced-air HVAC system. Outside air is also known as makeup air, since it replaces air exhausted from the air-handling unit. *Mixed air* is the blend of return air and outside air that is combined inside the air-handling unit and goes on to be conditioned. Exhaust air and supply air leave the air-handling unit. *Exhaust air* is air that is ejected from the forced-air HVAC system. Exhaust air is also known as relief air in some geographic regions. *Supply air* is newly conditioned mixed air that is distributed to the building space.

One or more fans draw air through the unit, and several sets of dampers control the relative proportions and mixing of the outside air and return air in the mixed air plenum (duct section). Incorporating outside air is a way to supply fresh air into the building to prevent the air from becoming stale, while permitting as much natural heating and cooling from the outside air as possible. Mixed air is filtered and passes across heating coils and cooling coils, as well as humidifiers, dehumidifiers, and/or sensors, and enters the building spaces as conditioned supply air. By controlling the fans, dampers, coils, and other devices within an air-handling unit, the unit can be used to heat, cool, humidify, dehumidify, or simply ventilate a building space.

After being distributed from the air-handling unit, air enters the building space through a terminal unit. A *terminal unit* is the end point in an HVAC distribution system where the conditioned medium (air, water, or steam) is added to or directly influences the environment of the conditioned building space. Forced-air terminal units include dampers to modulate the amount of conditioned supply air into the space, and may include other devices to further condition the supply air.

The devices within an air-handling unit can be controlled individually, but it is more common in building automation systems to use a controller that includes most or all the functions of the air-handling unit. The controller includes input connections for temperature, pressure, and other sensors and output connections for fan and damper control devices. The logic of how to change the outputs based on the different air conditions indicated from the input information may already be programmed into the controller. This is a much simpler way to interface with all of these devices, but the controller must be compatible with the specific air-handling unit and its functions. For example, not all air-handling units include heating functions, and some supply multiple building spaces with different climate requirements.

There are two ways that air-handling units add conditioned air to the indoor building spaces in order to achieve climate control: the constant-volume method and the variable-volume method. **See Figure 4-3.**

Constant-Air-Volume Air-Handling Units. A *constant-air-volume air-handling unit* is an air-handling unit that provides a steady supply of air and varies the heating, cooling, or other conditioning functions as necessary to maintain the desired setpoints within the building zone. Constant-air-volume air-handling units operate at their rated airflow capacity (in cubic foot per minute) at all times. Many air-handling units found in existing light commercial buildings are constant-volume air-handling units. The disadvantage of constant-air-volume air-handling units is that the fans operate at maximum power 100% of the time, which uses a great deal of energy.

Variable-Air-Volume Air-Handling Units. A *variable-air-volume air-handling unit* is an air-handling unit that provides air at a constant air temperature but varies the amount of supplied air in order to maintain the desired setpoints within the building zone. Many variable-air-volume air-handling units are used for cooling only.

When heating is provided, it is by reheat coils in terminal units. Variable-air-volume air-handling units are the most common air-handling units installed in new commercial buildings. Also, they generally use less energy and are quieter than constant-air-volume types because the fans operate at lower speeds.

Supply airflow is modulated either by controlling supply fan output or managing airflow within the air-handling unit. For example, bypass dampers (common for small rooftop units) can reduce the supply airflow by allowing excess supply air to flow directly into the return duct.

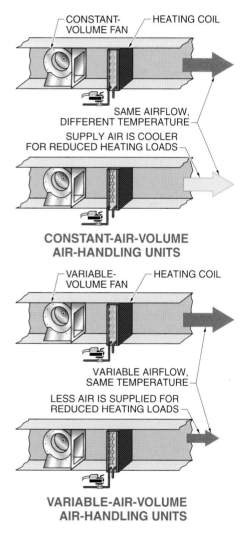

CONSTANT-AIR-VOLUME AIR-HANDLING UNITS

VARIABLE-AIR-VOLUME AIR-HANDLING UNITS

Figure 4-3. *Instead of changing the conditioning of the air, variable-air-volume air-handling units change the airflow volume of consistently conditioned air added to the building space.*

Variable-air-volume terminal boxes are individually controlled to further heat and/or modulate the flow of supply air into a single building zone.

systems require a pump to circulate the water throughout the building. Steam heating systems do not require circulating pumps because steam flows easily due to the difference in pressures between the boiler and the terminal units.

Variable-Air-Volume Terminal Boxes

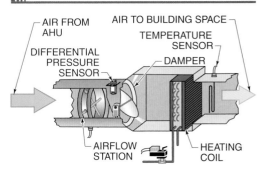

Figure 4-4. The controller for a variable-air-volume terminal box provides an individual zone with air at the optimal temperature and airflow.

Buildings may employ a small group of centralized variable-air-volume air-handling units, or they may use a system that incorporates many smaller units located throughout the building. A *variable-air-volume (VAV) terminal box* is a device located at the building zone that provides heating and airflow as needed in order to maintain the desired setpoints within the building zone. **See Figure 4-4.** As the zone temperature drops, heat is provided by a heating device within the VAV terminal box. When the zone temperature reaches the setpoint, the heating coil is modulated to maintain the zone setpoint. Most VAV terminal boxes have a differential pressure switch that de-energizes the heating device if airflow falls below a minimum setting, which can cause damage from overheating.

Hydronic HVAC Systems

A *hydronic HVAC system* is a system that distributes water or steam throughout a building as the heat-transfer medium for heating and cooling systems. Hydronic HVAC systems use pipes, valves, and pumps to distribute water throughout the building. **See Figure 4-5.** Hot water or steam is used for heating, while chilled water is used for cooling. These are typically distributed in two separate piping loops. Water

The water or steam can be used in two different ways. First, it can be piped directly into the building spaces, where it flows through a hydronic terminal unit that conditions the indoor air without forced airflow. In this case, the terminal unit is a heat exchanger that transfers heat between the water or steam and the indoor air through radiative and/or convective means. A *heat exchanger* is a device that transfers heat from one fluid to another fluid without allowing the fluids to mix. **See Figure 4-6.** This type of system is primarily used for heating. Some of the oldest and most common examples of terminal units are steam or hot-water radiators.

Alternatively, the water or steam can flow through a coil within the ductwork of a forced-air HVAC system, transferring heat between the air and the water or steam. Moving air provides a more efficient method of heat transfer and this arrangement can be used for both heating and cooling. Many commercial HVAC systems incorporate aspects of both forced-air and hydronic systems in this way, such as air-handling units that use steam from a boiler in the heating coils and chilled water from a chiller in the cooling coils.

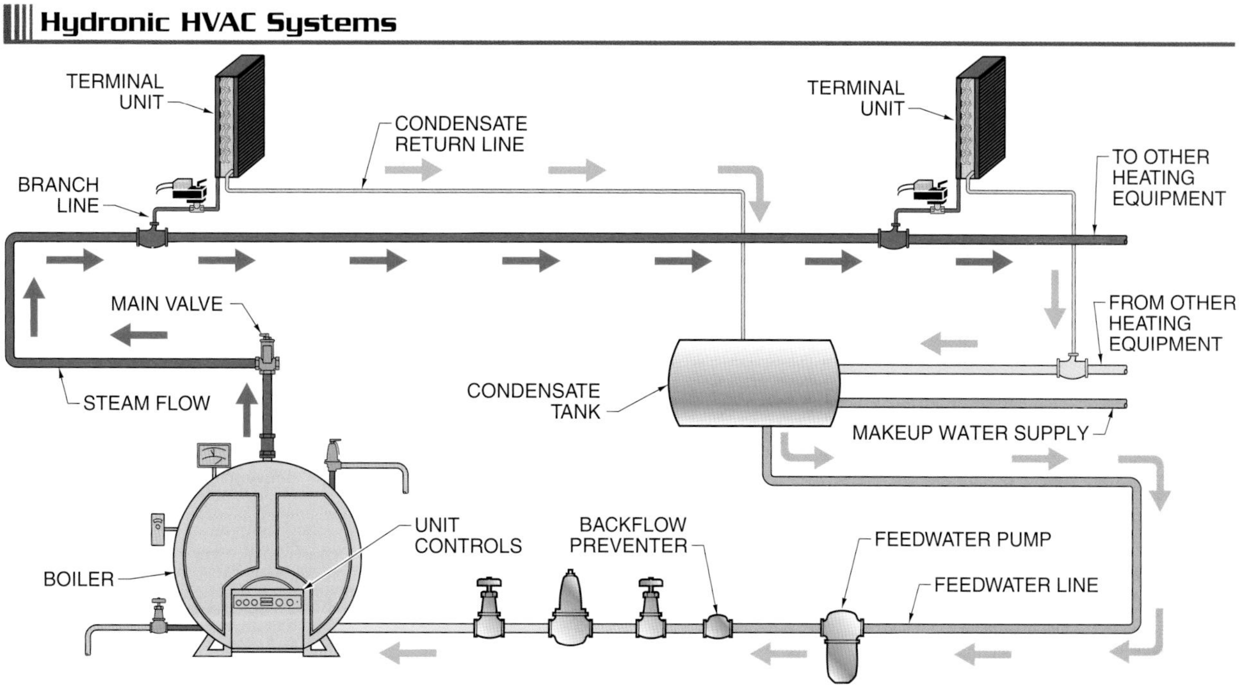

Hydronic HVAC Systems

Figure 4-5. Hydronic HVAC systems consist of a device to heat or chill water and the piping and equipment to distribute this water throughout the building for heating and cooling.

Heat Exchangers

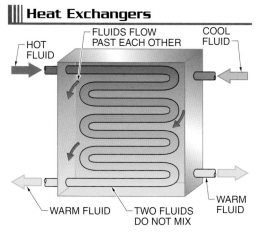

Figure 4-6. A heat exchanger allows heat to be transferred between two fluids without the fluids mixing.

Heating and Cooling Coils. Heating and cooling coils add or remove heat from indoor air. A *heating coil* is a heat exchanger that adds heat to the air surrounding or flowing through it. Heating coils use hot water or steam from boilers to heat the air. A *cooling coil* is a heat exchanger that removes heat from the air surrounding or flowing through it. Cooling coils use chilled water or refrigerant vapor to cool the air. Air-handling units may contain one or more of each type of coil to provide the required heating or cooling capacity.

The two types of coils are constructed in much the same way. Most consist of a long length of tubing bent and shaped to maximize the surface area exposed to the air. **See Figure 4-7.** Fins of sheet aluminum may be attached to the tubing to add more surface area for heat conduction. Either hot or cold fluid flows through the tubing, which acts as a heat exchanger, transferring heat to or from the air.

Heating coils may also be made from electric heating elements. These are shaped similarly to the fluid versions in order to maximize the heat transfer to the air, but they require only an electrical connection and no fluid piping.

The fluid or electrical current flow through a heating and cooling coil can be controlled with valves or relays to regulate the temperature of the coil. As the temperature in the building space changes, controllers allow more or less heating or cooling medium to flow. A faster flow provides greater heat transfer, which has a greater heating or cooling effect.

Heat Exchanger Coils

The Trane Company

Figure 4-7. Heating and cooling coils are heat exchangers made from folded lengths of pipe with tabs of sheet metal added to increase the surface area for heat conduction.

Alternatively, the supply air temperature can be controlled by changing the airflow volume over the coils, while the coil temperature remains constant. Similarly, faster airflow transfers more heat between the air and the coils.

Building Zones

Different areas within a building have different HVAC requirements. For simplicity, older HVAC systems attempted to provide an acceptable middle ground for the entire building. However, this can result in some areas receiving unnecessary climate control, which wastes energy, and other areas not being adequately controlled. HVAC systems with separate control over individual zones are becoming increasingly common, even in residential buildings.

Building codes may require HVAC zones in commercial buildings for energy efficiency. Zones must also be carefully defined to manage the number of subsystems and provide the proper conditioning for all spaces. For example, all rooms facing west may be combined into one zone since they will experience similar effects from solar heating in the afternoon.

A *zone* is an area within a building that shares the same HVAC requirements. A building may have many zones, divided by the use of the space and to create manageable units of HVAC conditioned spaces. **See Figure 4-8.** For example, a commercial building with a mixture of office and storage areas may divide the HVAC system functions into zones by these different areas. With many occupants, the office space climate should be carefully controlled for comfort. If this space is very large, it may be further divided into multiple zones. However, since fewer occupants are in the storage area, it may allow less stringent control over its indoor air climate, and can thus be a separate zone. This multizone approach makes efficient use of energy while maximizing the comfort of the building occupants.

Single-Zone HVAC Systems. A single-zone HVAC system conditions air to only one building zone. **See Figure 4-9.** The size of the zone can vary widely but cannot be so large that the air conditions vary significantly throughout the space. Single-zone HVAC systems are identified by the heating and cooling coils that are in series and located in the central air-handling unit. This is the simplest forced-air HVAC system configuration.

HVAC Zones

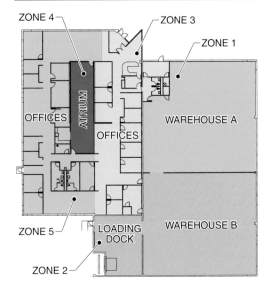

Figure 4-8. Buildings may be divided into multiple zones according to HVAC requirements, size, and location.

||| Single-Zone Air-Handling Units

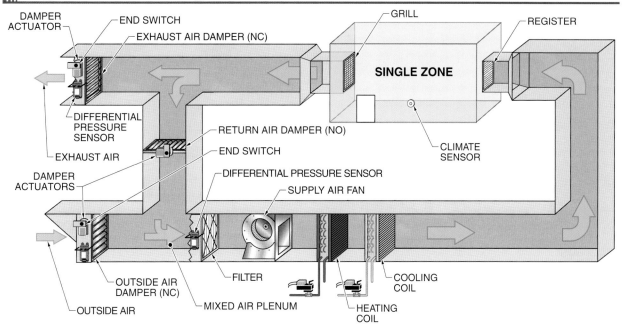

Figure 4-9. A single-zone air-handling unit manages all of the HVAC functions for a single area of the building.

Multizone HVAC Systems. A multizone system conditions air for more than one building zone. The mixing of outside air and the filtration of the supply air occurs at the main unit, which is located centrally. However, the devices managing the other functions, primarily heating and cooling, may be located elsewhere in the system, depending on the design. There are a few variations of multizone air-handling units, which are distinguished by the locations of these devices.

The simplest versions of multizone air-handling units retain all of the HVAC functions with the central unit, which outputs conditioned supply air to all of the zones. Heating and cooling coils may be located in series within the same duct or in separate parallel ducts. Either way, the supply air is fully conditioned or mixed before being distributed. Dampers at each zone control how much of this conditioned air is admitted to each zone, affecting the indoor climate of each zone differently. **See Figure 4-10.** This design is effectively a variable-air-volume system for each zone, since the zone controller

can only manage the airflow into the zone. The air is conditioned to the aggregate requirements of all the zones. This system does not control every zone efficiently since the conditioned air is not customized for each zone's climate requirements.

Fluke Corporation

Commercial air-handling units are usually purchased as packaged units and installed on the building rooftop.

||| Multizone Air-Handling Units (with Centralized Mixing)

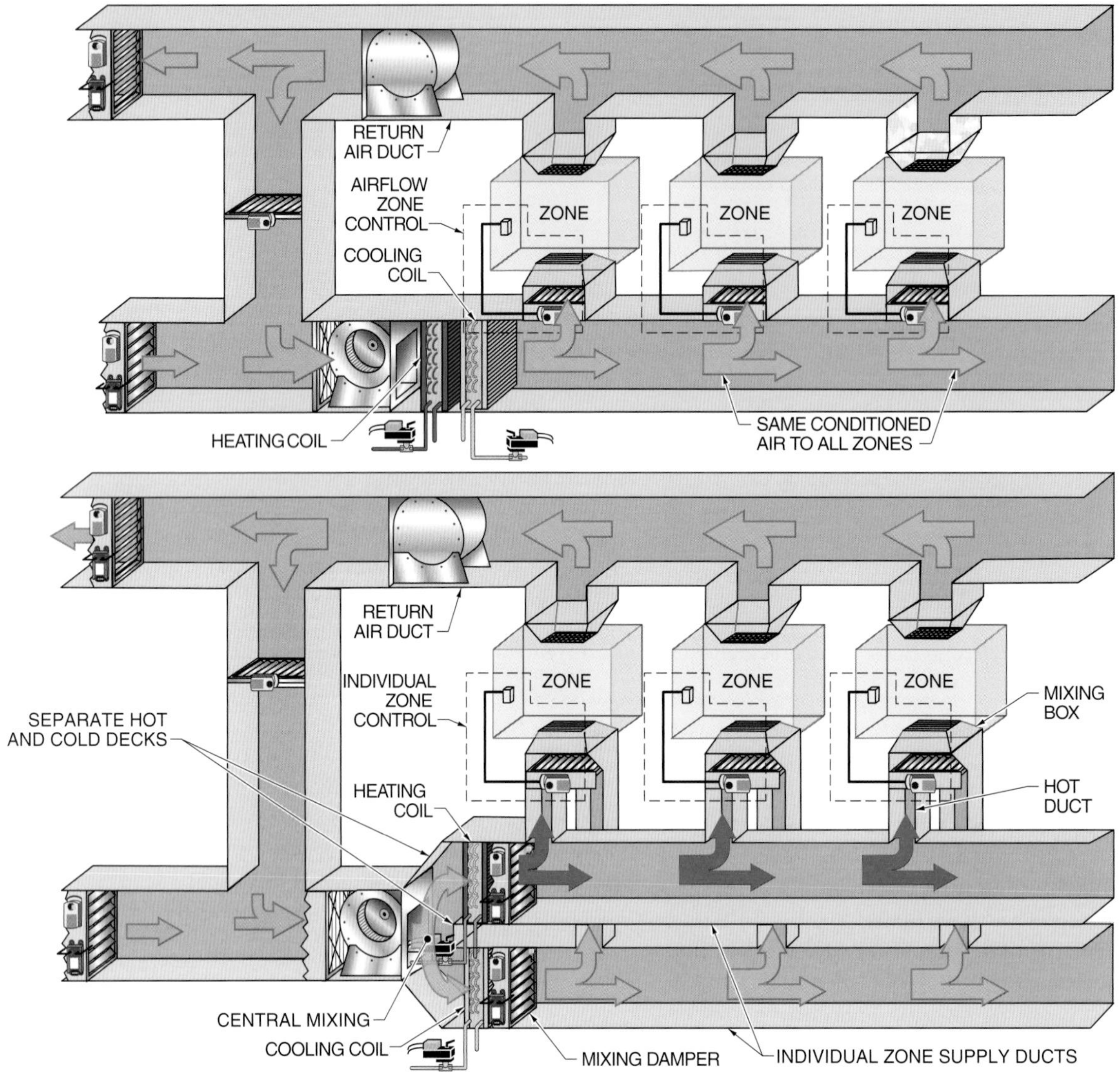

Figure 4-10. Air-handling units can be used to supply air for multiple zones, but each zone receives the same conditioned air, regardless of the comfort needs for the area.

There are several designs for forced-air HVAC systems that can control the supply air temperature to multiple zones individually. Some multizone air-handling units can work with as many as 50 zones. The primary differences among multizone air-handling unit types are the locations of the heating and cooling coils and how supply air is mixed. **See Figure 4-11.**

Dual-duct air-handling units distribute hot and cold supply air separately to the building zones. The heating and cooling devices are located centrally, but in two separate ducts. The pair of ducts distributes both hot and cold air throughout the building. Dampers at or near the individual zones then mix the hot and cold air for the desired supply air temperature. This provides greater control for individual zones, but is not very energy efficient, since air is both heated and cooled and the long lengths of ductwork can affect the supply air's outlet temperature (e.g., hot air cools and cold air warms).

⦀ Multizone Air-Handling Units (with Zone Mixing)

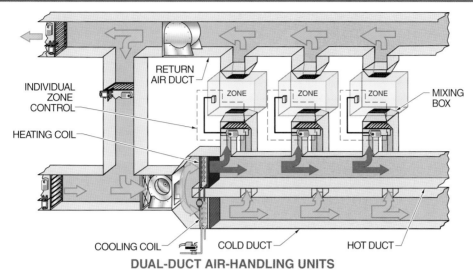

DUAL-DUCT AIR-HANDLING UNITS

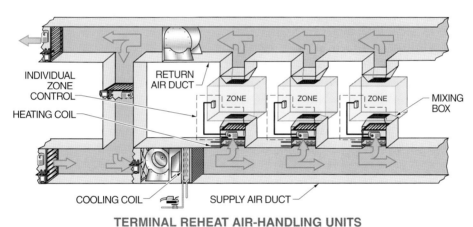

TERMINAL REHEAT AIR-HANDLING UNITS

Figure 4-11. *Various designs for air-handling units address the need for providing customized conditioned supply air to multiple zones from a single central unit.*

Alternatively, a terminal reheat air-handling unit places the cooling function with the central air-handling unit and distributes supply air at a constant 55°F temperature to all building zones. Each building zone has a heating coil in the nearby ductwork that warms the supply air, as needed, to the required temperature. When cooling is required at a zone, its heating coil is shut off and the cool air flows into the zone.

COMFORT

HVAC systems are used in buildings to provide comfort to occupants. *Comfort* is the condition of a person not being able to sense a difference between themselves and the surrounding air. The five requirements for comfort are proper temperature, humidity, filtration, circulation, and ventilation. **See Figure 4-12.** Comfortable air conditions can vary among different people, but generally fall within common ranges. People become uncomfortable when any of the conditions are outside their acceptable range.

> In large buildings, it is more energy efficient to control heating at the zones, which avoids heat loss in a large supply-air distribution system. Individual thermostats in the zones control valves for local heating coils and control dampers for the flow of heated air into the space.

Comfort Requirements

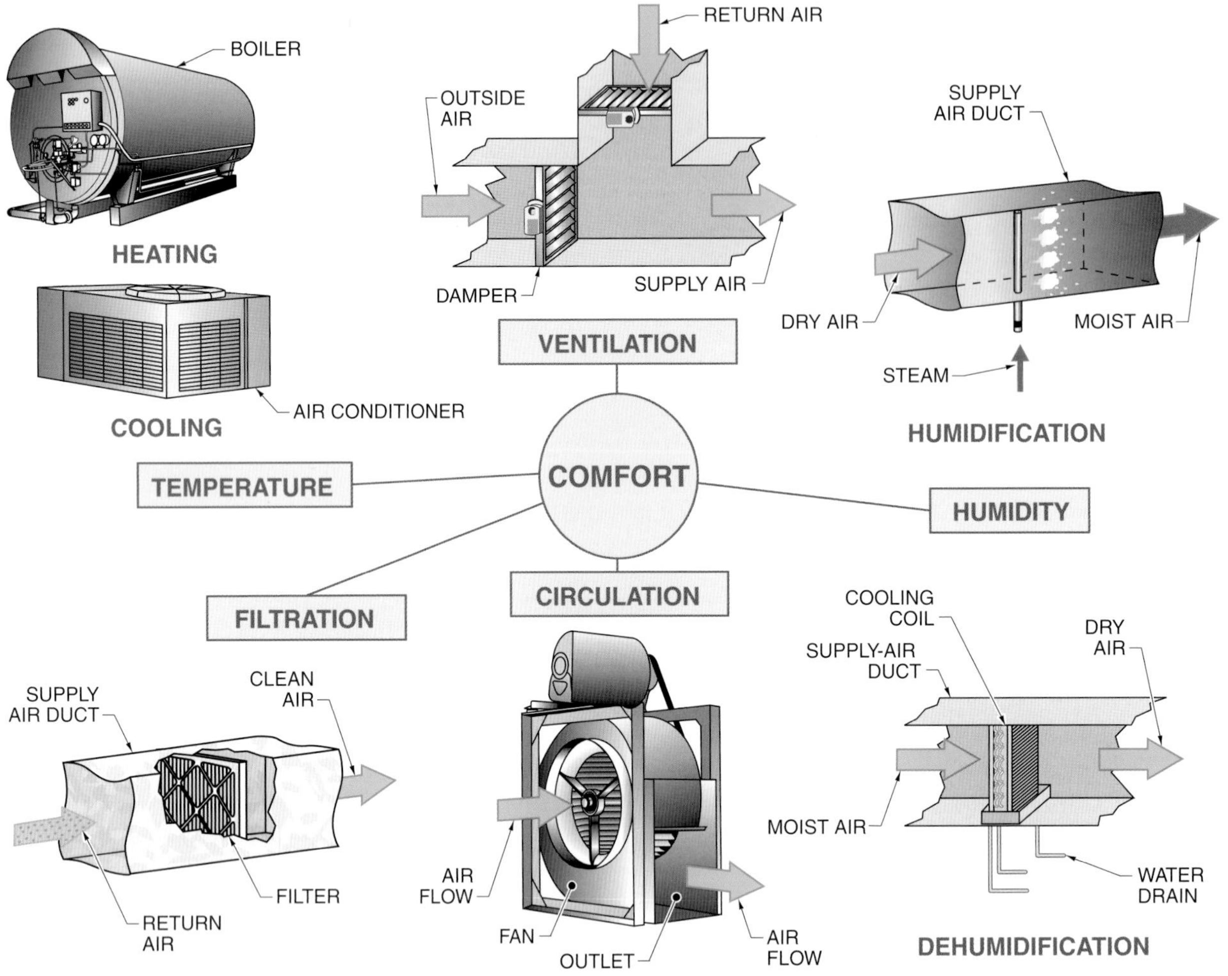

Figure 4-12. The comfort of building occupants relies on the temperature, humidity, circulation, filtration, and ventilation of the indoor climate.

Temperature

Temperature is the most important property of air that is controlled by an HVAC system. *Temperature* is the measurement of the intensity of the heat of a substance. When clothed, the human body is comfortable at an air temperature of approximately 70°F to 75°F. If the air temperature varies much above or below this range, the body begins to feel uncomfortably warm or uncomfortably cool. The HVAC system must condition indoor air to these temperatures, regardless of the outside temperature.

Temperature is quantified to a scale based on the freezing and boiling points of water. In the United States, the Fahrenheit temperature scale is the most common scale, though the Celsius scale may also be used. In the study of the properties of air, two different temperature measurements may be used: dry-bulb temperature and wet-bulb temperature. **See Figure 4-13.**

Dry-Bulb Temperature. The *dry-bulb temperature* is the temperature of air measured by a thermometer freely exposed to the air but shielded from radiation and moisture. Most temperature measurements are taken with dry bulb thermometers. If a temperature is not distinguished as dry-bulb or wet-bulb, it can be assumed to be a dry-bulb temperature. However, for a more detailed measurement of air conditions, both the dry-bulb temperature and wet-bulb temperature are considered.

Temperature

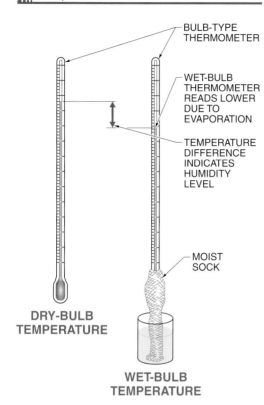

BULB-TYPE THERMOMETER

WET-BULB THERMOMETER READS LOWER DUE TO EVAPORATION

TEMPERATURE DIFFERENCE INDICATES HUMIDITY LEVEL

MOIST SOCK

DRY-BULB TEMPERATURE

WET-BULB TEMPERATURE

Figure 4-13. The difference between dry-bulb temperature and wet-bulb temperature is indicative of the amount of moisture in the air.

Wet-Bulb Temperature. The *wet-bulb temperature* is the temperature measured by a thermometer that has its bulb kept in contact with moisture. The bulb is wrapped in cloth that is wetted with water via wicking action. At a relative humidity below 100%, water evaporates from the cloth around the bulb, which cools the bulb to below ambient (dry-bulb) temperature. *Evaporation* is the process of a liquid changing to a vapor by absorbing heat. This lower temperature can be used to determine the relative humidity. A greater difference between the dry-bulb and wet-bulb temperatures, meaning that water is evaporating quickly from around the wet bulb, indicates a lower relative humidity. The precise relative humidity is determined with a psychrometric chart or by calculations.

Actually, temperatures for automated systems are rarely measured with bulb-type thermometers. Specialized electronic sensors are used to measure some temperature and humidity properties, and the related properties are calculated from those results. However, the "bulb" names for these properties remain, since they are descriptive of the relationships between temperature and humidity.

Humidity

Humidity is the amount of moisture present in the air. Some moisture is always present in air. A low humidity level indicates dry air that contains little moisture. A high humidity level indicates damp air that contains a significant amount of moisture. *Absolute humidity* is the amount of water vapor in a particular volume of air. The most common units of absolute humidity are pounds of water per pound of dry air or grams of water per cubic meter of dry air.

Warmer air has a greater capacity to be saturated with moisture than cooler air. Therefore, humidity is typically quantified as relative humidity, which relates the moisture content of the air within the context of current air temperature. *Relative humidity* is the ratio of the amount of water vapor in the air to the maximum moisture capacity of the air at a certain temperature. Relative humidity is represented as a percentage. A relative humidity of 50% means that the air is 50% saturated with moisture. A relative humidity of 100% represents the condition wherein the air is saturated with moisture for its temperature, and dew begins to condense. *Dewpoint* is the air temperature below which moisture begins to condense.

Even if the amount of water in the air remains the same, the relative humidity changes as the air temperature changes. **See Figure 4-14.** Since air has a greater moisture capacity at higher temperatures, the same amount of moisture in the air represents a smaller fraction at a higher temperature. For example, if air with 50% relative humidity is heated, the relative humidity may fall to 30%, even though the amount of moisture in the air remains the same.

Temperature and Humidity Relationship

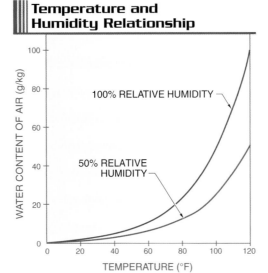

Figure 4-14. Relative humidity must be used in the context of temperature, since the two properties are closely related.

Humidity affects comfort because it determines the rate at which perspiration evaporates from the skin. Evaporation of perspiration cools the body. Higher humidity slows the evaporation rate, and lower humidity increases the evaporation rate. Because of this, humidity levels can affect the perception of temperature, even with no actual change in temperature. For example, at a constant temperature, an increase in humidity causes a person to feel warmer, while a decrease in humidity causes a person to feel cooler.

At normal room temperatures, comfortable relative humidity levels are between about 30% and 50%. The lower end of the range is ideal for winter, while the higher end is ideal for summer. If the humidity is too low, a higher air temperature is required to feel comfortable. If the humidity is too high, a lower air temperature is required for the same feeling of comfort.

The mounting location can affect the accuracy of the temperature-sensor measurement. For example, air-temperature sensors must be protected from exposure to direct sunlight, which can artificially raise measured ambient temperatures through radiation heat transfer.

Besides controlling humidity for human comfort, it is often important to control humidity within a building for inventory or process reasons. High humidity promotes corrosion and mold growth, and low humidity promotes the buildup of electrostatic charge. Items such as sensitive electronics, food, and natural materials (e.g., paper and wood) can be damaged by excessively high or low humidity.

Circulation

Air in a building must be circulated continuously for maximum comfort. *Circulation* is the continuous movement of air through a building and its HVAC system. Ductwork and registers are used to move air throughout the building spaces. A *register* is a cover for the opening of ductwork into a building space. Registers include dampers to control the amount and direction of the airflow. **See Figure 4-15.** The air is blown from the supply air registers and is drawn back to the HVAC system through the return registers. When sized and placed appropriately, the air flows through the entire space.

Registers

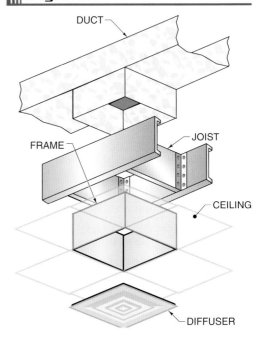

Figure 4-15. Registers are the interface between a building space and HVAC ductwork and can control the amount of airflow admitted to the space.

Psychrometrics

Psychrometrics is the scientific study of the properties of air and the relationships between them. Psychrometric charts are used to show the relationships between the various properties of air in any condition. The properties of air found on a psychrometric chart are dry-bulb temperature, relative humidity, humidity ratio, wet-bulb temperature, enthalpy, and specific volume. Dewpoint temperature is also sometimes included. When air is conditioned, one or more of the properties of air change. When one property changes, the others are affected. If any two properties of a sample of air are known, the others can be found by using a psychrometric chart.

The two properties used most often for identifying specific conditions of the air are temperature and humidity. In general, a dry-bulb temperature of 70°F to 75°F and a relative humidity of approximately 30% to 50% are considered comfortable. Therefore, comfort is defined as this area on the psychrometric chart.

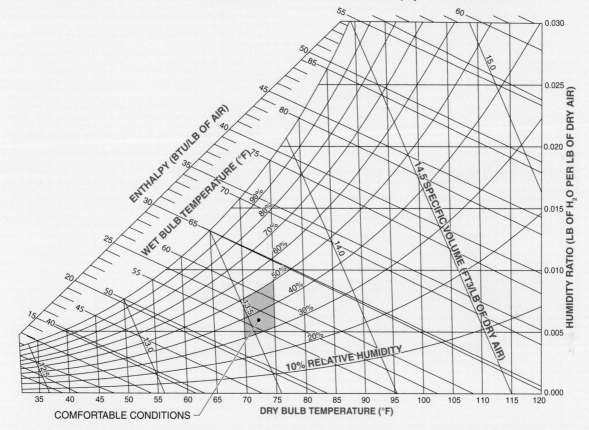

COMFORTABLE CONDITIONS

Any material with a temperature above absolute zero contains heat energy. Enthalpy is the total heat energy, meaning the sensible heat energy plus the latent heat energy, contained in a material. Sensible heat is related to temperature and is indicated by dry-bulb temperature. (Sensible heat is the heat that can be directly "sensed.") Latent heat is related to phase change and indicated by wet-bulb temperature. ("Latent" is from the Latin word "latere," meaning "to lie hidden.") Enthalpy is expressed in British thermal units per pound of air.

A psychrometric chart illustrates how interrelated the properties of air are. It also shows how important it is to consider all of these properties when conditioning air. For example, both dry-bulb and wet-bulb temperatures must be taken into account when considering the effect of changes in temperature on the other properties of air. A change in wet-bulb temperature indicates that moisture has been added to or removed from the air, which changes the humidity ratio and the relative humidity of the air.

A change in dry-bulb temperature also affects the specific volume of the air. If the dry-bulb temperature increases, the specific volume increases. If the dry-bulb temperature decreases, the specific volume will decrease.

When considering the effect of changes in humidity, the humidity ratio must be considered. As the humidity ratio increases, the latent heat content of the air also increases. As the humidity ratio decreases, the latent heat content of the air also decreases.

In a building where there is improper circulation, air rapidly becomes stale and uncomfortable. Proper air circulation also prevents temperature stratification. *Temperature stratification* is an undesirable variation of air temperature between the top and bottom of a space. **See Figure 4-16.** This is common in a building space, where warm air rises to the ceiling and cool air falls to the floor. Circulating air ensures that the characteristics of indoor air are consistent throughout the space. Circulation also cycles the air through the HVAC system, which cleans and conditions the air for the desired temperature and humidity level. Stratification can also occur within air-handling units, where it can cause control problems if temperature sensors are biased toward one extreme due to their location.

Temperature Stratification

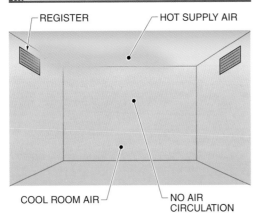

Figure 4-16. Poor air circulation can cause an uneven distribution of conditioned air throughout a building space.

Air velocity is the speed at which air moves from one point to another. Air velocity is measured in feet per second (fps). Air circulation helps cool the body by evaporating perspiration. An increase in air velocity increases the rate of evaporation of perspiration from the skin, causing a person to feel cool. A decrease in air velocity reduces the evaporation rate, causing a person to feel warm.

Circulation is also measured in the volume of air that is cycled through the system. Air movement of 40 cfm (cubic feet per minute) per person is considered ideal. The *air changes per hour (ACH)* is a measure of the number of times the entire volume of air within a building space is circulated through the HVAC system in one hour. The requirements for air movement and air changes per hour vary for different types of buildings and occupancy uses.

Filtration

Airborne particulate matter such as dirt, spores, and pollen creates an unhealthy environment for building occupants. Indoor air should be filtered to remove these potential health hazards. *Filtration* is the process of removing particulate matter from air.

The most common method of filtration is forcing the contaminated air through a filter. The filter contains tiny pores through which the molecules of air can pass, but the much larger particulates cannot. **See Figure 4-17.** They become trapped in the filter. Since forced-air HVAC systems already include the equipment to effectively circulate air, filtration involves only adding a filter within the ductwork.

Filters

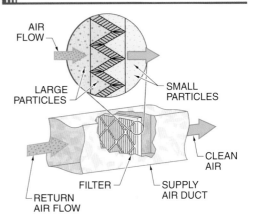

Figure 4-17. Filters trap particles in the airflow that are larger than the openings in the fabric-like filter material.

The primary consideration in air filter selection is the efficiency with which it removes particulates from the air. Its efficiency is related to the sizes of the particulates that are caught in the filter, compared to those that pass through. The size of airborne particles is measured in microns. A *micron* is a unit of measure equal to one-millionth of a meter, or 0.000039". Filters are available with different efficiencies. For example, prefilters, which are relatively inexpensive and are changed up to four times per year, have an efficiency of about 30%. This means that they capture about 30% of large particulate matter. They are changed so frequently because they are quickly clogged by the large particulates. Final filters, however, are changed every 12 to 18 months and have efficiencies of 60% to 90%. This means that they capture up to 90% of small particulates.

HVAC systems maximize the life of their filters by using two or more filters in series to capture progressively smaller particulates. For example, a prefilter captures the largest particulates, then a medium-efficiency filter captures many of the common-sized particulates, and a high-efficiency filter captures most of the remaining fine particulates.

Some filters are designed for disposal after use, while others can be cleaned and reused. Special filters are also available that remove organic and biological compounds or absorb specific types of gases and vapors from the air stream. The type of filter and size of the openings depends on the degree of filtration required.

Filters should be checked and changed periodically to maintain clean air and system performance. A dirty filter will excessively impede airflow and the HVAC system must work harder and use more energy to continue circulating the air.

Ventilation

Air within a building gradually becomes stale from exhaled carbon dioxide and chemical vapors, which are unhealthy for occupants. Indoor air is kept comfortable by diluting stale recirculated air with fresh outdoor air. **See Figure 4-18.** *Stale air* is air with high concentrations of carbon dioxide and/or other vapor pollutants. *Ventilation* is the process of introducing fresh outdoor air into a building. Outside air joins with return air recirculated from the building space to form mixed air, which is then conditioned as necessary and supplied back to the building space.

▌▌▌ Ventilation

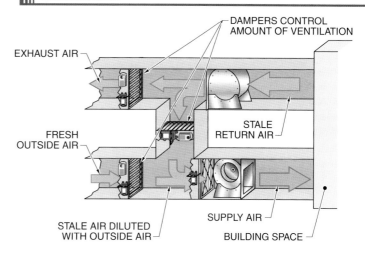

Figure 4-18. *The positions of damper blades control how much fresh outdoor air is added to the return air to ventilate the building spaces.*

Stale air is quantified by measuring the carbon dioxide (CO_2) level in the return air. Carbon dioxide is measured in parts per million (ppm) by volume. The amount of carbon dioxide in the atmosphere is about 390 ppm. Levels up to about 1000 ppm in indoor air are considered acceptable in many buildings.

Ventilation requirements are based on the building's occupancy and type of indoor activity, such as office work, classes, or manufacturing. For example, the ventilation requirement for a typical office building space is 20 cfm per person. Commonly, 5% to 30% of the air circulation for commercial buildings is composed of outside air. The position of the blades of air-handling unit dampers controls the amount of ventilation.

However, some industrial, research, and commercial buildings cannot recirculate any return air because of potentially dangerous air contaminants and must use 100% outside air for ventilation. If the indoor and outdoor climates are very different, such as during winter in northern locations, the amount of ventilation greatly affects the operation of the HVAC devices and the energy used.

HVAC CONTROL DEVICES

HVAC systems are composed of many devices and subsystems used together in some combination to accomplish the types of conditioning functions expected for the local climate. For example, an HVAC system in a warm climate includes devices for cooling, but may not include those for heating.

These devices are all involved in measuring or controlling the characteristics of the indoor climate that affect comfort. HVAC control devices include sensors to measure the current climate conditions and the presence of occupants with a zone, devices to control the temperature and humidity of the supply air, and devices to control the distribution of air, water, or steam throughout the HVAC system.

HVAC Control Devices

HVAC system documents, particularly those specifying the control devices and sequences, include schematic representations of the system and its components. These are not meant to be exact depictions of the physical size and shape of the equipment, but rather a representation of the basic operation of the system and the relationships between the components. Similar schematics may also appear on human-machine interfaces (HMIs) for monitoring system operation.

HVAC system schematics use symbols to represent many of the control devices involved in the system. These symbols are not standardized across the HVAC controls industry, but most are similar and recognizable as correlating to a certain physical component. The symbols may also include information about what type of information they share or receive from controllers or other devices, such as digital inputs (DIs), analog inputs (AIs), digital outputs (DOs), and analog outputs (AOs). Some devices with similar functions may be either analog or digital, depending on the needed information. For example, the valve controlling a heating coil may accept a digital or an analog signal, corresponding to a completely open/closed position or any position in between, depending on the type of valve or heating system.

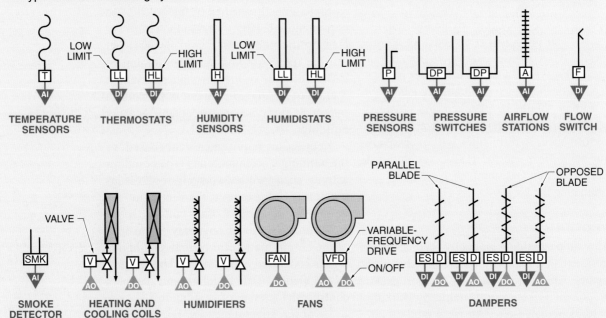

It is important to note that inputs and outputs are designated from the point of view of the HVAC system controller, such as an air-handling unit controller. For example, a temperature sensor outputs a signal, but that signal is received as an input to the controller. Therefore, the temperature sensor is considered an input device. Likewise, the controller outputs a signal to a damper actuator, which receives the signal as an input. Some devices interface with more than one signal, or even type of signal. For example, dampers receive an analog output signal to move to a different position, but may also include an end switch that provides a digital input to the controller if the damper closes completely.

Climate Sensors

Climate sensors measure the comfort-related properties of indoor air: temperature, humidity, circulation, filtration, and ventilation. There may be many climate sensors in a building, or even in an individual zone. Climate sensors are also installed within air-handling systems to monitor the mixing and conditioning of the supply air.

These electronic sensors communicate the values of these properties to local or central controllers, which then make decisions on HVAC system operation based on this information. The goal is to continually adjust the system functions until the sensors read values within an optimal range. This is an example of a closed-loop control system.

Temperature Sensor. A *temperature sensor* is a device that measures temperature. Temperature sensors are the most common inputs used in building automation systems and are used to measure temperatures in ducts, pipes, and rooms. Temperature sensors can be designed for measuring either air temperature or water temperature. **See Figure 4-19.** The technology for both types is the same, but the device packaging usually varies for different types of mounting or for protecting the sensor from the different environments.

Air-sensing temperature sensors can be mounted on walls or inside ducts. Wall-mount temperature sensors measure air temperatures within building spaces. Duct-mount temperature sensors are used to measure air temperatures inside ductwork. Since airflow through large ducts can actually have slightly different temperatures within the duct cross-section, small temperature sensors mounted in one spot can be inaccurate. Instead, an averaging temperature sensor is used. Averaging temperature sensors have a long, coiled sensing element that is mounted within a large duct to measures an overall air temperature. Temperature sensors for outside air temperatures are housed in weather-resistance enclosures.

Temperature sensors used for liquid measurements are known as immersion temperature sensors. The sensing element of an immersion temperature sensor is encased within a small, watertight, and thermally conductive casing, known as a thermowell, that is mounted inside the pipe, vessel, or fixture containing the water to be measured. Special fittings may be necessary to allow a leak-proof conduit for the conductors to pass outside the water-containing vessel.

There are several types of electrical temperature sensors that can measure and communicate temperature. These sensors are simple in design and relatively inexpensive, but offer significant capabilities when coupled with sensor electronics. The primary types of electrical temperature-sensing elements are thermocouples, thermistors, and resistance temperature detectors (RTD). Each of these can be coupled with electronics to convert their output to standard analog signals or structured network message information that corresponds to either the Fahrenheit or Celsius scales. **See Figure 4-20.**

A *thermocouple* is a temperature-sensing element consisting of two dissimilar metal wires joined at the sensing end. A unique property of this circuit is that the junction generates a very small voltage that varies in proportion to the temperature at the junction. With a sensitive voltmeter, the temperature can be accurately measured over a wide range.

Dwyer Instruments, Inc.

Temperature sensors typically include long probes to better measure an overall temperature of the fluid. Probes on duct-mounted averaging sensors are uncoiled and spread out inside the duct to measure the average temperature across the entire duct cross section.

Temperature Sensors

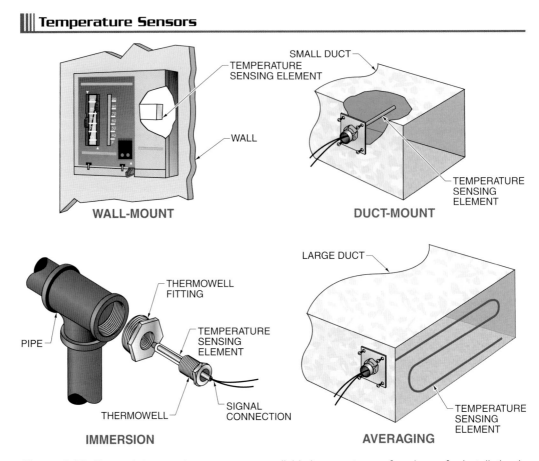

Figure 4-19. Electronic temperature sensors are available in many types of packages for installation in different parts of an HVAC system.

Temperature Sensors

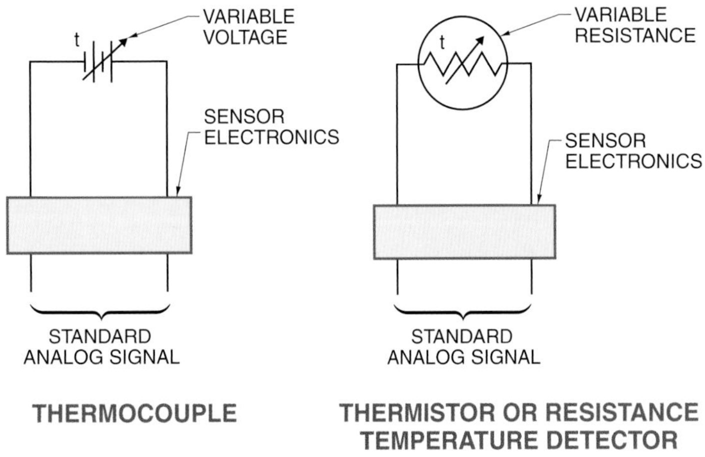

THERMOCOUPLE

THERMISTOR OR RESISTANCE
TEMPERATURE DETECTOR

Figure 4-20. The common types electronic temperature sensors include temperature-sensing elements that produce either a varying voltage or a varying resistance in response to temperature.

A *thermistor* is a temperature-sensing element made from a semiconductor material that changes resistance in response to changing temperatures. An external voltage source is connected to the thermistor and the voltage drop across its resistance is measured with a voltmeter. If the voltage source remains constant, the measured voltage drop changes with the thermistor's resistance and is proportional to the temperature.

Since the measured thermistor voltage can be much greater than thermocouple voltages for comparable temperatures, thermistor measurements are usually more precise. However, thermistors are susceptible to inaccuracies when the thermistor and the sensor electronics are connected with long conductors. Long conductors add resistance to a circuit, which affects the voltage measurements. If this is unavoidable, the sensor or system software may allow offset or calibration features to compensate for these inaccuracies.

A *resistance temperature detector (RTD)* is a temperature-sensing element made from a material with an electrical resistance that changes with temperature. Resistance temperature detectors are usually made from platinum, so they are relatively expensive. Resistance temperature detectors are similar to thermistors in operating principle, but are made from different materials and therefore have different electrical/temperature characteristics. They typically have a smaller resistance than thermistors, giving them a wider possible temperature range. They also have a slower response to temperature changes.

Thermostats. Most temperature inputs for automated HVAC systems require analog temperature values. However, there may be certain circumstances that require only a comparison of a temperature to a setpoint. Thermostats can be used in these applications. A *thermostat* is a switch that activates at temperatures either above or below a certain setpoint. For example, a thermostat may be used to close a set of contacts when a low temperature limit is reached, shutting down cooling processes to avoid ice forming in the HVAC ducts. When they are applicable, thermostats may be more desirable than temperature sensors because they are simpler to configure and may be less expensive.

Thermostats provide only contact closure outputs, which are normally open and/or normally closed. The temperature sensing may be done with either an electronic or electromechanical device. Electromechanical temperature-sensing devices include an element that moves in proportion to the ambient temperature. For example, a bimetallic strip is a temperature-sensing element composed of two dissimilar metals fused together, each with a different coefficient of thermal expansion. **See Figure 4-21.** As one material expands more than the other, the bonded strip bends toward one side. This movement can be used to open or close contacts at a certain temperature.

Thermostats

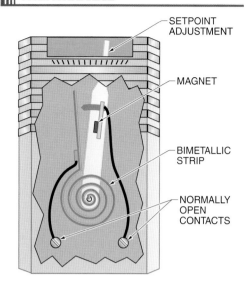

Figure 4-21. *Thermostats are simple devices that detect whether the ambient temperature around the device is either above or below a setpoint.*

The setpoint temperature is typically set manually with an adjustment screw in the device. In the simplest systems, the contact closure is used to switch another device ON or OFF through a relay, such as a fan. With electronic controllers, the contact closure is read as a digital input into the control program.

One of the most common applications of thermostats is as a low-temperature limit alarm. These thermostats include a manual reset that, once tripped, forces building personnel to physically investigate the reason for alarm. Low-limit thermostats have multiple connections to allow a direct connection to the fan safety circuit and also monitoring by controllers.

Hygrometers. A *hygrometer* is a device that measures the amount of moisture in the air. Hygrometers are also known as humidity sensors. The most common hygrometers measure percent relative humidity (% rh), while others measure dewpoint or absolute humidity. Humidity can also be calculated from the dry-bulb and wet-bulb temperature measurements.

Accurate humidity sensing is more difficult than temperature sensing. Since the sensing element must be directly exposed to the air, it is susceptible to contamination, which can affect readings. The accuracy of hygrometer measurements can drift over time, commonly 1% per year. Hygrometers should be checked annually for accuracy. Any instrument used to check the accuracy of hygrometers must be regularly calibrated against a known standard.

Dwyer Instruments, Inc.

Humidity sensors, or hygrometers, use the tiny changes in the length of a moisture-sensitive material to measure relative humidity.

Most hygrometers incorporate a hygroscopic element that absorbs moisture from the air. A *hygroscopic element* is a material that changes its physical or electrical characteristics as the humidity changes. **See Figure 4-22.** Electronic hygrometers apply a voltage to the hygroscopic element, measure its electrical characteristics to determine its moisture content, and then calculate the associated air humidity. Changes in either the capacitance or conductivity of the element indicate moisture content. Onboard electronics output the measurement as a standard analog signal or structured network message.

Since hygrometers and temperature sensors are often needed in similar locations, these two sensors are often combined in a single unit. In fact, some hygrometers require temperature information to convert absolute humidity values to relative humidity percentages.

Hygrometers

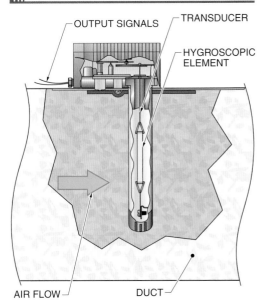

Figure 4-22. Hygrometers sense humidity levels by detecting very small changes in the length of an element that is sensitive to moisture.

Humidistats. Similar to thermostats, some humidity-sensing applications can be satisfied with the simpler humidistat. A *humidistat* is a switch that activates at humidity levels either above or below a certain setpoint. Humidistats include normally open and/or normally closed contacts, which are used to directly operate other devices, such as a humidifier, or as a digital signal to an electronic controller. Humidistats are normally less expensive than analog hygrometers.

Pressure Sensors. A *pressure sensor* is a sensor that measures the pressure exerted by a fluid, such as air or water. In HVAC systems, measured pressure is typically a differential pressure, rather than an absolute pressure. *Differential pressure* is the difference between two pressures. HVAC system differential pressure measurements are usually quantified in units of inches of water column (in. WC). Differential pressure sensors are often an integral part of air-, liquid-, and steam-flow measuring stations.

Differential pressure is used to compare the air pressure between two points within a duct, which is used to monitor fan operation or filter condition. Differential pressure sensors are connected to the two measuring points with a pair of tubes. **See Figure 4-23.** For example, a low differential pressure between the upstream and downstream sides of a filter means that the air is flowing freely, which indicates a clean filter. However, a dirty filter impedes the airflow significantly, which is indicated by a relatively large differential pressure. Therefore, differential air pressure sensors can be used to monitor the dirtiness of filters, indicating the level of air contamination within the building.

Differential Pressure

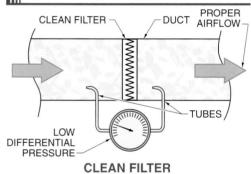

CLEAN FILTER

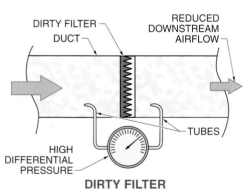

DIRTY FILTER

Figure 4-23. *The differential pressure between the upstream and downstream sides of a filter can be used to indicate the condition of the filter.*

Pressure sensors include some type of elastic deformation pressure element, which is a piece of material that flexes, expands, or contracts in proportion to the pressure applied to it. One of the most common types of elastic

deformation pressure elements is a diaphragm. One side of the flexible diaphragm is open to a pressure to be measured and the other is open to either another pressure (for differential pressure) or the atmosphere. The diaphragm moves into the space on the side with the lower pressure in response to the force of the higher-pressure side. The amount of the displacement is proportional to the pressure.

Motion of the pressure element is transferred to a transducer, where it is converted into an electrical signal. A *transducer* is a device that converts one form of energy into another form of energy. There are many different types of transducers that can produce electrical signals from some other type of signal, each taking advantage of a different electromechanical or material property. For example, piezoelectric elements are pressure-sensitive crystals that produce small voltages when squeezed. A pressure sensor may couple a diaphragm with a piezoelectric element. When the diaphragm moves to one side, it puts pressure on the piezoelectric element, which outputs a tiny voltage in proportion to the pressure. Additional electronics convert the transducer outputs into standard sensor output signals, such as 4 mA to 20 mA. **See Figure 4-24.**

Pressure Sensors

Dwyer Instruments, Inc.

Figure 4-24. *Pressure sensors use electronics to convert the small movement of a flexible diaphragm into a proportional electrical signal.*

Differential Pressure Switches. A *differential pressure switch* is a switch that activates at a differential pressure either above or below a certain value. Differential pressure switches are often used across a fan or filter and indicate conditions that are impeding airflow. Like other HVAC sensor switches, differential pressure switches include a setpoint adjustment and normally open and/or normally closed contacts.

Airflow Stations. An *airflow station* is a sensor that measures the velocity of the air in a duct system. The velocity multiplied by the duct cross-sectional area yields a volume expressed in cubic feet per minute (cfm). Airflow stations are installed either in a fan inlet or in ductwork. Airflow stations are very sensitive, so they must be located some distance downstream of anything causing turbulence, which can produce erroneous readings.

There are two types of airflow stations. The most common type is a pitot tube station, which is a differential pressure sensor comparing the airflow's total pressure (velocity pressure plus static pressure) and the static pressure. **See Figure 4-25.** The difference is proportional to the air velocity. The second type is a hot wire anemometer. In this type, a temperature sensor is placed on a wire and current is applied to the wire to keep it at a constant temperature as air moves across the device. As more air moves across the wire, more current is needed to keep the constant temperature. This current draw is proportional to the air velocity. In either type, the air velocity reading is converted into an airflow rate and output as a standard analog signal.

Flow Switches. A *flow switch* is a switch with a vane that moves from the force exerted by the water or air flowing within a duct or pipe. Flow switches are installed so that the vane is inside the duct or pipe and the rest of the device is outside. **See Figure 4-26.** The pressure of the air or water flow pushing on the vane causes it to move downstream slightly, which activates the switch. Switch connections include normally open and/or normally closed contacts.

Airflow Stations

Dwyer Instruments, Inc.

Figure 4-25. Airflow stations measure air velocity within a duct.

Flow Switches

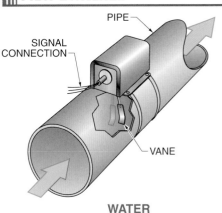

WATER

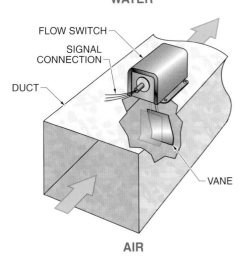

AIR

Figure 4-26. Both water flow switches and airflow switches use a vane deflected by the fluid stream to detect the movement of the fluid.

When the flow switch is activated, some of the force of the fluid flow must be removed before the switch returns to its previous state. This results in a deadband in the switch's response to changing flow. The vane deflection is a function of the vane area, pipe or duct size, and the characteristics of the flowing air or water. Setpoint adjustments determine how much vane deflection, and therefore flow rate, is required to actuate and deactivate the switch.

Like differential pressure switches, flow switches are used in HVAC systems to indicate the flow status of a pump or fan. However, due to fan or pump flow surges, flow switches are susceptible to false activations. For this reason, the use of differential pressure switches is more common.

Carbon Dioxide Sensors. The primary feature of stale air is an excess concentration of carbon dioxide. Ventilation incorporates outside air into the building space, lowering the carbon dioxide level. However, the ventilation system may not run continuously, especially if there are no heating or cooling requirements at the time. Therefore, carbon dioxide sensors are used to control ventilation systems so the system runs only when needed.

A *carbon dioxide sensor* is a sensor that detects the concentration of carbon dioxide (CO_2) in air. The reading is output as a standard analog signal. Controllers use measured carbon dioxide levels to activate and deactivate the ventilation system. They may also increase the percentage of outside air mixed into the supply air when the carbon dioxide level rises rapidly, such as when many people occupy a space, or decrease the percentage when spaces are unoccupied.

This ventilation system control, however, can be overridden by other HVAC functions. For example, in economizer mode, the system will incorporate a large amount of cool outside air into the building for cooling, regardless of the indoor carbon dioxide level.

Occupancy Sensors

An *occupancy sensor* is a sensor that detects whether an area is occupied by people. Occupancy sensors are currently more commonly used with lighting and security control systems than with HVAC systems, though they are becoming more popular for the control of ever-smaller HVAC zones, such as individual rooms and workstations. They are used to save energy by deactivating certain HVAC functions to a zone when no one is in the room or area.

Indoor Air Quality

Indoor air quality is a description of the concentration of contaminants present in the air that could affect the health and comfort of building occupants. Potential contaminants include microbes (such as mold and bacteria), chemicals (such as carbon monoxide and radon), allergens, or any other particulates that can induce health effects. Common building materials such as carpet fibers, adhesives, and ceiling tiles are sources of contaminants.

Good indoor air quality means the air is free of harmful particles or chemicals. Poor indoor air quality means that the air is contaminated. Recent studies have found that indoor air is often more polluted than outdoor air (though with different pollutants). In fact, indoor air is often a greater health hazard than the corresponding outdoor setting. The Environmental Protection Agency (EPA) ranks poor indoor air quality as one of the top five environmental threats to human health.

Ventilation, filtration, and source control are the primary methods for improving indoor air quality. As indoor air contaminants build up to unacceptable levels, outside air is introduced in greater quantities to dilute the contaminants.

Techniques for analyzing indoor air quality include the collection of air samples, the collection of samples on building surfaces, and computer modeling of airflow inside buildings. The resulting samples can be analyzed for mold, bacteria, chemicals, and other pollutants. Currently, no universal standard exists for measuring the amounts of contamination in a building space, though carbon dioxide levels in a building space are commonly used as an indicator of indoor air quality.

Occupancy sensors are mounted on the ceiling or high on a wall to maximize their field-of-view. **See Figure 4-27.** The sensors provide an ON/OFF signal that is used to control a relay or send a digital signal to a controller.

Occupancy Sensors

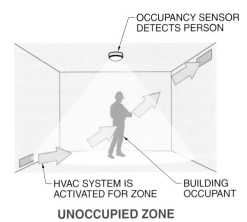

OCCUPANCY SENSOR — FIELD OF VIEW

UNOCCUPIED ZONE

OCCUPANCY SENSOR DETECTS PERSON

HVAC SYSTEM IS ACTIVATED FOR ZONE — BUILDING OCCUPANT

UNOCCUPIED ZONE

Figure 4-27. Occupancy sensors can save energy from the operation of HVAC systems by turning off some HVAC functions when no occupants are within the building zone.

Different technologies can be used to detect occupants. A *passive infrared (PIR) sensor* is a sensor that activates when it senses the heat energy of people within its field-of-view. Infrared occupancy sensors are line-of-sight devices and require the entire detection area to be within view. An *ultrasonic sensor* is a sensor that activates when it senses changes in the reflected sound waves caused by people moving within its detection area. The sensor emits low-power, high-frequency sound waves and listens for a phase shift in the reflected

sound returning to the sensor, which indicates a moving target. This even works to some degree around corners, so the detection area is larger than the line-of-sight field-of-view. A *microwave sensor* is a sensor that activates when it senses changes in the reflected microwave energy caused by people moving within its field-of-view. Dual-technology occupancy sensors include a combination of two types of occupancy sensing, such as a passive infrared sensor with an ultrasonic sensor.

Temperature Control Devices

The main aspect of controlling building occupant comfort is regulating indoor air temperature. Temperature control devices maintain a specified temperature range within the building zones and include furnaces, boilers, electric heat sources, air conditioners, chillers, and heat pumps. These devices deliver or absorb heat via heating or cooling coils.

Furnaces. A *furnace* is a self-contained heating unit for forced-air HVAC systems. Furnaces include a fan, heat source, and controls that handle the operation of all the valves, switches, regulators, and safety devices within the furnace. Typically, the only outside input needed to operate a furnace is an ON/OFF signal when the building space requires heating. The onboard controls manage the startup, operation, and shutdown of the furnace automatically.

Furnaces produce heat by either fuel combustion or electrical energy. The heat in a combustion furnace is produced by burning fuel, which may be coal, wood, fuel oil, natural gas, or propane. The combustion and the resulting gases are contained within pipes and ducts until they are vented to the outside. **See Figure 4-28.** These pipes absorb heat from the combustion, forming a heat exchanger. As cooler air flows past the heat exchanger, it gains heat energy, which is then added to the air within the heated building space.

Electric furnaces work in much the same way, except that there are no combustion gases to segregate or exhaust. Electricity flows through heating elements that become very hot due to their high resistance. Air is heated as it flows directly by the hot electric elements.

▌▌▌ Furnaces

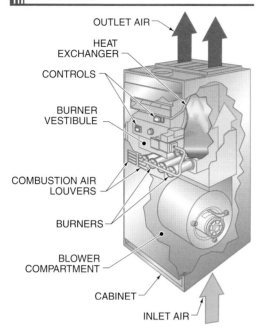

Figure 4-28. Furnaces are incorporated into forced-air HVAC systems to heat air quickly and efficiently.

Boilers. A *boiler* is a closed metal container that heats water to produce steam or hot water. The heat is produced primarily by fuel combustion, though small electric units are becoming more common for limited on-demand applications, such as radiant floor heating for a single zone. Similar to a furnace, the hot gases from combustion heat water in a heat exchanger. **See Figure 4-29.**

The resulting steam or hot water is used to carry heat through piping systems to other parts of the building where other heat exchangers transfer the heat to the building space. The high amount of heat energy contained in steam causes the air temperature in building spaces to rise quickly.

In some urban areas, it is possible for building owners to purchase steam or chilled water from a very large centralized heating and cooling plant. This eliminates for building owners the expense of purchasing, operating, and maintaining the boiler or chiller equipment themselves. These district heating and cooling plants distribute steam and chilled water to many separately owned buildings via a piping network and charge according to the consumption.

▌▌▌ Boilers

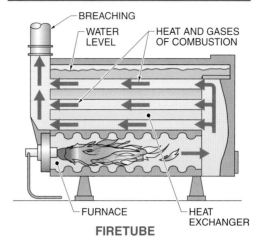

FIRETUBE

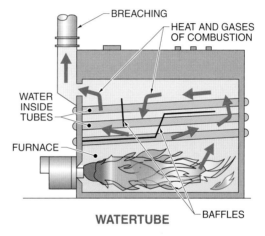

WATERTUBE

Figure 4-29. Boilers heat water to provide hot water or steam to the building for heating coils throughout the building.

The water or steam flows through the system in a continuous loop, propelled by water pumps or steam pressure. As the steam gives up its heat, some of it condenses back into water. Steam heating systems include steam traps attached to the discharge sides of heating units to separate the condensate and return it to the boiler. *Condensation* is the formation of liquid (condensate) as moisture or other vapor cools below its dewpoint.

Like furnaces, boilers include electronic controls to manage the operation of the boiler and its associated equipment. However, instead of simple ON/OFF commands from outside controllers, the boiler controls may receive analog signals corresponding to specific temperatures, within an acceptable range. For example, as the outside air temperature drops from 55°F to –10°F, an HVAC controller may change the boiler's hot water setpoint from 100°F to 190°F. The boiler controller modulates the heating function and a three-way mixing valve (incorporating cool water) to obtain the correct water temperature. Also, the supply of hot water or steam to the heat exchangers conditioning the building spaces can be modulated with valves. For example, the steam supply to a heating coil in an air-handling unit can be increased or decreased as needed by changing the position of the valve.

Unlike furnaces, which are relatively simple to turn on and off and produce heated air quickly, boiler control involves outlet water temperature control. In fact, since large boilers require much more complicated start-up and shut-down procedures, it is common to leave them operating, but at a lower water output temperature, when the heating demand is low.

Electric Heat Sources. Electricity can be used to create heat with electric resistance heating elements. Resistance to the flow of electric current causes the heating elements to become hot. The electrical power supply to the unit is simple to control for both ON/OFF and staged heating. Since they require only an electrical connection and no piping for water or steam distribution, electric heating units are relatively simple and inexpensive to install. **See Figure 4-30.** The disadvantages of electric heating systems include the high cost of electricity and the precautions that must be taken to ensure that the heating elements and building are not damaged due to excessive heat.

Electric heating elements can be used as a heating coil within an air-handling unit. This type of heating is most common in systems with many VAV terminal boxes, since the electrical supply is relatively simple to distribute to many heating coils, as opposed to additional piping for hot water or steam supplies. These units include fans to direct the airflow over the heating elements, thus warming the air.

Electric Resistance Heating Elements

Figure 4-30. Electric heat sources create heat from electrical current flowing through elements with high resistances.

Air Conditioners. Furnaces and boilers create heat energy from other types of energy, such as chemical fuel energy. However, heat energy cannot be destroyed to produce a cooling effect. Cooling appliances such as air conditioners can only move heat energy to where its effect is negligible, like outside the building. An *air conditioner* is a self-contained cooling unit for forced-air HVAC systems.

Air conditioners rely on a mechanical compression refrigeration cycle to move heat. This process uses the phase-change properties of a refrigerant to absorb heat in one part of the system and release it in another. A *refrigerant* is a fluid that is used for transferring heat. Refrigerants have low boiling points, so they vaporize at room temperatures.

Air conditioners are divided into two parts. The evaporator is located inside the building, where the refrigerant cools the air by absorbing heat. The condenser is located outside the building, where the refrigerant rejects heat from the indoor air into the outside air. The refrigerant flows in one direction through a loop of piping connecting the two parts. **See Figure 4-31.**

Air Conditioners

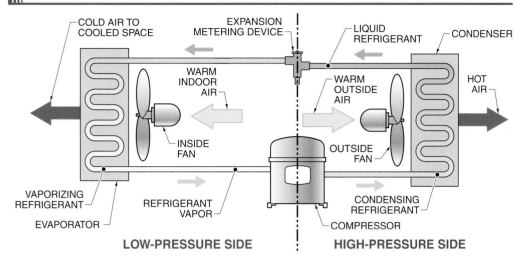

COLD AIR TO COOLED SPACE
EXPANSION METERING DEVICE
LIQUID REFRIGERANT
CONDENSER
WARM INDOOR AIR
WARM OUTSIDE AIR
HOT AIR
INSIDE FAN
OUTSIDE FAN
VAPORIZING REFRIGERANT
REFRIGERANT VAPOR
CONDENSING REFRIGERANT
EVAPORATOR
COMPRESSOR
LOW-PRESSURE SIDE
HIGH-PRESSURE SIDE

Figure 4-31. Air conditioners use the evaporating and condensing properties of a refrigerant to move heat from one area to another.

Liquid refrigerant is metered into the evaporator by an expansion device, which lowers the pressure and temperature of the refrigerant. An *evaporator* is a heat exchanger that adds heat to low-pressure refrigerant liquid. Warm return air from the building is blown over the evaporator coils, transferring heat to the refrigerant and causing it to vaporize. This cools the air. Refrigerant leaves the evaporator as a vapor and flows to the compressor.

The compressor increases the pressure of refrigerant vapor and circulates the refrigerant through the system. Because the refrigerant absorbs heat from the compression process, the refrigerant that leaves the compressor is hotter than the refrigerant in the rest of the refrigeration system. The hot refrigerant vapor discharged from the compressor flows to the condenser. A *condenser* is a heat exchanger that removes heat from high-pressure refrigerant vapor. A fan in the condenser blows warm outside air over the condenser coils. Heat transfers from the refrigerant to the air, lowering the temperature of the vapor until it condenses into a liquid. This liquid returns to the expansion device to begin the cycle again.

Air conditioners are simply switched ON or OFF to start the refrigeration cycle. However, the amount of refrigerant metered into the evaporator coils may be modulated by valves controlled by analog signals.

Chillers. Chillers are very similar to air conditioners. A *chiller* is a refrigeration system that cools water. Chillers use the same mechanical compression refrigeration cycle as air conditioners, but instead of the refrigerant absorbing or rejecting heat to the air in the heat exchangers, they exchange heat with loops of water, on both sides, which then exchange heat with air. Chiller compressors circulate the water and refrigerant throughout their closed-loop systems.

Commercial buildings with large cooling loads often use chillers. Water is cooled to about 45°F in the chiller evaporator and pumped throughout a building for cooling purposes. **See Figure 4-32.** Heat from building spaces is absorbed by the water within terminal devices located in the building spaces or cooling coils located in an air-handling unit.

Many large cities have a district system that provides steam and chilled water through underground piping. For a service fee, a building in the service district may be connected to the network, which saves the building's managers from having to maintain their own steam or chiller equipment.

⫴ Chillers

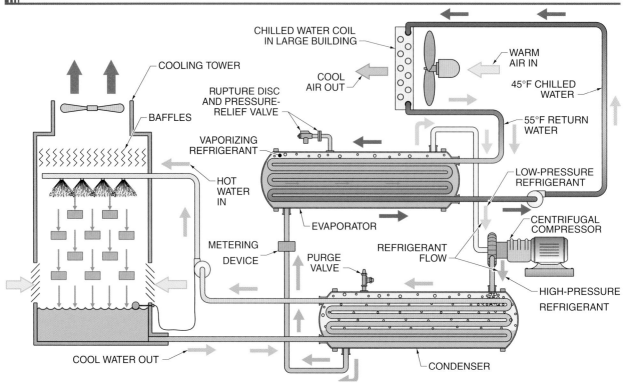

Figure 4-32. Water that is cooled in a chiller can then be distributed and used in cooling coils throughout the building.

Warmer return water (approximately 55°F) is pumped into the chiller and its heat is used to vaporize the refrigerant in the evaporator. Liquid refrigerant surrounds the coils. The resulting cooled water is sent back to the building to absorb heat and cool the building air. The heat taken in by the evaporator is rejected in the condenser.

The condenser is cooled by water that has been cooled in a cooling tower located outside. A *cooling tower* is a device that uses evaporation and airflow to cool water. A fan forces air upward through the tower as warm water is sprayed downward and drips through rows of tiles. Some of the water is evaporated and carried away by the upward airflow, removing heat from the water left behind. The cool water falls to the bottom of the tower for reuse in the condenser. Makeup water is added to replace the water lost to evaporation.

Like boilers, chillers are not turned ON or OFF for short periods of time, since it takes some time during startup for the unit to establish normal operating temperatures.

Rather, the chiller remains operating most of the time and analog-controlled water valves at the terminal devices or cooling coils provide temperature control.

Heat Pumps. A *heat pump* is a mechanical compression refrigeration system that moves heat from one area to another area. Heat pumps are nearly identical to air conditioning systems, except that heat pumps can reverse the flow of refrigerant to switch between heating and cooling modes. **See Figure 4-33.** Heat pumps are turned ON or OFF as needed to provide heating or cooling to the indoor supply air.

When a heat pump is in the cooling mode, it moves heat from inside the building to outside the building, just like an air conditioner. The indoor unit is the evaporator and the outdoor unit is the condenser. However, when a heat pump is in the heating mode, the flow of refrigerant is reversed, which moves heat from outside the building to inside the building. In heating mode, the indoor unit is the condenser and the outdoor unit is the evaporator.

⦀ Heat Pumps

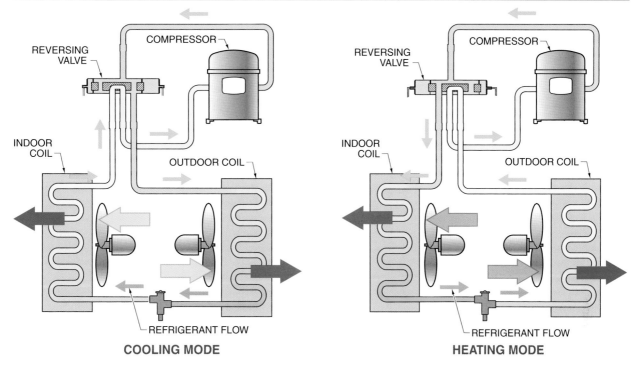

Figure 4-33. Heat pumps are nearly identical to air conditioners in operation, except that heat pumps can reverse the flow of heat for heating needs.

The heat is moved between the building air and either the outside air, water, or the ground. Air-to-air heat pumps use the outside air as the heat source (for heating) and heat sink (for cooling). Commercial heat pumps are commonly water-to-air heat pumps, which use water from lakes, streams, wells, or retention ponds as the heat source and heat sink. Water-to-air heat pumps have a coil heat exchanger with water as the heat transfer medium. A geothermal heat pump is a heat pump that uses the ground below the frost line as the heat source and heat sink.

Humidity Control Devices

Humidification control devices are needed to control the amount of moisture present in the air. These devices rely on forced-air HVAC systems to affect the humidity of the supply air and distribute it throughout the building space. Humidification is controlled by humidifiers and dehumidifiers. **See Figure 4-34.**

Humidifiers. Humidification requirements depend primarily on the local climate. Buildings in warm climates may not require humidification if temperatures are rarely low, though there are some regions with extremely dry and warm weather. Buildings in northern climates, however, require humidification during cold weather to increase the relative humidity, which falls during the heating process.

A *humidifier* is a device that adds moisture to the air by causing water to evaporate into the air. Humidifiers are typically installed as central units in an air-handling unit after the heating coil (if there is one). This ensures that the air can absorb as much water as possible, since warm air can hold more moisture than cool air. Humidifiers can also be installed downstream of air terminal units for special applications.

Because their cooling coils also provide passive dehumidification, air conditioners and heat pumps must collect and drain the accumulated water that condenses on the evaporator (cooling) coils. This prevents water damage and mold problems.

░ Humidity Control Devices

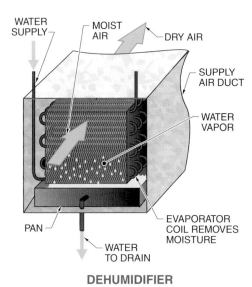

HUMIDIFIER

DEHUMIDIFIER

Figure 4-34. *Humidifiers and/or dehumidifiers are installed within an air-handling unit's supply duct to add moisture to or remove moisture from the conditioned air.*

Methods of humidification in commercial buildings include steam jets, flow-through metal filters, and drum-type humidifiers. Steam jets are the most common type and simply inject steam into the air flowing through the supply air duct. The humidification is easily modulated by controlling the steam flow with valves operated by analog signals. The latter two types rely on air being blown through a filter-like material that holds water, and the air picks up moisture by evaporating some of the water. However, these two types of humidifiers are increasingly uncommon due to indoor air quality concerns.

Dehumidifiers. Dehumidification is separate from humidification, and the two respective devices are not always included together, depending on the building's HVAC system requirements. A *dehumidifier* is a device that removes moisture from air by causing the moisture to condense. It does this by cooling the air until it reaches its dewpoint, or 100% relative humidity. Further cooling causes the moisture to condense into water. The water is either collected in a container to be emptied later or, more commonly, drains directly into the wastewater plumbing system.

Dehumidifiers incorporated into building HVAC systems are very similar to air conditioners. The moist air is blown over a cooling coil in the air-handling unit that is chilled with low-pressure refrigerant (though chilled water could also be used). The resulting air is both cooler and drier. If cooling the air is not desired, the air then flows over the warm condensing coils of the refrigeration system. This extra step does not affect the moisture content of the air, but the relative humidity falls again by warming the air temperature.

Dehumidifiers are not easily modulated to different levels of dehumidification, so they are generally only turned ON or OFF as needed. However, since temperature and relative humidity are closely related, with changes in one property often affecting the other, heating and cooling functions also have some control over humidification and dehumidification.

Airflow Distribution Control Devices

Airflow is essential to a forced-air system, which relies on the movement of air over temperature and humidity control devices to affect its properties and to distribute the conditioned air throughout the building space. Airflow is also important to the efficiency of the HVAC system. Conditioned air should flow to the areas where it is needed and not be wasted in areas where it is not needed. The design of a building and the installation of its HVAC system greatly influence airflow. However, fans and dampers within the system can be used to further adjust the way air flows through the ducts and the building spaces.

Fans. Fans are primary components in forced-air HVAC systems. Separate supply air and return air fans are used to move air within different parts of the duct system. Fans are also included with some unit devices, such as furnaces. A *fan* is a mechanical device with spinning blades that move air. Fans are used within air ducts and in open environments such as air conditioners and heat pumps to force outside air over the coils located outside of the building.

Fan designs fall within two main types: centrifugal fans and axial-flow fans. **See Figure 4-35.** Air enters a centrifugal fan through the center of the impeller. An *impeller* is the bladed, spinning hub of a fan or pump that forces fluid to its perimeter. As the impeller rotates, the blades move rapidly through the air, forcing the air to move outward. The housing of the fan, the scroll, directs this airflow to the fan outlet. Because of this design, centrifugal fans change the direction of the airflow by 90°. Axial-flow fans, however, do not change the direction of airflow. Fan hubs for either type may be mounted directly on the shaft of a motor or may be turned by a motor with a pulley and belt arrangement.

Fan airflow output is modulated in variable-air-volume air-handling units in a variety of ways. The velocity of the airflow is affected by the pitch (angle) of the blades, restriction of the fan inlet, and the fan speed. **See Figure 4-36.**

Some fans can be adjusted for different blade pitches, within an allowable range. A small, geared motor in the hub rotates the attachment points of each blade, changing its angle relative to the direction of airflow. Within limits, steeper pitches increase the air velocity and shallower pitches decrease the air velocity. This allows the air velocity to be controlled without changing the motor speed.

Alternatively, vortex dampers control the air volume output of a fan by restricting the air flowing into the fan. A *vortex damper* is a pie-shaped damper at the inlet of a centrifugal fan that reduces the ability of the fan to grip and move air. Like other dampers, vortex dampers can be set at various positions to allow a percentage of air to pass through.

Fans

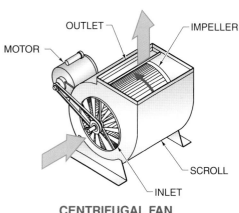

CENTRIFUGAL FAN

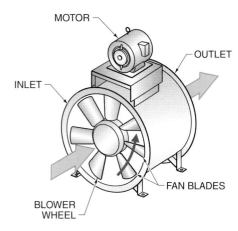

AXIAL-FLOW FAN

Figure 4-35. Centrifugal fans change the direction of the airflow by 90°, while axial-flow fans have no effect on the airflow direction.

Most commonly, though, the speed of the electric motor is controlled with a variable-frequency drive. A *variable-frequency drive* is a motor controller that is used to change the speed of AC motors by changing the frequency of the supply voltage. This drive changes the characteristics of the electrical power to the motor so that the motor operates at different speeds. The drive can even change the direction of rotation, causing the air to flow in the opposite direction. All three methods are very effective at providing airflow control, though variable-frequency electric motor drives are the most energy efficient and the most common in new installations.

||| Variable Airflow Control

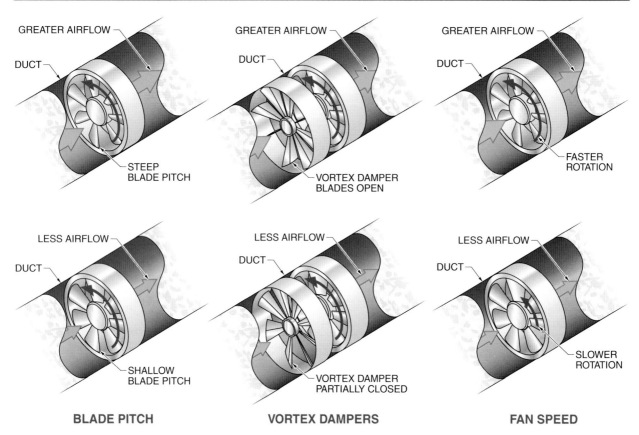

Figure 4-36. The volume and velocity of airflow can be controlled by the blade pitch and vortex dampers but are most commonly controlled by changing the fan speed.

Fans and dampers within an air-distribution system are analogous to the pumps and valves within a water- or steam-distribution system. Both systems are used to drive and control the distribution of conditioned air (or the means to condition air) from a central location to terminal units throughout a building.

Fan airflow output control requires an analog input to produce the desired airflow. Depending on the fan, inputs may be either a standard analog signal, such as 0 VDC to 10 VDC, or a structured network message that includes the desired settings.

Dampers. A *damper* is a set of adjustable metal blades used to control the amount of airflow between two spaces. Damper positions range from fully open, which has minimal effect on airflow through the damper, to fully closed, which almost completely seals a section or space off from the airflow on the other side (very little air gets through). Dampers can also be set to any position in between to modulate the volume of airflow through the damper. **See Figure 4-37.**

Dampers are used throughout the air-handling units and ductwork to control outside air, return air, and exhaust airflow. Dampers are also used at the ends of ductwork, in air terminal units, where the air flows into the building spaces. Depending on the type and application of the air-handling unit, dampers may also be used to control airflow across heating or cooling coils. There may also be manually operated dampers used to balance the HVAC system, though these are not involved in automation.

Damper

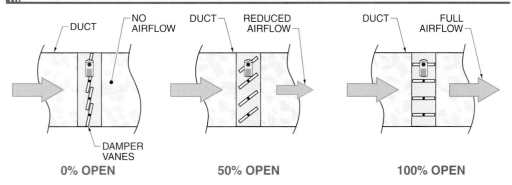

Figure 4-37. The position of dampers within a duct affects the amount of airflow allowed to flow through the section.

The angle of the damper blades is controlled by an actuator containing a small, geared electric motor. The blades are connected together so that they all move simultaneously. The actuator responds to signals from a controller to set the damper position. Digital (ON/OFF) signals correspond to completely open or completely closed. Tristate actuators use two digital signals to drive the actuator open or closed. When a signal stops, the actuator stalls in its current position. Alternatively, many actuators accept standard analog signals, which correspond to any position between 0% open (100% closed) and 100% open, allowing the damper opening to be positioned precisely.

End switches are commonly installed on many dampers. An *end switch* is a switch that indicates the fully actuated damper positions. End switches are similar to limit switches. Its contacts are actuated by the position of the damper, providing a digital (ON/OFF) output signal. Some end switches are installed in the damper actuator, proving that the actuator moved to a full position in response to a control signal. Other end switches are installed near one of the damper blades, proving that the damper blades actually moved to a full position. **See Figure 4-38.** This distinction may be important if an actuator linkage breaks; in this case, the actuator may move, but the blades may not.

Dampers vary in their response to a control system problem or power failure. Spring return dampers return to a default position, either fully open or fully closed, without the actuator's influence. Others stall in their current position. The choice between the two damper design defaults depends on the damper's function and placement within the HVAC system. For example, exhaust air and outside air dampers are normally closed dampers, which means that they automatically close if the damper motor control loses power. The return air damper, however, is normally open.

End Switches

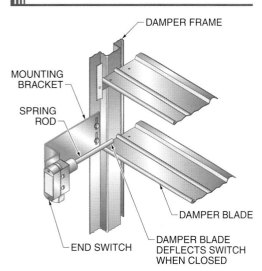

Figure 4-38. End switches are installed on dampers to verify damper position.

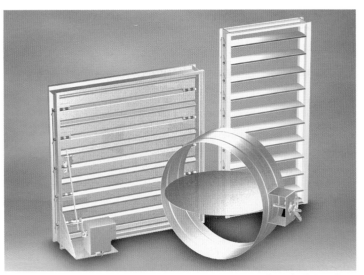

Greenheck Fan Corp.

Dampers are available in a variety of sizes, shapes, and blade configurations for different pressure, velocity, and airflow requirements.

Some dampers are used exclusively during a building fire by the fire alarm system. These dampers help control the spread of fires and smoke or purge smoke outside the building. However, because they provide life safety functions, the control of smoke dampers is heavily regulated. Engineered smoke control system applications may be controlled only by the fire protection system, which is the only building system UL-listed for reliability.

Dampers can also be used to improve HVAC system control by avoiding temperature stratification within an air-handling unit. The outside air and return air damper blades can be arranged so their air streams collide, thoroughly mixing the air. Also, a two-section outside air damper, with one stage opening before the other, keeps the outside air velocity fast in order to promote mixing. The first stage of the damper is called the minimum outside air damper, while the second stage is an economizer damper.

Water Distribution Control Devices

Systems that use either hot or chilled water require pumps and valves to circulate and control the water through the pipes and heat exchangers. A pump and several valves are usually integral to the water-conditioning device, such as a boiler or chiller, and controlled by that device's on-board controller. Additional pumps and valves may be controlled separately by an HVAC system controller, particularly those involved in the supply of the water to terminal devices.

Pumps. A *pump* is a device that moves water through a piping system. A pump moves water from a lower pressure to a higher pressure, overcoming this pressure difference by adding energy to the water. Water pumps are normally driven by electric motors. The motor for smaller pumps is typically integrated into the pump housing. For pumps with a pumping rate above 100 gpm (gallons per minute), the motor is normally connected by a coupling between the shaft and motor.

The motors can be turned ON or OFF with relays or motor starters. With a variable-frequency drive, the speed of the electric motor can be controlled to achieve the desired pumping rate.

Centrifugal pumps are the most widely used type of pump in HVAC systems. **See Figure 4-39.** A *centrifugal pump* is a pump with a rotating impeller that uses centrifugal force to move water. Centrifugal pumps are very similar to centrifugal fans, even though they move different fluids. Water enters the pump through the center of the impeller. The impeller rotates at a relatively high speed, imparting a centrifugal force to the water, which moves it outward. The shape of the casing directs this water to the outlet.

Centrifugal Pumps

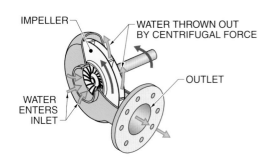

Figure 4-39. Pumps circulate water from a water heating or chilling device, such as a boiler, throughout a hydronic system.

Valves. Indoor air temperature control in a hydronic system may be maintained with a valve, which controls the flow of chilled water, heated water, or steam to a heat exchanger. The amount of flow has a direct impact on the rate of heat transfer in the exchanger. A *valve* is a fitting that regulates the flow of water within a piping system.

A *full-way valve* is a valve designed to operate in only the fully open or fully closed positions. Full-way valves are also called shutoff valves. When open, the internal design permits a straight and unrestricted fluid flow through the valve, resulting in very little pressure loss due to friction.

A *throttling valve* is a valve designed to control water flow rate by partially opening or closing. Due to the internal configuration, water flowing through the valve changes direction several times, resulting in flow resistance and a pressure drop. This makes them ideal for reducing flow and pressure. Since they can be used for any position between fully open and fully closed, the actuators for these valves accept analog inputs.

Valves used in automation systems are essentially the same as manually actuated valves except that they include electronically controlled actuators, usually a small geared electric motor mounted to the rotating valve stem. **See Figure 4-40.** However, these motors require onboard controls to allow for operation in both directions (for both opening and closing) and any position in between (for throttling).

Hydronic System Valves

Dwyer Instruments, Inc.

Figure 4-40. Many types of water valves are used to control the flow of water for a HVAC system. The actuators attached the valves receive signals from controllers to open or close the valves.

The electronic valve packages include the necessary electronics to operate the motor and accept input signals that indicate the desired valve position. For example, full-way valves may require only digital inputs, which correspond to completely open and completely closed. Throttling valves require analog values to determine intermediate positions. Either type may be available for use with structured network message systems.

Refer to Quick Quiz® on CD-ROM

Summary

- Precise control of HVAC systems maintains a comfortable indoor climate for the building occupants, optimizes the indoor conditions for operating equipment or stored inventory, and does these functions with minimum energy use during all times of the year.
- The hub of a forced-air HVAC system is an air-handling unit.
- Air-handling units manage four types of air: return air, outside air, exhaust air, and supply air.
- By controlling the fans, dampers, coils, and other devices within an air-handling unit, the unit can be used to heat, cool, humidify, dehumidify, or simply ventilate a building space.
- Hydronic HVAC systems use pipes, valves, and pumps to distribute water throughout the building.
- Many commercial HVAC systems incorporate aspects of both forced-air and hydronic systems, such as air-handling units that use steam from a boiler in the heating coils and chilled water from a chiller in the cooling coils.

- The fluid or electrical current flow through a heating and cooling coil can be controlled to regulate the temperature of the coil.
- A building may be divided into many zones, depending on the use of the space and to create manageable units of HVAC-conditioned spaces.
- The five requirements for comfort are proper temperature, humidity, filtration, circulation, and ventilation.
- Warmer air has a greater capacity to be saturated with moisture than cooler air.
- Relative humidity relates the moisture content of the air within the context of the current air temperature.
- Ventilation prevents indoor air from becoming stale by diluting stale, recirculated air with fresh, outdoor air.
- Climate sensors measure and communicate the comfort-related properties of indoor air and the HVAC system is adjusted until the sensors read values within an optimal range.
- Some sensors measure the exact value of the property and output the analog signal, while others indicate only whether the measured property is either above or below a certain setpoint.
- Occupancy sensors are used to save energy by deactivating certain HVAC functions to a zone when no one is in the room or area.
- Temperature control devices deliver or absorb heat via heating or cooling coils.
- Furnaces, electric heat sources, air conditioners, and heat pumps can easily be turned on or off, but the output from boilers and chillers is usually modulated in some way instead because of their complicated start-up requirements.
- Heat pumps are nearly identical to air conditioning systems, except that heat pumps can reverse the flow of refrigerant to switch between heating and cooling modes.
- Humidification control devices are needed to control the amount of moisture present in the air.
- Fans and dampers within the HVAC system alter the way air flows through ducts and building spaces.
- The velocity of the airflow output from a fan depends on the pitch (angle) of the blades and the speed of their rotation.
- Dampers can be positioned from fully open to fully closed, and any position in between, to modulate the volume of airflow through the damper.
- In the case of a control system problem or power failure, each damper is designed to default to either a fully open or fully closed position.
- Systems that use hot or chilled water require pumps and valves to circulate and control the water through the pipes and heat exchangers.
- Centrifugal pumps are very similar to centrifugal fans, even though they move different fluids.
- Valves are classified as either full-way or throttling valves, which accept digital or analog inputs, respectively.

Definitions

- An **HVAC system** is a building system that controls a building's indoor climate.
- A **forced-air HVAC system** is a system that distributes conditioned air throughout a building in order to maintain the desired conditions.
- **Conditioned air** is indoor air that has been given desirable qualities by the HVAC system.

- An *air-handling unit (AHU)* is a forced-air HVAC system device consisting of some combination of fans, ductwork, filters, dampers, heating coils, cooling coils, humidifiers, dehumidifiers, sensors, and controls to condition and distribute supply air.

- *Return air* is the air from within the building space that is drawn back into the forced-air HVAC system to be exhausted or reconditioned.

- *Outside air* is fresh air from outside the building that is incorporated into the forced-air HVAC system.

- *Mixed air* is the blend of return air and outside air that is combined inside the air-handling unit and goes on to be conditioned.

- *Exhaust air* is air that is ejected from the forced-air HVAC system.

- *Supply air* is newly conditioned mixed air that is distributed to the building space.

- A *terminal unit* is the end point in an HVAC distribution system where the conditioned medium (air, water, or steam) is added to or directly influences the environment of the conditioned building space.

- A *constant-air-volume air-handling unit* is an air-handling unit that provides a steady supply of air and varies the heating, cooling, or other conditioning functions as necessary to maintain the desired setpoints within the building zone.

- A *variable- air-volume air-handling unit* is an air-handling unit that provides air at a constant air temperature but varies the amount of supplied air in order to maintain the desired setpoints within the building zone.

- A *variable-air-volume (VAV) terminal box* is a device located at the building zone that provides heating and airflow as needed in order to maintain the desired setpoints within the building zone.

- A *hydronic HVAC system* is a system that distributes water or steam throughout a building as the heat-transfer medium for heating and cooling systems.

- A *heat exchanger* is a device that transfers heat from one fluid to another fluid without allowing the fluids to mix.

- A *heating coil* is a heat exchanger that adds heat to the air surrounding or flowing through it.

- A *cooling coil* is a heat exchanger that removes heat from the air surrounding or flowing through it.

- A *zone* is an area within a building that shares the same HVAC requirements.

- *Comfort* is the condition of a person not being able to sense a difference between themselves and the surrounding air.

- *Temperature* is the measurement of the intensity of the heat of a substance.

- The *dry-bulb temperature* is the temperature of air measured by a thermometer freely exposed to the air but shielded from radiation and moisture.

- The *wet-bulb temperature* is the temperature measured by a thermometer that has its bulb kept in contact with moisture.

- *Evaporation* is the process of a liquid changing to a vapor by absorbing heat.

- *Humidity* is the amount of moisture present in the air.

- *Absolute humidity* is the amount of water vapor in a particular volume of air.

- *Relative humidity* is the ratio of the amount of water vapor in the air to the maximum moisture capacity of the air at a certain temperature.

- *Dewpoint* is the air temperature below which moisture begins to condense.

- *Circulation* is the continuous movement of air through a building and its HVAC system.

- A *register* is a cover for the opening of ductwork into a building space.
- *Temperature stratification* is an undesirable variation of air temperature between the top and bottom of a space.
- *Air velocity* is the speed at which air moves from one point to another.
- The *air changes per hour (ACH)* is a measure of the number of times the entire volume of air within a building space is circulated through the HVAC system in one hour.
- *Filtration* is the process of removing particulate matter from air.
- A *micron* is a unit of measure equal to one-millionth of a meter, or 0.000039″.
- *Stale air* is air with high concentrations of carbon dioxide and/or other vapor pollutants.
- *Ventilation* is the process of introducing fresh outdoor air into a building.
- A *temperature sensor* is a device that measures temperature.
- A *thermocouple* is a temperature-sensing element consisting of two dissimilar metal wires joined at the sensing end.
- A *thermistor* is a temperature-sensing element made from a semiconductor material that changes resistance in response to changing temperatures.
- A *resistance temperature detector (RTD)* is a temperature-sensing element made from a material with an electrical resistance that changes with temperature.
- A *thermostat* is a switch that activates at temperatures either above or below a certain setpoint.
- A *hygrometer* is a device that measures the amount of moisture in the air.
- A *hygroscopic element* is a material that changes its physical or electrical characteristics as the humidity changes.
- A *humidistat* is a switch that activates at humidity levels either above or below a certain setpoint.
- A *pressure sensor* is a sensor that measures the pressure exerted by a fluid, such as air or water.
- *Differential pressure* is the difference between two pressures.
- A *transducer* is a device that converts one form of energy into another form of energy.
- A *differential pressure switch* is a switch that activates at differential pressures either above or below a certain value.
- An *airflow station* is a sensor that measures the velocity of the air in a duct system.
- A *flow switch* is a switch with a vane that moves from the force exerted by the water or air flowing within a duct or pipe
- A *carbon dioxide sensor* is a sensor that detects the concentration of carbon dioxide (CO_2) in air.
- An *occupancy sensor* is a sensor that detects whether an area is occupied by people.
- A *passive infrared (PIR) sensor* is a sensor that activates when it senses the heat energy of people within its field-of-view.
- An *ultrasonic sensor* is a sensor that activates when it senses changes in the reflected sound waves caused by people moving within its detection area.
- A *microwave sensor* is a sensor that activates when it senses changes in the reflected microwave energy caused by people moving within its field-of-view.
- A *furnace* is a self-contained heating unit for forced-air HVAC systems.
- A *boiler* is a closed metal container that heats water to produce steam or hot water.

- *Condensation* is the formation of liquid (condensate) as moisture or other vapor cools below its dewpoint.
- An *air conditioner* is a self-contained cooling unit for forced-air HVAC systems.
- A *refrigerant* is a fluid that is used for transferring heat.
- An *evaporator* is a heat exchanger that adds heat to low-pressure refrigerant liquid.
- A *condenser* is a heat exchanger that removes heat from high-pressure refrigerant vapor.
- A *chiller* is a refrigeration system that cools water.
- A *cooling tower* is a device that uses evaporation and airflow to cool water.
- A *heat pump* is a mechanical compression refrigeration system that moves heat from one area to another area.
- A *humidifier* is a device that adds moisture to the air by causing water to evaporate into the air.
- A *dehumidifier* is a device that removes moisture from air by causing the moisture to condense.
- A *fan* is a mechanical device with spinning blades that move air.
- An *impeller* is the bladed, spinning hub of a fan or pump that forces fluid to its perimeter.
- A *vortex damper* is a pie-shaped damper at the inlet of a centrifugal fan that reduces the ability of the fan to grip and move air.
- A *variable-frequency drive* is a motor controller that is used to change the speed of AC motors by changing the frequency of the supply voltage.
- A *damper* is a set of adjustable metal blades used to control the amount of airflow between two spaces.
- An *end switch* is a switch that indicates the fully actuated damper positions.
- A *pump* is a device that moves water through a piping system.
- A *centrifugal pump* is a pump with a rotating impeller that uses centrifugal force to move water.
- A *valve* is a fitting that regulates the flow of water within a piping system.
- A *full-way valve* is a valve designed to operate in only the fully open or fully closed positions.
- A *throttling valve* is a valve designed to control water flow rate by partially opening or closing the valve.

Review Questions

1. What is the primary difference between a forced-air HVAC system and a hydronic HVAC system?
2. How can establishing separate HVAC zones within a building improve energy efficiency and occupant comfort?
3. Why are relative humidity measurements dependent on the air temperature?
4. Why is ventilation important?
5. What is the primary difference between a temperature sensor and a thermostat?
6. How can differential pressure sensors be used to indicate filter condition?
7. How can occupancy sensors be used to save energy in the HVAC system?
8. How does a heat pump change between being a heating device to a cooling device?
9. How do some dehumidifiers operate similarly to air conditioners?
10. How are airflow distribution control device and water distribution control device functions similar?

Chapter Five

HVAC System Applications

HVAC system control is the most common application of building automation but can be quite complex. Multiple zones with different requirements, constantly changing variables, and the interrelationships between conditioned air characteristics make HVAC system control seem particularly complicated. However, the key to understanding HVAC system controls is to study each application individually. Most applications can be broken down into a simple sequence of inputs, decision making, and outputs. Additional considerations of optional controls, safeties, alarms, and energy-saving measures only build further upon this basic sequence.

Chapter Objectives

- Describe the interactions between common HVAC control devices.
- Identify the common applications in air-handling units, air terminal units, and hydronic systems.
- Describe system-wide applications with the zone-specific applications of terminal units.
- Identify the regulations and limitations for automation of smoke control applications.
- Describe the heating and cooling applications using water or steam distribution.

HVAC SYSTEM APPLICATIONS

Each HVAC system sequence specifically addresses a single application with a desired outcome. Many control sequences occur simultaneously within the HVAC system to manage different aspects of comfort, such as temperature, humidity, and airflow. The sequences are managed locally by controllers that collect input signals from HVAC sensors and other building automation devices, make decisions on how to operate the unit to maintain the desired setpoints, and generate output signals to the necessary devices to change their operation in accordance with these decisions. The controllers also share HVAC operating information with other devices on the building automation network.

However, some sequences also affect each other. **See Figure 5-1.** The output from one sequence may become the input of another. Or, the change of a setpoint may affect the calculations of another. Also, many sequences involve some of the same devices. Some devices send out control signals that are used in more than one sequence, and some devices receive control signals from more than one sequence. Priority information may be necessary to ensure that a device receiving multiple control inputs behaves appropriately.

All of these interactions can make HVAC control applications especially tricky to balance. The consulting-specifying engineer and the controls contractor must pay careful attention to sequences and controller programming in order to appropriately manage the system as a whole.

Control applications in HVAC systems may also include special considerations for safety functions. While some failures in an HVAC control application cause relatively minor discomfort for the building occupants, other failures or adverse conditions may cause equipment damage and even life safety issues. Therefore, some applications require special safety functions. The most vital safeties are hardwired as separate circuits that are independent of the automation system, which ensures reliable operation regardless of the status of the building automation system. **See Figure 5-2.** The opening of the circuit by any safety control device on the circuit immediately shuts down part of the system by de-energizing the variable-frequency drives (VFDs) or motor starters, regardless of the current mode of operation. Safety circuits also typically provide an input to controllers to generate alarms. Other safeties may be present only as a feature that is programmed into the controller.

||| HVAC Control Sequence Interaction

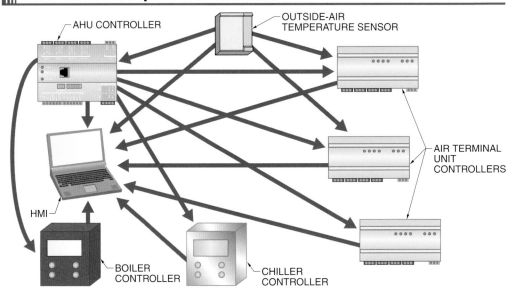

Figure 5-1. *Many sensor inputs and controller outputs are shared between some of the common HVAC control sequences.*

▌▌▌ Safety Circuits

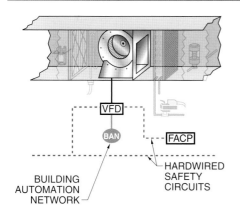

Figure 5-2. Hardwired safety circuits ensure reliable shutdown of the system for a major problem, regardless of the status of the automation system.

Since an HVAC system is one of the largest consumers of electrical and fuel energy in a building, some control applications allow variations in the sequence in order to reduce energy consumption. Depending on the application, energy-saving measures may involve a small sacrifice in comfort. The decision to implement these measures requires input from the building owner to balance the energy savings with less-than-optimal occupant comfort.

HVAC control applications can be divided by the primary unit or controller involved in executing the sequence. Air-handling units are the most common central HVAC unit used in commercial buildings. Their controllers manage all the control devices found in the unit in order to properly condition the air supplied to building spaces. Located at the building spaces, air terminal units are typically managed by a separate controller to further condition the supply air for the precise requirements of an individual building zone. Other HVAC control applications involve more specific heating and cooling control with hydronic and refrigeration systems.

AIR-HANDLING UNIT CONTROL APPLICATIONS

The hub of a forced-air HVAC system is an air-handling unit (AHU). This device combines many HVAC control devices together to fully condition the air in a central location before it is distributed throughout the building. An *air-handling unit* is a forced-air HVAC system device consisting of some combination of fans, ductwork, filters, dampers, heating coils, cooling coils, humidifiers, dehumidifiers, sensors, and controls to condition and distribute supply air. Depending on the application, air-handling units may vary in size and configuration, but most are arranged similarly.

Supply-Air Fan Control

The supply-air fan moves conditioned air from the AHU to the building spaces via ductwork. There must be enough supply airflow to adequately distribute air through the ducts. Control of the supply-air fan is managed by the AHU controller and is based primarily on the static air pressure within the supply-air duct, which should be maintained at a setpoint. **See Figure 5-3.** The air pressure is raised or lowered as necessary by changing the speed of the variable-frequency drive (VFD) operating the supply-air fan (or energizing a motor starter if the AHU is a constant volume-type).

Control Sequence. A typical sequence for controlling the supply-air fan in an AHU includes the following:

1. The AHU controller monitors inputs from the building automation system.
   - The AHU controller receives a digital input indicating desired occupied/ unoccupied fan operation based on a preprogrammed schedule.
   - The AHU controller receives digital input indicating an override of a scheduled unoccupied period in order to prepare it in anticipation of occupancy.
2. The AHU controller monitors inputs from physical sensors.
   - A static air pressure sensor provides an analog input indicating the supply-air duct static air pressure. (If there are multiple supply-air duct static pressure sensors, such as one on every floor in a multistory building, the sensor with the lowest static pressure [greatest differential from setpoint] is utilized for control.)

||||Supply-Air Fan Control

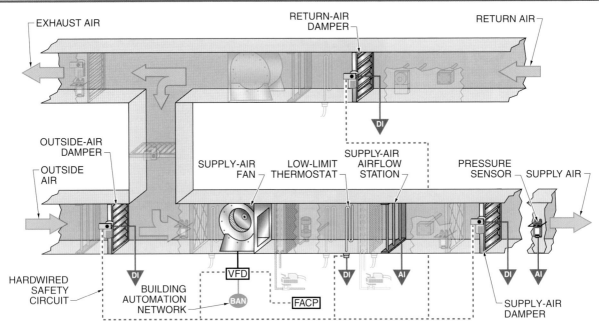

Figure 5-3. Supply-air fan control maintains an adequate static air pressure in the supply-air duct.

- A control device provides a digital input indicating an override of the occupancy schedule, such as an occupancy sensor that indicates unscheduled use of the building space.

3. If the digital inputs indicate that the supply-air fan should be ON, the controller compares the measured static air pressure to the setpoint.
   - If the measured static air pressure is lower than the setpoint, the supply-air fan speed must be increased.
   - If the measured static air pressure is higher than the setpoint, the supply-air fan speed must be decreased.

4. The AHU controller generates an analog signal to the supply-air fan variable-frequency drive (VFD) to modulate the speed of the supply-air fan. If a larger air-handling unit has multiple fans, each receives individual but identical signals during normal operation.

Optional Controls. The AHU controller may monitor inputs from other sensors for control. An airflow station just upstream of the supply-air fan provides an analog input that indicates the supply-air fan airflow. This airflow station is also used extensively for control of the return-air fan. While it is possible to configure a fan control system without airflow stations, many engineers decide to add these components to increase system stability.

The VFD provides an analog input that indicates the motor speed (as a percentage of the maximum rated motor speed). This information is used to generate an alarm if the fan is at maximum speed and the pressure setpoint is not being met.

Safeties. Safety circuits are hardwired to a digital input on the supply-air fan VFD (or motor starter) to shut the fan down immediately upon the opening of a safety circuit. Three conditions require shutdown of the supply-air fan: reaching the low-temperature limit, closed isolation dampers, or a shut-down command from the fire alarm system. Indications of these conditions are provided by the low-limit thermostat, damper end switches, and a relay from the fire alarm control panel (FACP), respectively. These devices are hardwired into two safety circuits. One circuit provides the digital input to the VFD as the safety control and the other circuit provides an input to the building automation system to generate alarms.

Alarms. Besides the alarm for the activation of the safety circuit, the controller transmits an alarm for either of the following conditions:

• The supply-duct static air pressure falls below a preset limit.

• The supply-air fan is at 100% rated speed and the supply-duct static air pressure is below the setpoint.

Energy-Saving Measures. The most obvious energy-saving measure is to simply shut the fans OFF when they are not needed. This is accomplished by using the software inputs and other applicable external inputs to determine when the building space is being used or when the space conditions must be maintained in anticipation of use. However, this may not be possible in some systems due to climate extremes, the quality of the building envelope, or other factors.

Also, a fixed static pressure setpoint represents a design or worst-case load, plus a safety factor, which is established in the balancing process by the balancing contractor. This conservative value sacrifices energy efficiency to ensure that the system always satisfies the desired conditions. However, the static air pressure setpoint can be dynamically adjusted for peak-load and low-load conditions. First, all of the air terminal units on the AHU must be polled to confirm that all of the airflow setpoints are being satisfied. Second, the damper positions of all the air terminal units are checked to ensure that no dampers are wide open. It is important to check both because the sensors used on an air terminal unit may not be very accurate. When both conditions are satisfied, the setpoint can be lowered, which allows the fan to slow down and saves energy. Conversely, if the air terminal units are not maintaining flow or are 100% open, then the supply-air fan must be sped up to make sure all systems are satisfied.

Return-Air Fan Control

The return-air fan assists the supply-air fan in moving conditioned air into the building spaces by drawing air back out of the spaces. There must be enough return airflow to adequately circulate air through the building space. Control of the return-air fan in the first sequence is based on the static air pressure within the building, similar to the supply-air fan sequence, and involves many of the same AHU components. **See Figure 5-4.** The air pressure is raised or lowered as necessary by changing the speed of the variable-frequency drive (VFD) operating the return-air fan (or energizing a motor starter if the AHU is a constant volume-type).

Control Sequence. A typical sequence for controlling the return-air fan in an AHU includes the following:

1. The AHU controller monitors inputs from the building automation system.
   • The AHU controller receives a digital input indicating desired ON/OFF fan operation based on a preprogrammed schedule.
   • The AHU controller receives a digital input indicating an override of a scheduled unoccupied period in order to prepare it in anticipation of occupancy.

2. The AHU controller monitors inputs from physical sensors.
   • A differential pressure sensor provides an analog input indicating the difference in air pressure between the building space and outside.
   • A control device provides a digital input indicating an override of the occupied/unoccupied schedule, such as an occupancy sensor that indicates unscheduled use of the building space.

3. If the digital inputs indicate that the return-air fan should be in occupied mode, then the controller compares the measured differential air pressure to the setpoint.
   • If the measured differential pressure is below the setpoint (outside pressure is greater than inside pressure), the return-air fan speed must be decreased.
   • If the measured differential pressure is above the setpoint (outside pressure is less than inside pressure), the return-air fan speed must be increased.

4. The AHU controller generates an analog signal to the return-air fan VFD to modulate the speed of the return-air fan.

Return-Air Fan Control

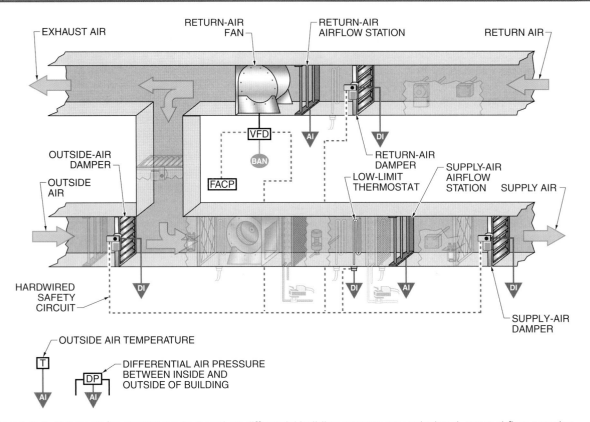

Figure 5-4. Return-air fan control can be based on differential building pressure or a calculated return airflow setpoint.

While this sequence is simple to install, it may be unstable. Building pressure is influenced by wind gusts, wind direction, and turbulence around the building itself. Instead, an alternate sequence relies on airflow measurements. These measurements are then used to calculate a return airflow setpoint with the following formula:

$$RA = SA - EA - p$$

where

RA = return airflow (in cfm)

SA = supply airflow (in cfm)

EA = exhaust airflow (in cfm)

p = pressurization factor (in cfm)

The exhaust airflow is a fixed value from the fan schedule and should include only the fans that are ON. The pressurization factor is an airflow that represents 2% to 3% (adjustable) of the supply airflow.

Control Sequence. An alternative sequence for controlling the return-air fan in an AHU includes the following:

1. The AHU controller monitors inputs from the building automation system.
   - The AHU controller receives a digital input indicating desired occupied/unoccupied fan operation based on a pre-programmed schedule.
   - The AHU controller receives a digital input indicating an override of a scheduled unoccupied period in order to prepare it in anticipation of occupancy.
2. The AHU controller monitors inputs from physical sensors.
   - A differential pressure sensor provides an analog input indicating the difference in air pressure between the building space and outside.
   - An airflow station provides an analog input indicating the supply airflow.

- An airflow station provides an analog input indicating the return airflow.
- A control device provides a digital input indicating an override of the occupancy schedule, such as an occupancy sensor that indicates unscheduled use of the building space.
3. If the digital inputs indicate that the return-air fan should be in occupied mode, then the controller compares the measured differential air pressure to the calculated setpoint.
    - If the differential pressure is below the setpoint (outside pressure is greater than inside pressure), the pressurization factor must be increased in order to reduce return airflow.
    - If the differential pressure is above the setpoint (outside pressure is less than inside pressure), the pressurization factor must be decreased in order to increase return airflow.
4. The AHU controller generates an analog signal to the return-air fan VFD to modulate the speed of the return-air fan to maintain the new return airflow setpoint.

Optional Controls. The AHU controller may also monitor inputs from other sensors for control. The VFD provides an analog input that indicates the motor speed (as a percentage of the maximum rated motor speed). This information is used to generate an alarm if the fan is at maximum speed and the pressure setpoint is not being met.

Safeties. The hardwired safety circuits for the return-air fan are the same as those on the supply-air fan. The return-air fan shuts down upon the opening of the safety control circuit, and the alarm safety circuit provides an alarm signal to the building automation system if a safety shutdown occurs.

Alarms. Besides the alarm for the activation of the safety circuit, the controller transmits an alarm for any of the following conditions:
- The return-air fan is at 100% rated speed and the differential pressure is negative (for the first sequence).
- The airflow is below the minimum setpoint (for the second sequence).

- The return-air fan falls to a minimum speed in a variable-volume system. This protects the fan motor, which has a minimum speed of 10% of full speed, while at the same time maintaining building pressurization.

Energy-Saving Measures. The energy-saving measures for the return-air fan are the same as the methods used by the supply-air fan.

Outside-Air and Return-Air Damper Control

The outside-air and return-air dampers control the relative proportions of the two components of supply air: outside air and return air from the building spaces. Together, these dampers serve two purposes. The first is to satisfy the ventilation requirements, thus maintaining good indoor air quality. The second is to take advantage of the outside air conditions for reducing building energy use. This function opens the outside-air damper even more than the minimum outside air setting required for adequate ventilation.

A building space may require cooling even when the outdoor weather is cool. This is because building occupants and electrical equipment, such as lighting, can significantly warm the air within a space. In this situation, the building space can be cooled very efficiently by ventilating with the cool outside air. *Economizing* is a cooling strategy that adds cool outside air to the supply air. This is a very energy-efficient method of cooling but can only be used when the outside-air temperature falls below the supply-air temperature. Economizers will work as the outside-air temperature continues to fall until the economizer airflow is equal to the minimum ventilation airflow. When the minimum ventilation airflow exceeds the economizer airflow, the AHU activates the preheat coil.

Control of the outside- and return-air dampers for economizing is based primarily on outside temperature and airflow. **See Figure 5-5.** The mixed-air temperature is raised or lowered as necessary by changing the positions of the outside-air and return-air dampers. The two dampers maintain equal but opposite positions.

Outside-Air and Return-Air Damper Control

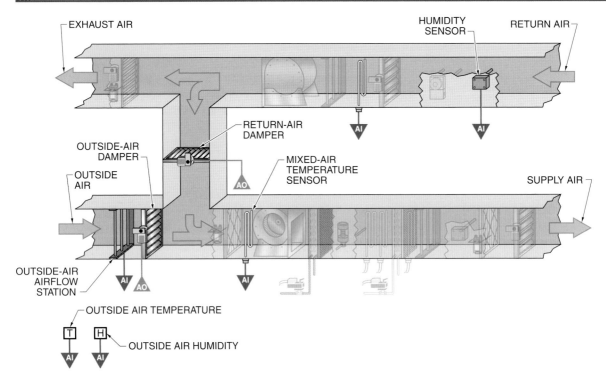

Figure 5-5. Economizing involves controlling the relative positions of the outside-air and return-air dampers.

The return airflow setpoint is calculated with the following formula:

$$OA = EA + p$$

where

OA = outside airflow (in cfm)

EA = exhaust airflow (in cfm)

p = pressurization factor (in cfm)

The exhaust airflow is a fixed value from the fan schedule and should include only the fans that are ON. The pressurization factor is an airflow that represents 2% to 3% (adjustable) of the supply airflow.

The requirements for air movement and air changes per hour for different types of buildings and occupancy uses are specified in the standards ANSI/ASHRAE 62.1, *Ventilation for Acceptable Indoor Air Quality,* and 62.2, *Ventilation and Acceptable Indoor Air Quality in Low-Rise Residential Buildings.*

Control Sequence. A typical sequence for controlling the outside-air damper and the return-air damper in an AHU includes the following:

1. The AHU controller monitors inputs from the building automation system.
   - The AHU controller receives a digital input indicating desired occupied/unoccupied fan operation based on a preprogrammed schedule.

2. The AHU controller monitors inputs from physical sensors.
   - A temperature sensor provides an analog input indicating the outside-air temperature.
   - A temperature sensor provides an analog input indicating the mixed-air temperature.
   - An airflow station provides an analog input indicating the outside airflow.

3. The controller compares the measured outside airflow to the calculated outside airflow setpoint.
   • If the outside airflow is below the setpoint, the outside-air damper must be opened.
   • If the outside airflow is above the setpoint, the outside-air damper must be closed.
4. If the outside-air temperature is below the supply-air discharge temperature, the economizer is enabled. The controller compares the mixed-air temperature measurement with a setpoint.
   • If the mixed-air temperature is above the setpoint, the outside-air damper must be opened.
   • If the mixed-air temperature is below the setpoint, the outside-air damper must be closed.
5. The AHU controller generates analog signals to the two damper actuators. The outside-air damper is slowly opened and the return-air damper is slowly closed. However, any direction to close the damper can be overridden to maintain the minimum ventilation requirements. The dampers are stopped in their current position when the mixed-air temperature reaches the setpoint.

Optional Controls. The AHU controller may instead control the outside-air damper and return-air damper based on other inputs. Relative humidity sensors in both the return air and outside air in conjunction with outside- and return-air temperatures provide analog inputs to enable damper control based on enthalpy (total heat energy) instead of just dry-bulb temperatures. Enthalpy economizing allows the AHU to take advantage of the low-cost cooling when the outside-air temperature is between the return-air temperature and the AHU supply-air temperature. Measuring enthalpy determines the amount of energy used in dehumidification. Enthalpy control is considered optional because of the difficulty associated with maintaining the accuracy of the humidity sensors with frequent maintenance.

Safeties. Most AHU control damper actuators include a spring return so that upon loss of power or loss of a control signal, the dampers return to their default positions. The outside-air damper is normally closed and the return-air damper is normally open.

Exhaust-Air Damper Control

There are many possible sequences for controlling building pressure (relative to outside air pressure), though modulation of the exhaust-air damper is one of the more common and stable methods. While it is possible to control the exhaust-air dampers with the mixed-air damper signal (to the outside-air and return-air dampers), controlling it independently is more stable. This sequence works in conjunction with the return airflow setpoint calculation to control building pressure. The exhaust-air damper only opens if the return-air fan discharge plenum is positive, which prevents outside air from entering through the exhaust-air damper.

Control of the exhaust dampers is managed by the AHU controller and is based primarily on differential air pressure across the exhaust-air damper. **See Figure 5-6.** The damper position is modulated as necessary to maintain the differential air pressure setpoint.

Dampers and fans control the movement of air in HVAC systems.

Exhaust-Air Damper Control

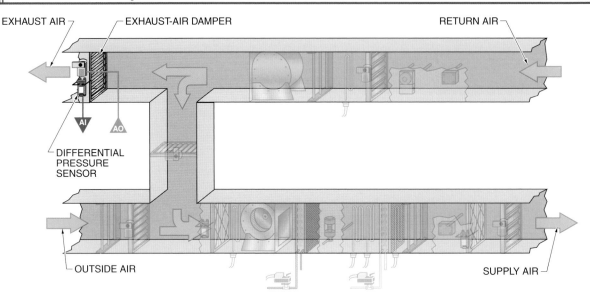

EXHAUST AIR   EXHAUST-AIR DAMPER   RETURN AIR

AI   AO

DIFFERENTIAL PRESSURE SENSOR

OUTSIDE AIR   SUPPLY AIR

Figure 5-6. The position of the exhaust damper is controlled to ensure that outside air does not enter the system through the exhaust-air duct.

Control Sequence. A typical sequence for controlling the exhaust-air damper in an AHU includes the following:

1. The AHU controller monitors inputs from the building automation system.
   - The AHU controller receives a digital input indicating desired occupied/unoccupied fan operation based on a preprogrammed schedule.
2. The AHU controller monitors an input from a physical sensor.
   - A differential pressure sensor provides an analog input indicating the differential air pressure across the exhaust-air damper.
3. The controller compares the measured differential air pressure to the setpoint.
   - If the differential pressure is above the setpoint, the exhaust-air damper must be opened.
   - If the differential pressure is below the setpoint, the exhaust-air damper must be closed.
4. The AHU controller generates a signal to modulate the position of the exhaust-air damper actuator.

Safeties. Most AHU control damper actuators include a spring return so that upon loss of power or loss of a control signal, the dampers return to their default positions. The exhaust-air damper is normally closed.

Cooling Control

The AHU cooling coil both chills (as measured in dry-bulb temperature) and dehumidifies the supply air. Both of these actions are accomplished by controlling the cooling coil to a temperature setpoint. Typically, dehumidification is not controlled via humidity sensors. Instead, temperature sensors, which are less expensive and more accurate, can control humidity by taking advantage of the relationships between temperature and humidity.

Some AHU configurations place the supply-air fan upstream of the cooling coil (blow-through AHUs) and some place the supply-air fan downstream of the cooling coil (draw-through AHUs). In either case, the supply-air temperature sensor must be downstream of both the cooling coil and the supply-air fan. The fan motor generates heat that can raise the temperature of the supply air, which must be accounted for in the temperature of the conditioned air supplied to the building space.

The control of the cooling coil is managed by the AHU controller based on the supply-air temperature. **See Figure 5-7.** The cooling function is raised or lowered as necessary by opening or closing the valve supplying the coil with chilled water. If a cooling coil is on a chilled water system with variable-speed pumping, the chilled water valve is a two-way control valve. If it is a constant-speed pumping system, the valve is a three-way valve, which bypasses excess chilled water back to the chiller.

Control Sequence. A typical sequence for controlling the cooling coil in an AHU includes the following:

1. The AHU controller monitors inputs from the building automation system.
   - The AHU controller receives a digital input indicating desired occupied/ unoccupied fan operation based on a preprogrammed schedule.
2. The AHU controller monitors inputs from physical sensors.
   - A temperature sensor provides an analog input indicating the supply-air temperature.
   - A humidity sensor provides an analog input indicating the return-air humidity.
3. The controller compares the measured supply-air temperature to the setpoint.
   - If the supply-air temperature is above the setpoint, the chilled water valve must be opened.
   - If the supply-air temperature is below the setpoint, the chilled water valve must be closed.
4. The AHU controller generates an analog signal to modulate the actuator controlling the chilled water valve.

Safeties. There is a significant difference of opinion as to whether a cooling coil should have a spring return on the valve actuator upon loss of power or control signal. One opinion contends that the cooling coil should close in order to stabilize the operation of the chiller. Another opinion states that cooling coils should be open in case the owner does not drain the cooling coil in winter. Yet another theory claims that the valve should fail in its last known position, thereby leaving at least some cooling capacity. The ultimate decision is left to the specifying engineer.

Cooling Control

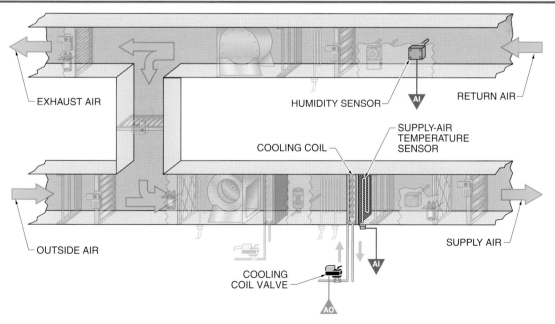

Figure 5-7. Cooling provided by an air-handling unit is modulated by opening or closing the chilled water valve serving the cooling coil.

Alarms. The controller transmits an alarm for the following conditions:

- The supply-air temperature falls below or rises above the setpoint by a predetermined amount.
- The return-air humidity rises above a setpoint.

Heating Control

The heating coil heats (as measured in dry-bulb temperature) the supply air by controlling the heating coil to a temperature setpoint. When the heating coil is upstream of the supply fan, the heating coil discharge temperature sensor setpoint may be reset by the supply-air temperature sensor downstream of the fan. The fan motor generates heat that can raise the temperature of the supply air, which must be accounted for in the temperature of the conditioned air supplied to the building space.

If the heating coil is supplied by a glycol heating system with variable-speed pumping, the valve controlling fluid to the coil is a two-way control valve. If it is a constant-speed pumping system, the valve is a three-way valve that bypasses the excess hot water back to the heat exchanger.

Control of the heating coil is managed by the AHU controller based primarily on temperature measurements. A blow-through AHU uses the supply-air temperature sensor. **See Figure 5-8.** A draw-through AHU requires an additional temperature sensor at the discharge of the heating coil. **See Figure 5-9.** The heating coil is then modulated by controlling the heating coil valve actuator.

Control Sequence. A typical sequence for controlling a glycol heating coil in an AHU includes the following:

1. The AHU controller monitors inputs from the building automation system.
   - The AHU controller receives a digital input indicating desired the occupied/ unoccupied fan operation based on a preprogrammed schedule.
2. The AHU controller monitors inputs from physical sensors.
   - A temperature sensor provides an analog input indicating the supply-air temperature.
   - A temperature sensor (for the draw-through AHU) provides an analog input indicating the heating coil discharge temperature.

Heating Coil Control (Blow-Through AHU)

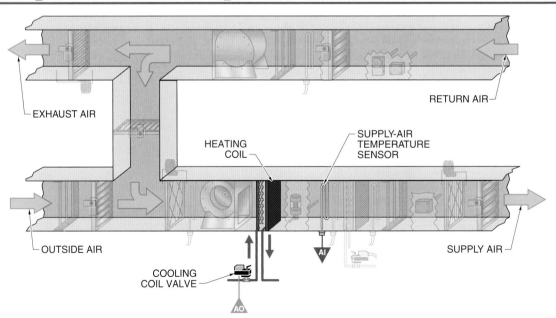

EXHAUST AIR

RETURN AIR

HEATING COIL

SUPPLY-AIR TEMPERATURE SENSOR

OUTSIDE AIR

SUPPLY AIR

COOLING COIL VALVE

AI

AO

Figure 5-8. *Heating from glycol heating coils is modulated by opening or closing the fluid valve.*

Heating Coil Control (Draw-Through AHU)

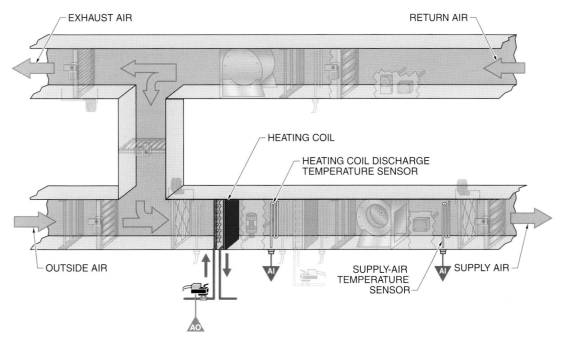

EXHAUST AIR

RETURN AIR

HEATING COIL

HEATING COIL DISCHARGE
TEMPERATURE SENSOR

OUTSIDE AIR

SUPPLY-AIR
TEMPERATURE
SENSOR

SUPPLY AIR

AI

AI

AO

Figure 5-9. Heating with a draw-through AHU requires an additional temperature measurement from a sensor near the heating coil.

3. In a draw-through AHU, the controller compares the measured supply-air temperature to the supply-air temperature setpoint.
   - If the supply-air temperature is above the setpoint, the heating coil discharge temperature setpoint is decreased.
   - If the supply-air temperature is below the setpoint, the heating coil discharge temperature setpoint is increased.
4. The controller compares the measured heating coil discharge temperature to the setpoint.
   - If the discharge temperature is above the setpoint, the heating coil valve must be closed.
   - If the discharge temperature is below the setpoint, the heating coil valve must be opened.
5. The AHU controller generates an analog signal to the actuator controlling the glycol heating water valve.

If a heating coil is supplied by steam instead, there is an airflow control damper in addition to the fluid (steam) control valve. This is known as a face-and-bypass heating coil. In this configuration, the face (cross-sectional area) of the AHU is divided into two components. The first is the coil component and the second is the bypass component that allows air to move around the heating coil without picking up heat. A bypass damper diverts air from one component to another. **See Figure 5-10.**

When the outside-air temperature is above freezing and heating is required, the bypass damper moves air over the coil and the steam valve modulates to maintain the setpoint. When the outside-air temperature is below freezing and heating is required, the steam valve is 100% open and the bypass damper modulates to push air over the coil or through the bypass. This avoids condensate from freezing in the piping when exposed to subfreezing air. Full-pressure steam inside the heating coil ensures that there is enough pressure to push condensate through the steam trap serving the coil and out of the system.

Steam Heating Control

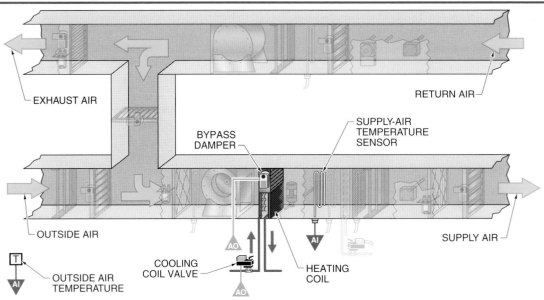

EXHAUST AIR

RETURN AIR

BYPASS DAMPER

SUPPLY-AIR TEMPERATURE SENSOR

OUTSIDE AIR

SUPPLY AIR

OUTSIDE AIR TEMPERATURE

COOLING COIL VALVE

HEATING COIL

Figure 5-10. *Steam heating coils require control of both the steam valve and a bypass damper that directs air away from the coil when the outside-air temperature is below freezing.*

Control Sequence. A typical sequence for controlling a steam-heating coil in an AHU includes the following:

1. The AHU controller monitors inputs from the building automation system.
   - The AHU controller receives a digital input indicating desired occupied/ unoccupied fan operation based on a preprogrammed schedule.
2. The AHU controller monitors inputs from physical sensors.
   - A temperature sensor provides an analog input indicating the supply-air temperature.
   - A temperature sensor (for a draw-through AHU) provides an analog input indicating the heating coil dis-charge temperature.
   - A temperature sensor provides an analog input indicating the outside-air temperature.
3. The controller compares the measured supply-air temperature to the supply-air temperature setpoint.
   - If the supply-air temperature is above the setpoint, the heating coil discharge temperature setpoint is raised.

   - If the supply-air temperature is below the setpoint, the heating coil discharge temperature setpoint is increased.
4. If the outside-air temperature is above freezing and heating is required, the bypass damper is positioned to move air across the heating coil.
   - If the supply-air temperature is above the setpoint, the heating coil valve must be closed.
   - If the supply-air temperature is below the setpoint, the heating coil valve must be opened.
5. If the outside-air temperature is below freezing and heating is required, the steam valve is opened fully.
   - If the supply-air temperature is above the setpoint, the bypass damper must allow more air through the bypass.
   - If the supply-air temperature is below the setpoint, the bypass damper must allow more air across the heating coil.
6. The AHU controller outputs analog signals to the steam valve actuator and bypass damper actuator.

Safeties. A heating coil valve actuator and bypass damper should have a spring return on the value and damper actuator to open upon loss of power or control signal.

Alarms. The controller transmits an alarm for the following conditions.

- The supply-air temperature falls below or rises above the setpoint by a predetermined amount.

Humidification Control

While dehumidification is often accomplished with cooling coils, humidification requires a separate device. The most common method of humidification is the addition of steam into the supply air. The steam is piped to the AHU from a boiler system.

Control of the humidifier is managed by the AHU controller based primarily on the humidity sensor. **See Figure 5-11.** The amount of moisture added to the air can be modulated by controlling the position of a valve in the humidifier.

Control Sequence. A typical sequence for controlling humidification in an AHU includes the following:

1. The AHU controller monitors inputs from the building automation system.
   - The AHU controller receives a digital input indicating the desired occupied/ unoccupied fan operation based on a preprogrammed schedule.
2. The AHU controller monitors an input from a physical sensor.
   - A humidity sensor provides an analog input indicating return-air humidity.
3. The controller compares the measured return-air humidity to the setpoint.
   - If the return-air humidity is above the setpoint, the humidifier valve must be closed.
   - If the return-air humidity is below the setpoint, the humidifier valve must be opened.
4. The AHU controller generates an analog signal to the actuator controlling the humidifier steam valve.

Optional Controls. A humidity sensor in the return airstream gives an average value for all spaces served by the AHU. However, humidity sensors can be installed in the occupied zones as well. The AHU controller can then be configured to respond differently to the multiple humidity level measurements.

Some types of humidifiers can only be turned either fully ON or fully OFF, which compromises fine control of the supply air's humidity. In this case, the controller would cycle the humidifier ON and OFF as necessary (with a digital signal) to stay close to the setpoint.

▌▌▌ Humidification

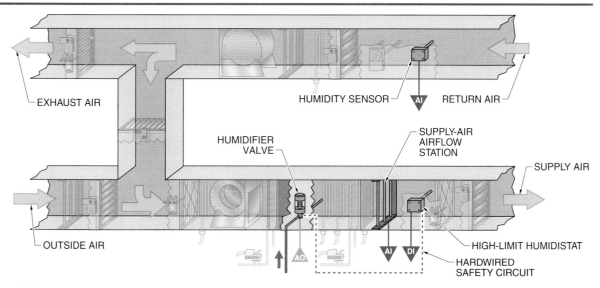

Figure 5-11. Humidification is modulated by opening or closing the steam valve serving the humidifier.

Safeties. Humidifier safeties help avoid excessive moisture in the AHU, which can lead to corrosion and mold inside the ductwork. A humidistat downstream of the humidifier is set to activate if the humidity level reaches a high-limit setpoint, typically 85% rh. The opening of the high-limit humidistat circuit will close the steam valve.

Also, if the supply-air airflow station measures insufficient airflow, the steam valve is closed. This safety circuit is a signal from the supply-fan VFD or positive feedback.

Alarms. The AHU controller transmits an alarm for either of the following conditions:
• The supply-air humidistat indicates the high-limit setpoint.
• The return-air humidity deviates from the setpoint by a certain amount.

Energy-Saving Measures. Most humidifiers require significant amounts of energy. The only available energy-savings measure is to lower the return-air humidity setpoint and sacrifice comfort. However, very low humidity can cause problems. At approximately 15% rh, copiers and printers start having problems. At approximately 10% rh, static

electricity becomes a risk to electronic equipment. In fact, codes may require a minimum humidity for certain critical applications, such as healthcare facilities.

Filter Pressure Drop Monitoring

The monitoring of the filter pressure drop is managed by the AHU controller based on differential pressure across the filter. **See Figure 5-12.** Dirty filters cause a greater pressure drop and can have a significant effect on fan energy consumption. Filters cannot be controlled by the automation system, so this sequence only provides the information to the operator workstation.

The output must be the highest recorded value over the previous three days. The filter pressure drop will be higher on a summer afternoon than a summer morning because more air is being forced through the filter on a VAV system when it is hot. The sliding window to cancel out the effect that unoccupied weekend days may have on the readings is three days.

⦀ Filter Pressure Drop Monitoring

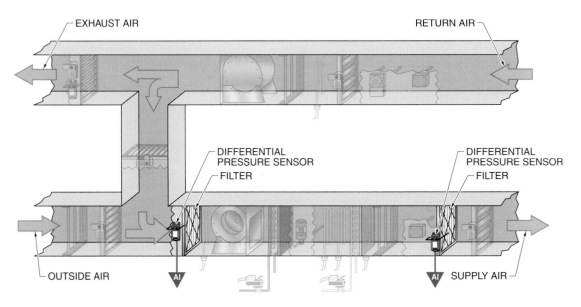

Figure 5-12. The condition of the filter is monitored by a differential pressure sensor and communicated to an operator workstation.

Control Sequence. A typical sequence for monitoring the filter bank's pressure drop includes the following:

1. The AHU controller monitors an input from a physical sensor.
   - A differential pressure sensor provides an analog input indicating pressure drop across a filter.
2. The AHU controller generates an analog signal to the operator workstation.

Alarms. The AHU controller transmits an alarm for the following condition:

- The differential pressure exceeds the high-limit setpoint by a predefined amount. This alarm point can be found in the bid document's filter schedule.

SMOKE CONTROL APPLICATIONS

There are many regulations governing the installation and control of smoke control systems in buildings. NFPA standards, other applicable fire protection codes, and even an owner's fire insurer all affect the operation of smoke control sequences in air-handling units. Smoke control sequences must be tightly regulated because failures can easily cause smoke to be spread throughout the building, endangering all occupants.

Smoke purge sequences are controlled to a default fire response position by the fire protection system. For example, the fire alarm control panel (FACP) automatically closes smoke dampers to isolate areas and contain the smoke. Then, a manual command by responding fire personnel may override this action and control the fans and dampers to instead purge the smoke, but only if a subsequent failure returns them to the default fire response position. Therefore, if there is a failure in the purge sequence, the smoke is at least again contained.

Alternatively, an engineered smoke control sequence is executed entirely automatically, but may only be controlled by the building's fire protection system, which is the only building system that must be UL-listed for reliability. For example, an engineered smoke control system, via the fire protection system, opens the exhaust-smoke dampers and turns on the fans automatically.

Smoke Control Modes

In a smoke control system, the signals controlling smoke dampers can be put into one of three different modes. **See Figure 5-13.** The three modes are the smoke evacuation, smoke purge, and combination evacuation/purge modes.

Smoke Control Modes

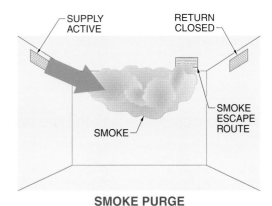

SMOKE EVACUATION

SMOKE PURGE

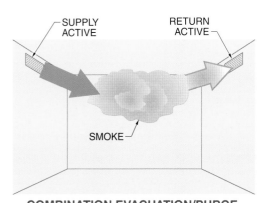

COMBINATION EVACUATION/PURGE

Figure 5-13. Control and smoke dampers can be put into one of three different modes for smoke control.

The smoke evacuation mode creates a negative pressure in the smoke area by running only the return-air fan as an exhaust-air fan. This mode is common in buildings without much perimeter wall area in comparison to floor area, and when there is a minimal risk of contaminating adjacent zones.

The smoke purge mode creates a positive pressure in the smoke area by running only the supply-air fan. This carries a significant risk of contaminating adjacent zones, but it is the best option in certain circumstances. For example, in a chemistry lab, a negative smoke control sequence may reverse the flow in a safety fume hood, pulling chemical fumes back into a fire. Instead, a positive pressure helps push contaminated air out of the area, though there must be adequate outside exposure so that the contaminated air can escape.

The combination evacuation/purge mode runs both the supply-air fan and the return-air fan at the same time. The supply-air fan brings in 100% outside air and the return-air fan exhausts 100% of the return airflow to outside. This completely ventilates the area with fresh, outside air. However, this function requires an oversized mechanical system to ensure the AHU can deliver 100% outside air even in the middle of winter.

Smoke Control with the Air-Handling Unit Dampers

AHU isolation smoke dampers prevent the AHU from recirculating air contaminated with smoke back into the building spaces. The smoke detector in the return-air duct closes the return-air isolation damper and the smoke detector in the supply-air duct closes the supply-air isolation damper. **See Figure 5-14.**

Control Sequence. A typical sequence for controlling smoke with the AHU includes the following:

1. Upon detection of smoke within the air-handling unit, the fire alarm control panel (FACP) automatically closes the smoke dampers within the unit.
2. Upon a manual command, the AHU controller initiates one of the three smoke control modes. It sends analog and/or digital output signals to the AHU

fans and dampers according to the operating mode.

- If the smoke evacuation mode is indicated, the AHU control dampers are positioned so that the outside-air damper is closed, the return-air damper is open, and the exhaust-air damper is open. The AHU smoke-isolation dampers are positioned so that the supply-air damper is closed and the return-air damper is open. The return-air fan operates at a preset fixed speed and the exhaust-air fan turns ON.

- If the smoke purge mode is indicated, the AHU control dampers are positioned so that the outside-air damper is open, the return-air damper is closed, and the exhaust-air damper is closed. The AHU smoke-isolation dampers are positioned so that the supply-air damper is open and the return-air damper is closed. The exhaust-air fan turns ON. The supply-air fan operates utilizing standard control based on supply-air duct static pressure. If this signal is not available, then the fan operates at a preset fixed speed.

- If the combination evacuation/purge mode is indicated, the AHU control dampers are positioned so that the outside-air damper is open, the return-air damper is closed, and the exhaust-air damper is open. The AHU smoke-isolation dampers are positioned so that both the supply-air and return-air dampers are open. The supply-air fan operates by standard control. The return-air fan operates with fan-tracking control. The exhaust-air fan turns ON.

Smoke Control with Zone Dampers

Zone smoke dampers can also be used to isolate specific floors or partial floors of a building. The smoke detector located in the building space or in the air-distribution ducts controls the zone smoke dampers. **See Figure 5-15.**

Smoke Control with Air-Handling Unit Dampers

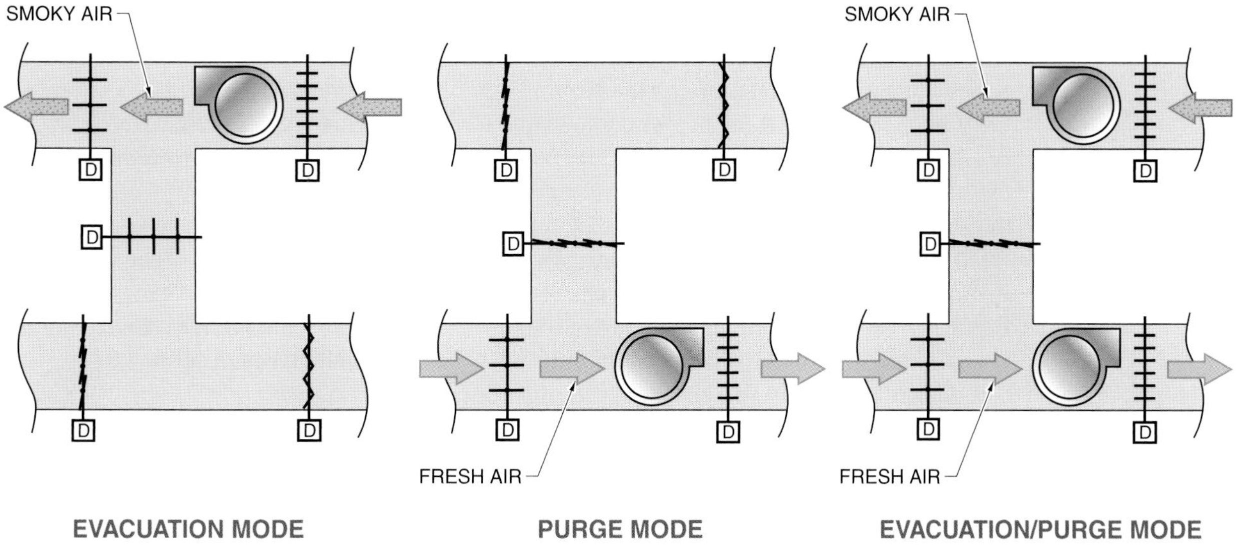

SMOKY AIR

SMOKY AIR

FRESH AIR

FRESH AIR

EVACUATION MODE   **PURGE MODE**   **EVACUATION/PURGE MODE**

Figure 5-14. *Isolation dampers prevent the AHU from circulating smoke-contaminated air throughout the building.*

Control Sequence. A typical sequence for controlling smoke in a certain zone served by an AHU includes the following:

1. Upon detection of smoke within a certain zone, the fire alarm control panel (FACP) automatically closes the smoke dampers for that zone.

2. Upon a manual command, the AHU controller initiates one of the three smoke control modes. It sends analog and/or digital output signals to the AHU fans and dampers according to the operating mode.

   • If the smoke evacuation mode is indicated, the zone damper serving the supply-air duct remains closed, the zone damper serving the return-air duct is opened, and the zone damper serving the exhaust-air duct is opened.

   • If the smoke purge mode is indicated, the zone damper serving the supply-air duct is opened, the zone damper serving the return-air duct is closed, and the zone damper serving the exhaust-air duct is opened.

   • If the combination evacuation/purge mode is indicated, the zone damper serving the supply-air duct is opened,

the zone damper serving the return-air duct is opened, and the zone damper serving the exhaust-air duct is opened.

3. The AHU operates in parallel mode.

Smoke Control with Zone Dampers

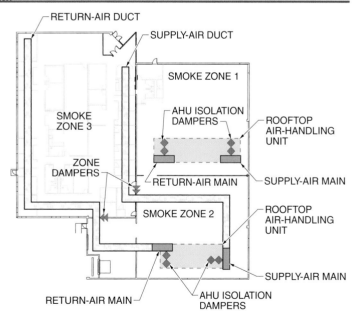

RETURN-AIR DUCT

SUPPLY-AIR DUCT

SMOKE ZONE 1

SMOKE ZONE 3

AHU ISOLATION DAMPERS

ROOFTOP AIR-HANDLING UNIT

ZONE DAMPERS

RETURN-AIR MAIN

SUPPLY-AIR MAIN

SMOKE ZONE 2

ROOFTOP AIR-HANDLING UNIT

SUPPLY-AIR MAIN

RETURN-AIR MAIN

AHU ISOLATION DAMPERS

Figure 5-15. *Zone smoke dampers are used to isolate certain areas of the building to prevent the spread of smoke-contaminated air.*

Stairwell Pressurization

Another major smoke control sequence is to control the pressurization of the escape stairwells. This sequence is not controlled nor overridden by the building automation system, but it is important to understand this element of smoke control.

A fan brings in outside air to pressurize the stairwell so that if someone exits a zone filled with smoke via the stairwell, the smoke is not introduced into the escape route. **See Figure 5-16.** Control of the pressurization fan is hardwired directly into two circuits. The differential pressure sensor measuring the stairwell pressure is tied directly into the fan's VFD, bypassing the building automation system. The other circuit provides an input to the building automation system to generate alarms.

Stairwell Pressurization

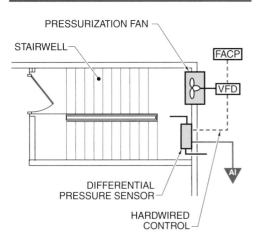

Figure 5-16. *Controlling stairwells to a pressure higher than the rest of the building keeps smokes from entering this escape route.*

Control Sequence. A typical sequence for controlling the stair pressurization fan includes the following:
1. Upon activation by the fire alarm system, the pressurization fan VFD initiates stair pressurization mode.
2. The VFD monitors an input from a physical sensor.

* A differential air pressure sensor provides an analog input indicating the stairwell static air pressure.
3. The VFD compares the static air pressure measurement to a setpoint.
   * If the static air pressure is below the setpoint, the pressurization fan speed must be increased.
   * If the static air pressure is above the setpoint, the pressurization fan speed must be decreased.
4. The VFD modulates the speed of the fan to maintain the static air pressure .

Alarms. Besides the alarm for the activation of the safety circuit, the controller transmits an alarm for either of the following conditions:
* The stairwell static air pressure falls below a preset limit.

TERMINAL UNIT CONTROL APPLICATIONS

After being distributed from the air-handling unit, air enters the building space through a terminal unit. A *terminal unit* is the end point in an HVAC distribution system where the conditioned medium (air, water, or steam) is added to or directly influences the environment of the conditioned building space. Air terminal units (ATUs) include dampers to modulate the amount of conditioned supply air into the space, and may include other devices to further condition the supply air. A single AHU can serve multiple ATUs.

Air Terminal Unit Control

The two common designs of air terminal units are single-duct units and dual-duct units. Single-duct units are provided with cool air from the AHU, which the air terminal units then heat as needed. **See Figure 5-17.** In older buildings, it is common to have air terminal units with reheat coils on the exterior zones and air terminal units without reheat coils on the interior zones. However, with the evolution of the ventilation code, it is often necessary to include reheat coils on every ATU to maintain both comfort and indoor air quality.

Single-Duct Air Terminal Unit Control

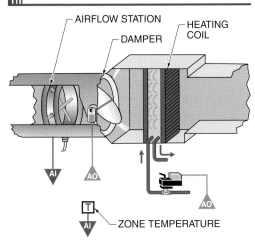

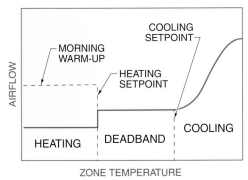

Figure 5-17. *Single-duct ATUs control cooling via airflow and heating via a reheat device.*

The ATU controller uses temperature to determine the airflow setpoint. The change in airflow rate into the room causes the temperature to change. The temperature is also raised or lowered as necessary by modulating the heating function.

Control Sequence. A typical sequence for controlling conditioned air to a certain zone with a single-duct ATU includes the following:

1. The ATU controller monitors an input from the building automation system.
   - The AHU controller receives a digital input indicating the desired operation based on a preprogrammed schedule. This schedule also contains minimum and maximum airflow setpoints as well as temperature setpoints for various situations.

2. The ATU controller monitors inputs from physical sensors.
   - A temperature sensor provides an analog input indicating zone temperature.
   - An airflow station provides an analog input indicating airflow at the inlet of the ATU.
3. The controller compares the zone temperature measurement with the zone cooling setpoint.
   - If the zone temperature is above the setpoint, the airflow must be increased.
   - If the zone temperature is below the setpoint, the airflow must be decreased.
4. The controller generates a signal to modulate the damper to maintain the new airflow setpoint.
5. The controller compares the zone temperature measurement with the zone heating setpoint.
   - If the zone temperature is above the setpoint, the heating function must be decreased.
   - If the zone temperature is below the setpoint, the heating function must be increased.
6. The controller generates an analog signal to modulate the heating function to maintain the temperature setpoint. This involves either opening or closing a hot water valve or turning electric heating elements ON or OFF.

A dual-duct air terminal unit is essentially two single-duct air terminal units operating in parallel and the airstreams mixing before going to the diffuser. Dual-duct systems were popular 25 years ago because of their stability in constant-volume systems. However, with the advent of more sophisticated and more accurate control systems, implementation in new buildings has decreased. However, many existing buildings continue to use these systems.

Control of the air terminal unit is managed by the ATU controller and based primarily on zone temperature. **See Figure 5-18.** The temperature is also raised or lowered as necessary by modulating the relative contribution of airflow from each duct.

Dual-Duct Air Terminal Unit

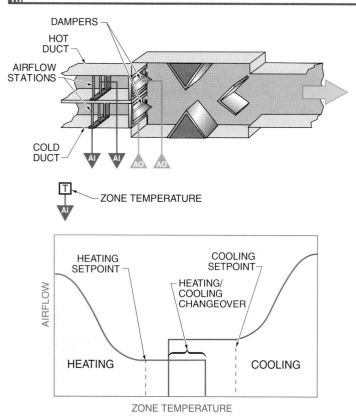

Figure 5-18. Dual-duct ATUs control heating and cooling by modulating the relative position of dampers in the hot and cold ducts respectively.

Control Sequence. A typical sequence for controlling conditioned air to a certain zone with a dual-duct ATU includes the following:

1. The ATU controller monitors an input from the building automation system.
   - The AHU controller receives a digital input indicating the desired occupied/unoccupied fan operation based on a preprogrammed schedule. This schedule also contains minimum and maximum airflow setpoints as well as temperature setpoints for various situations.
2. The ATU controller monitors inputs from physical sensors.
   - A temperature sensor provides an analog input indicating zone temperature.
   - An airflow station provides an analog input indicating airflow in the hot duct.

- An airflow station provides an analog input indicating airflow in the cold duct.
3. The controller compares the zone temperature measurement with the zone cooling setpoint.
   - If the zone temperature is above the setpoint, the cold-duct airflow must be increased.
   - If the zone temperature is below the setpoint, the hot-duct airflow must be increased.
4. The controller generates two analog signals to modulate the positions of the duct dampers: one for the cold-duct damper and one for the hot-duct damper.

Many air terminal units can control airflow only by modulating their integral dampers, but some include a circulation fan that operates in parallel with the airflow from the central air-handling unit to assist with heating. During cooling, the fan is OFF and cool supply air is served from the central AHU. A backdraft damper prevents reverse flow through the fan. During heating, though, the air terminal unit may turn on its circulating fan to take advantage of the warm air that has risen to near the ceiling. Recirculating some of that warm air reduces the heating load.

The control of the air terminal unit is managed by the ATU controller and is based primarily on zone temperature. **See Figure 5-19.** The necessary output signals to change the operation of the ATU are sent to the heating function actuators.

Control Sequence. A typical sequence for controlling conditioned air to a certain zone with a fan-powered ATU includes the following:

1. The ATU controller monitors an input from the building automation system.
   - The AHU controller receives a digital input indicating the desired occupied/unoccupied fan operation based on a preprogrammed schedule. This schedule also contains minimum and maximum airflow setpoints as well as temperature setpoints for various situations.
2. The ATU controller monitors inputs from physical sensors.

- A temperature sensor provides an analog input indicating zone temperature.
- An airflow station provides an analog input indicating airflow in the primary inlet of the air terminal unit.

3. The controller compares the zone temperature measurement with the zone cooling setpoint.
   - If the zone temperature is above the setpoint, the airflow must be increased.
   - If the zone temperature is below the setpoint, the airflow must be decreased.

4. The controller generates an analog signal to the duct damper to modulate the airflow.

5. The controller compares the zone temperature measurement with the zone heating setpoint.
   - If the zone temperature is above the setpoint, the heating function must be decreased.
   - If the zone temperature is below the setpoint, the heating function must be increased.

6. The controller generates an analog signal to modulate the heating function to maintain the temperature setpoint. This involves either opening or closing a hot water valve or turning electric heating elements ON or OFF.

7. If the zone is unoccupied and requires heating, the controller generates a digital signal to turn ON the circulation fan. Heat is picked up as the recirculated air is drawn from near the ceiling.

Optional Controls. An air terminal unit controller, single-duct, dual-duct, or circulating-fan type, may also monitor inputs from other sensors.

An occupancy sensor provides a digital input indicating occupancy outside scheduled hours of operation. Or, an occupancy override switch provides a digital input indicating that the zone must be changed to occupied mode, regardless of the schedule or occupancy-sensor reading. This override switch is often integrated with the zone temperature sensor unit and requires no additional wiring or input points on the controller. Activation of either device causes the air terminal unit to use the "occupied" setpoints for a preset amount of time.

The supply-air fan sequence provides a digital input indicating a maximum supply airflow. This input tells all ATUs to modulate to their dampers to their maximum airflow setpoint. This function facilitates testing and balancing of the air terminal units.

Safeties. Air terminal unit safeties monitor the unit to detect or avoid operating problems. These functions require additional optional inputs.

Fan-Powered Air Terminal Unit

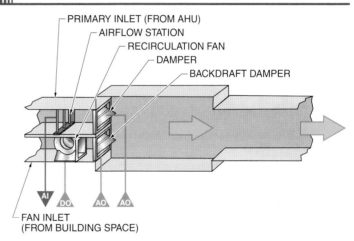

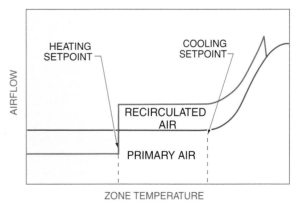

Figure 5-19. *Fan-powered ATUs operate similarly to conventional single-duct units but include a fan to further control airflow for cooling.*

The return-air fan sequence provides a digital input indicating a minimum return airflow. If a return-air fan must slow down in order to maintain the differential airflow setpoint with the supply-air fan, but is already running at its minimum speed, it can create a severely negative building pressure. Instead, this input tells all ATUs to "bias" their airflow settings upward to ensure that the return-air fan speed does not fall below minimum speed.

A temperature sensor located downstream of a reheat coil provides an analog input indicating the discharge air temperature. The controller monitors this sensor and checks for conflicts with the operating sequence. For example, if a space is calling for cooling but the ATU discharge is more than a degree above the AHU supply-air temperature, then the reheat valve must be stuck.

> Air terminal units, also known as variable-air-volume (VAV) boxes, include their own controllers for managing the integral damper and control devices such as heating coils and sensors.

Alarms. The controller transmits an alarm for either of the following conditions:
* The ATU is unable to maintain temperature setpoint.
* The ATU is unable to maintain airflow setpoint.

These alarms are separate because airflow for ventilation may be needed even if airflow for temperature control is not needed. It is possible to meet the temperature criteria of a space but not the airflow, which would not be noticed without a separate alarm.

Energy-Saving Measures. One of the major ways to save energy costs is to reduce electric demand during times of peak usage. It cannot be emphasized enough that demand limiting must begin at the air terminal unit level and work back to the larger equipment in order for all zones to be treated equally. Upon a trigger from the overall building energy monitor, all the air terminal units on an AHU "bias" their airflow settings downward in order to use less energy. This reduced flow raises zone temperatures but also reduces the energy used by the AHU fan and the cooling system.

Isolation Rooms

Two different types of isolation rooms are often used in hospitals to help control the spread of airborne contagions. An infectious isolation room has a negative relative pressure, which keeps infectious agents from easily spreading into surrounding areas. This is used to quarantine infectious patients. Alternatively, a protective isolation room has a positive relative pressure, which keeps infectious agents from easily entering the room. This is used to protect patients with compromised immune systems.

Most isolation rooms have an anteroom or airlock to help seal the room and to allow at least one door to be closed at all times. The anteroom also provides a space for the medical staff to don protective clothing or wash their hands before coming into the room.

The operation of an infectious isolation room is simple. The ATU provides a constant airflow into the room, which is determined by the number of air changes per hour required by code. Room exhaust is equal to the supply airflow plus approximately 80 cfm, which is exhausted via the patient toilet exhaust fan. This creates a negative pressure, even with typical air leakage from around the door. A pressure monitoring alarm measures the differential pressure between the patient room and the hallway by measuring the velocity of the air moving through a tube with an airflow sensor. A pressure alarm panel mounted on the hallway wall indicates to the medical staff if the room is operating properly.

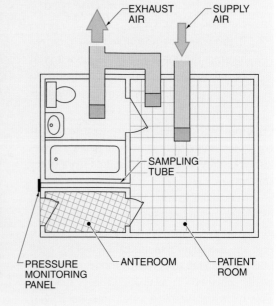

Perimeter Heating System Control

The perimeter heating system counteracts the loss of heat through the building envelope. Perimeter heating systems generally include either radiant heating panels, which heat with infrared energy, or fintube units, which use hot water or steam to heat via convection airflow. Each type has advantages and disadvantages. For example, fintube heaters prevent occupants from placing furniture against an outside wall because it would block airflow, but radiant heating panels complicate the mounting of nearby window coverings. However, both are controlled in the same manner.

Perimeter heating systems are generally used in northern states in facilities with large amounts of exterior glass or relatively high humidity levels throughout the year, such as healthcare facilities. Perimeter heating systems are usually controlled by extra connection points on an ATU controller. This is a complementary relationship because the perimeter heat must be coordinated with the ATU reheat system.

The perimeter heating system is activated in relationship to when the heating device inside the air terminal unit is activated. The two systems can operate at the same time, but also the perimeter heating system can be programmed to activate either before or after the overall zone temperature measurements call for heating. All three options are valid, and any one option may be implemented by the specifying engineer.

Control of the perimeter heating system is managed by the ATU controller based on temperature. The temperature is raised or lowered as necessary by modulating the heating function. Depending on the application, the control of perimeter heating systems may be influenced by the fact that radiant heat and convective air loops have very long response times. This also means that perimeter heat must never be controlled by ON/OFF or triac control techniques.

Control Sequence. A typical sequence for controlling perimeter heating systems includes the following:

1. The ATU controller monitors an input from a physical sensor.

- A temperature sensor provides an analog input indicating overall zone temperature. (This is the same sensor used by the air terminal unit for other sequences.)

2. The controller compares the zone temperature measurement with the zone heating setpoint.
    - If the zone temperature is above the setpoint, the heating function must be decreased.
    - If the zone temperature is below the setpoint, the heating function must be increased.

3. The controller generates an analog signal to modulate the hot water valve or electric heating element to maintain the temperature setpoint.

Optional Controls. An alternative is to base the control of the perimeter heating system on a schedule of outside-air temperature so that certain outside-air temperatures require perimeter heating. Additional required temperature measurements include the outside-air temperature and zone temperature at the perimeter of the building. When determining the values in the reset schedule, the schedule must reflect the presence of the hot water heating system being reset as well.

Energy-Saving Measures. Perimeter heating systems are themselves an energy-savings feature. Since the systems do not rely on fans to circulate air, which consumes energy, the inclusion of such systems is an energy-savings measure.

Unit Heater Control

A unit heater is a relatively small device used for spot heating, meaning that there is no air distribution system. Unit heaters typically serve utility spaces such as penthouses, loading docks, exterior stairwells, and entrance vestibules. Unit heaters are often controlled with stand-alone controls, but it is possible to integrate them into a building automation system. Unit heaters may use either hot water coils or electric heating elements.

If the hydronic unit is supplied by a heating system with variable-speed pumping, the valve controlling the heating water to the coil is a two-way control valve. If it is a constant-speed pumping system, the valve is a three-way valve that bypasses the excess hot water back to the heat exchanger.

On a hydronic unit heater, when the outside-air temperature is above freezing and heating is required, the fan cycles ON and the valve modulates to maintain the setpoint. When the outside-air temperature is below freezing, the heating valve is 100% open and the fan cycles to maintain setpoint. Constant circulation prevents the water from being exposed to freezing temperatures, even though it requires constant pumping.

Control of the unit heater is managed by the unit heater controller based on temperature. **See Figure 5-20.** Ideally, the zone temperature sensor should be mounted far from the unit heater so that convective air loops do not influence the sensor when the fan is OFF.

Unit Heater Control

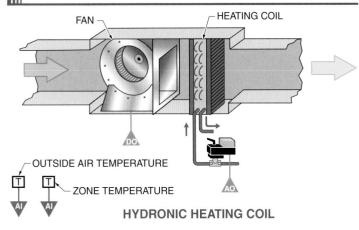

HYDRONIC HEATING COIL

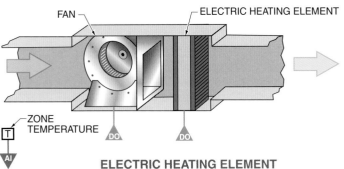

ELECTRIC HEATING ELEMENT

Figure 5-20. Unit heaters are small, individual units with their own controls.

Control Sequence. A typical sequence for controlling hydronic unit heaters includes the following:

1. The unitary controller monitors inputs from physical sensors.
   - A temperature sensor provides an analog input indicating the zone temperature.
   - A temperature sensor provides an analog input indicating the outside-air temperature.
2. The controller compares the measured outside-air temperature with a programmed setpoint (typically 38°F).
   - If the outside-air temperature is below the setpoint, the heating coil valve must be opened.
   - If the outside-air temperature is above the setpoint, the heating coil valve must be closed.
3. The controller generates an analog signal to modulate the actuator controlling the hot water valve.
4. The controller compares the measured zone temperature with the setpoint.
   - If the zone temperature is below the setpoint, the fan must be turned ON.
   - If the zone temperature is above the setpoint, the fan must be turned OFF.
5. The controller outputs a digital signal to turn the fan ON or OFF.

Safeties. A heating coil valve actuator should have a spring return on the valve to open upon the loss of power or control signal.

Control Sequence. A typical sequence for controlling electric unit heaters includes the following:

1. The unit controller monitors an input from a physical sensor.
   - A temperature sensor provides an analog input indicating the zone temperature.
2. The controller compares the measured zone temperature with the setpoint.
   - If the zone temperature is below the setpoint, the fan and heating element must be cycled ON.
   - If the zone temperature is above the setpoint, the fan and heating element must be cycled OFF.

3. The controller outputs a digital signal to turn the fan and heating element ON or OFF.

Alarms. The controller transmits an alarm for the following condition on either unit heater configuration.

- The supply-air temperature falls below or rises above the setpoint by a predetermined amount.

Fan Coil Control

Fan coils are similar to unit heaters. They are installed for similar applications and the heating function is controlled in the same way. The difference is that a fan coil also includes a cooling function. **See Figure 5-21.** Again, the zone temperature sensor should be mounted far from the unit heater so convective air loops do not influence the sensor when the fan is OFF.

Control Sequence. A typical sequence for controlling fan coils includes the following:

1. The unit controller monitors inputs from physical sensors.
   - A temperature sensor provides an analog input indicating the zone temperature.
   - A temperature sensor provides an analog input indicating the outside-air temperature.
2. The controller compares the measured zone temperature with the zone cooling setpoint.
   - If the zone temperature is below the setpoint, the cooling coil valve must be closed.
   - If the zone temperature is above the setpoint, the cooling coil valve must be opened.
3. The controller generates an analog signal to modulate the cooling coil valve to maintain the temperature setpoint. However, if the heating function is operating, the cooling valve must be closed.
4. The controller compares the measured zone temperature with the zone heating setpoint.
   - If the zone temperature is above the setpoint, the fan must be turned ON.

Fan Coil Control

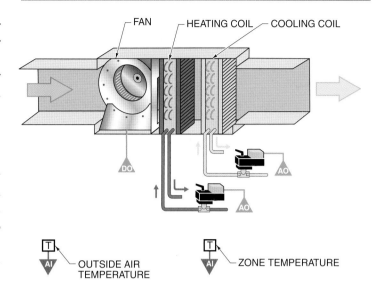

Figure 5-21. Fan coil units are similar to unit heaters, but they also include a cooling function.

- If the zone temperature is below the setpoint, the fan must be turned OFF.
5. The controller generates a digital signal to turn the fan ON or OFF.

Exhaust-Air Fan Control

An exhaust-air fan removes contaminated air from the building. **See Figure 5-22.** Exhaust requirements can range from relatively safe to highly toxic, which affects the physical configuration of exhaust system components, as well as the operating sequence. For example, toilet exhaust in an office building can be turned OFF at night, but fume hood exhaust must be ON 24 hours per day, 7 days a week.

Control Sequence. A typical sequence for controlling the exhaust-air fan includes the following:

1. The controller monitors an input from the building automation system.
   - The AHU controller receives a digital input indicating desired occupied/ unoccupied fan operation based on a preprogrammed schedule.
2. The controller monitors an input from a

2. The controller monitors an input from a physical sensor.
   • A digital input from the fan indicates its ON/OFF status.
3. The controller generates a digital signal to turn the exhaust fan ON or OFF based on the schedule.

Exhaust-Air Fan Control

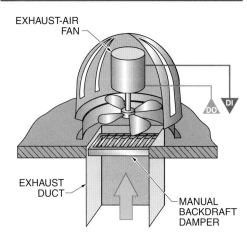

EXHAUST-AIR FAN

DO DI

EXHAUST DUCT

MANUAL BACKDRAFT DAMPER

Figure 5-22. Exhaust fans operate according to a preprogrammed schedule.

Alarms. The controller transmits an alarm for the following condition:
• The fan is scheduled to be ON but does not indicate ON status.

Energy-Saving Measures. The most obvious energy-saving measure is to simply shut the fans OFF when they are not needed. This is accomplished by using the software inputs and other applicable external inputs to determine when the building space is being used or when the space conditions must be maintained in anticipation of use. However, this may not be possible in some systems due to climate extremes, the quality of the building envelope, or other factors.

Boilers manage a significant amount of heat energy. They contain water and steam under pressure, so they require many safety measures. For example, if the pressure in a steam boiler exceeds a certain rating, a mechanical safety valve opens to relieve the excess pressure.

HYDRONIC AND STEAM HEATING CONTROL APPLICATIONS

A *hydronic HVAC system* is a system that distributes water or steam throughout a building as the heat-transfer medium for heating and cooling systems. Hydronic HVAC systems use pipes, valves, and pumps to distribute water throughout the building. Hot water or steam is generated by boilers and is distributed in a piping loop for heating use. Water systems require a pump to circulate the hot water throughout the building. Steam heating systems do not require circulating pumps because steam flows easily due to the difference in pressures between the boiler and the terminal units.

The water or steam is piped into a heat exchanger that transfers heat between the water or steam and the indoor air. A *heat exchanger* is a device that transfers heat from one fluid to another fluid without allowing the fluids to mix.

Package Boiler System Control

The package boiler system produces hot water or steam for hydronic heating. Control within package boiler units is managed by the factory-installed controller. A package boiler is only enabled or disabled as needed by the building automation system.

Control Sequence. A typical sequence for controlling package boiler systems includes the following:
1. The unit controller monitors inputs from physical sensors.
   • A temperature sensor provides an analog input indicating the outside-air temperature.
   • A temperature sensor provides an analog input indicating the boiler header supply temperature.
   • A temperature sensor provides an analog input indicating the boiler header return temperature.
   • The boiler control panel provides a digital input indicating its ON/OFF status.
   • The boiler control panel provides a digital input indicating a trouble alarm.

2. The controller compares the measured outside-air temperature with a setpoint.
   • If heating is required, the boiler must be enabled.
   • If heating is not required, the boiler must be disabled.
3. The controller generates a digital signal to the boiler control panel to enable or disable the boiler.

Optional Controls. An HVAC controller may also monitor other inputs from various sensors related to boiler systems. Steam heating systems, in particular, may require these additional controls. A pressure sensor can provide an analog input indicating the steam header pressure. Two additional digital inputs indicate trouble alarms from the boiler feedwater pump and de-aerator tank package and the boiler transfer pump/surge tank package, which both use factory-installed controls.

Safeties. All boiler safety features are part of the boiler control package that comes shipped from the factory. The package boiler manufacturer is responsible for conformance with all the applicable boiler regulations. The boiler control panel monitors all the boiler system control devices and generates a digital alarm signal for operating problems.

Alarms. An alarm is triggered for any of the following conditions:
• The boiler is enabled and the hot water supply header is indicating a low temperature.
• The boiler activates its trouble alarm.

For steam heating systems, additional alarms may be necessary for the following conditions:
• The header pressure falls to a preset setpoint.
• The boiler feedwater pump/de-aerator tank sounds a trouble alarm.
• The boiler transfer pump/surge tank sounds a trouble alarm.

Energy-Saving Measures. Since the boilers are not directly controlled by the building automation system, there are few if any opportunities on the control side to save energy. Energy efficiency is a result of the boiler having additional features such as a boiler economizer or better burners.

Heat Exchanger Control

A heat exchanger transfers heat between two fluids. A common application of heat exchangers is the transfer of heat energy from a steam system to a hydronic (hot water) heating system. **See Figure 5-23.** A hydronic heating system then serves hydronic reheat coils in air terminal units and perimeter heating systems. Substituting a glycol mixture in the tube side of the heat exchanger creates the glycol heating system used in AHU preheat coils.

Heat exchangers can experience extreme load swings, from a 100% load to a load less than 5%, which complicates the selection of a steam valve for all circumstances. Valves sized for full load can become unstable when turned down. One solution is to use two valves on the steam inlet. The first valve is sized at approximately one-third full load and the second is sized at approximately two-thirds load. Both valve actuators receive the same controller output signal, but the smaller valve is set to open first, thereby giving good control at both ends of the spectrum.

Control Sequence. A typical sequence for controlling a heat exchanger includes the following:
1. The controller monitors inputs from physical sensors.
   • A temperature sensor provides an analog input indicating the water inlet temperature.
   • A temperature sensor provides an analog input indicating the water outlet temperature.
2. The controller compares the measured outlet temperature to the heat exchanger setpoint.
   • If the outlet temperature is below the setpoint, the heat exchanger valve must be opened.
   • If the outlet temperature is above the setpoint, the heat exchanger valve must be closed.
3. The controller generates an analog signal to modulate the heat exchanger valves to maintain the setpoint temperature.

Heat Exchanger Control

Figure 5-23. Heat exchangers control the transfer of heat between fluids by modulating valves.

Boiler Room Air-Handling Units

Boiler rooms usually have a dedicated AHU because they include some unique hazards that affect air-handling unit safeties, such as the presence of open flames in the boiler burner.

Before electronic controls, it was common to supply the combustion air for a boiler by opening a door to allow in outside air. Now, room conditions must be much more carefully controlled in order to protect the electronics prevalent in these areas.

A differential pressure sensor provides an analog input to the controller indicating the boiler room static air pressure in relation to an outside reference. This is to ensure that the building is positive and that the burners have enough air. The forced draft fans are sized only to accommodate the back pressure in the boiler and stack.

If the differential pressure measurement is above the setpoint, the return airflow must be increased. If the differential pressure measurement is below the setpoint, the return airflow must be decreased. If the return-air fan is at minimum speed, the ATUs and supply-air fan supply more air to satisfy both the differential pressure requirement and the return-air fan minimum speed. Any attempt to turn off the return-air fan should be undertaken with caution because this is a significant change to the system.

All other AHU control functions are controlled per standard AHU sequences.

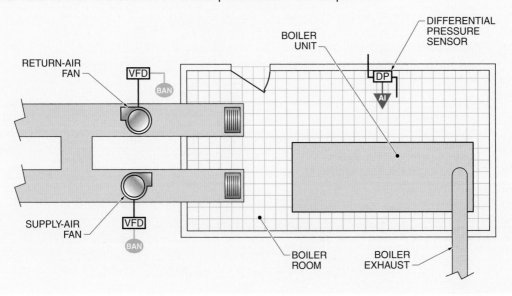

Optional Controls. When there are multiple heat exchangers, the header downstream of the last heat exchanger should also have the supply and return temperatures monitored. Some heat exchanger control sequences may also use outside-air temperature measurements.

Safeties. The heat exchange valve may include a spring return so that upon loss of power or loss of a control signal, the valve returns to a default position.

Alarms. An alarm is triggered for either of the following conditions:
- The individual heat exchanger output falls below the setpoint.
- The supply header temperature falls below the setpoint.

Energy-Saving Measures. One energy-saving technique is to "reset" the hot water supply setpoint based on the outside-air temperature. It is debatable as to whether the reduced temperature in the pipes results in less heat loss and that the heat loss is not offset by increased pump energy necessary to move more fluid with a lower temperature. But this does stabilize the control loops on the reheat coil air terminal units and perimeter heating system valves. Besides multiple valves, another way to handle extreme load variation is to lower the heat-carrying capacity so that the valve requires moderate fluid flow always operating in its sweet spot.

Hot Water Pumping Control

A hydronic heating system distributes hot water to various terminal devices such as hydronic heating coils in air terminal units, unit heaters, and perimeter heating systems. A pump circulates hot water throughout the hot water distribution system and must provide adequate flow. The differential water pressure between the hot water supply pipe and the hot water return pipe indicates if there is enough energy to push water through the heating coils. **See Figure 5-24.**

Control Sequence. A typical sequence for controlling hot water pumping includes the following:
1. The controller monitors an input from a physical sensor.

- A differential water pressure sensor provides an analog input indicating the differential water pressure between the hot water supply and hot water return lines.
2. The controller compares the measured differential pressure to the pressure setpoint.
   - If the differential pressure is below the setpoint, the pump speed must be increased.
   - If the differential pressure is above the setpoint, the pump speed must be decreased.
3. The controller generates an analog signal to the pump motor VFD to modulate the speed of the pump.

Alarms. An alarm is triggered for the following condition:
- The differential static pressure sensor is not satisfied and the pump is already running at 100% rated speed.

COOLING CONTROL APPLICATIONS

HVAC systems use refrigerant or chilled water for cooling air in an AHU. Refrigerant is controlled with a direct expansion (DX) system and chilled water is controlled with a water chiller system. Either cold fluid flows through a cooling coil within the ductwork of a forced-air HVAC system, transferring heat from the air to the fluid. Moving air provides a more efficient method of heat transfer.

Direct Expansion Cooling Coil Control

Direct expansion (DX) cooling coils use a refrigeration cooling cycle. These units are also known as air conditioners, though this term is more commonly applied to residential and light commercial installations. These are typically installed as package units that manage their own internal controls and devices for efficient operation. External signals enable and modulate the cooling effect of the direct expansion cooling coil units and receive feedback on their status.

Hot Water Pumping Control

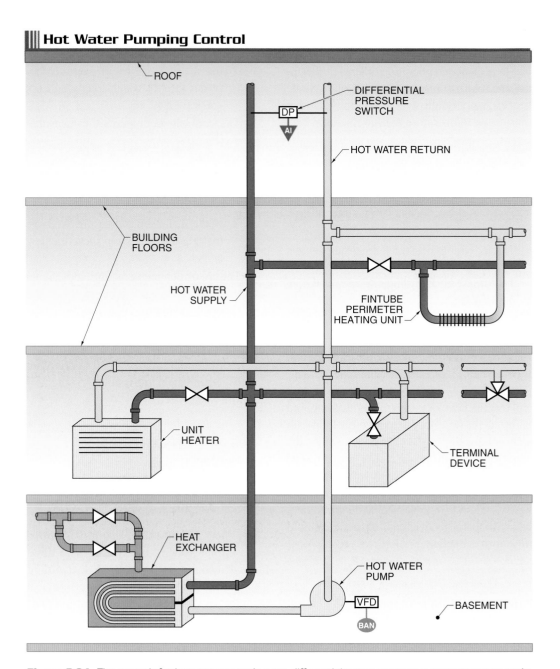

Figure 5-24. The controls for hot water pumping use differential water pressure measurements to maintain adequate water flow.

Control Sequence. A typical sequence for controlling a direct expansion (DX) cooling coil in an air-handling unit includes the following:

1. The controller monitors inputs from physical sensors.
   - A temperature sensor provides an analog input indicating the outside-air temperature.

   - A temperature sensor provides an analog input indicating the supply-air temperature.
   - The cooling coil unit control panel provides a digital input indicating its ON/OFF status.
   - The cooling coil unit control panel provides a digital input indicating a trouble alarm.

2. The controller compares the measured outside-air temperature with a cooling setpoint.
   • If the outside-air temperature is below the setpoint, the cooling coil must be disabled.
   • If the outside-air temperature is above the setpoint, the cooling coil must be enabled.
3. The controller generates a digital signal to the cooling coil control panel to enable or disable the unit.

4. The controller compares the measured supply-air temperature with a AHU discharge setpoint.
   • If the supply-air temperature is below the setpoint, the cooling function must be decreased.
   • If the supply-air temperature is above the setpoint, the cooling function must be increased.
5. The controller generates an analog signal to the direct expansion controller to modulate the cooling coil.

Computer Room Air Conditioning (CRAC) Units

The racks of equipment in computer data centers produce a significant amount of heat, requiring large cooling systems. The need for cooling has grown dramatically over the last decade. In the late 1990s, power consumption or heat density in a typical data center was 35 W/ft² to 50 W/ft². In 2007, power consumption was about 125 W/ft², with capacities of 150 W/ft² to 175 W/ft² being planned for the near future.

Special computer room air conditioning (CRAC) units are used in these applications to cool the equipment. A computer room air conditioning (CRAC) unit is a hybrid device. It is similar to a cooling-only fan coil unit, with a cooling coil, fan, and filter being the primary components. It is also similar to an AHU, with high-quality fans, a variety of cooling coil types, and high cooling capacity.

The physical configuration of data center equipment effects the operation of a cooling system in extreme loading. The cold aisle on the left is being compromised by system effects while the cold aisle on the right is performing satisfactorily. No amount of control modifications will remedy the situation until the core problem of poor air distribution is addressed.

Instead, a common configuration is to have the CRAC discharge supply air into a raised floor acting as a plenum. The air is then forced out into the computer space through a series of perforated tiles in the floor. These tiles are located in a cold aisle. The rack-mounted servers then pull the air from the cold aisle across the server and discharges it into the hot aisle where it is then pulled back to the CRAC for cooling and to start the cycle over.

CRAC units typically have factory-installed controls, but many data centers integrate the CRAC controls into the building automation system for single-point monitoring.

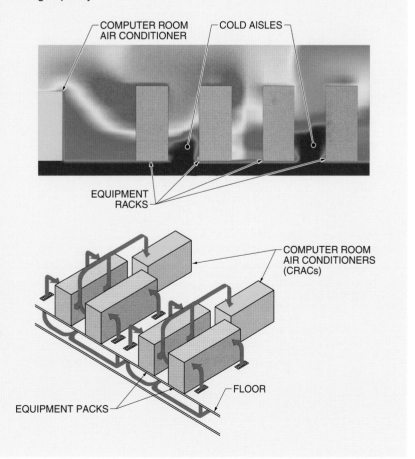

A chiller removes heat from water that circulates through a building for cooling purposes. The chilled water is distributed to the cooling coils of a building at about 45°F. It increases about 10°F through the cooling coils and is returned to the chiller at about 55°F to be cooled again. The three basic types of chillers are high-pressure chillers, low-pressure chillers, and absorption chilled-water systems.

Safeties. All direct expansion cooling coil safety features are part of the control package that comes shipped from the factory. The direct expansion cooling coil control panel monitors all the system control devices and generates a digital alarm signal for operating problems.

Alarms. An alarm is triggered for any of the following conditions:

- The direct expansion cooling coil is enabled, the AHU discharge temper sensor is indicating a high temperature, and the direct expansion cooling unit is OFF.
- The AHU discharge temperature sensor temperature rises above a preset setpoint.
- The direct expansion cooling unit activates its trouble alarm.

Energy-Saving Measures. Since the direct expansion cooling unit is not directly controlled by the building automation system, there are few if any opportunities on the control side to save energy.

Water Chiller Control

Chilled water is distributed to cooling coils in various terminal devices such as air-handling units, fan coil units, and computer room air conditioners. Control of the chiller is managed by the factory-installed controller. A chiller is only enabled/disabled by the building automation system.

Control Sequence. A typical sequence for controlling a water chiller includes the following:

1. The controller monitors inputs from physical sensors.
   - A temperature sensor provides an analog input indicating the outside-air temperature.
   - A temperature sensor provides an analog input indicating the chilled water supply temperature.
   - A temperature sensor provides an analog input indicating the chilled water return temperature.
   - The chiller unit control panel provides a digital input indicating its ON/OFF status.
   - The chiller unit control panel provides a digital input indicating a trouble alarm.
2. The controller compares the measured outside-air temperature with a setpoint.
   - If the outside-air temperature is below the setpoint, the chiller must be disabled.
   - If the outside-air temperature is above the setpoint, the chiller must be enabled.
3. The controller generates a digital signal to the chiller control panel to enable or disable the unit.

Safeties. All chiller safety features are part of the control package that comes shipped from the factory. The chiller control panel monitors all the system control devices and generates a digital alarm signal for operating problems.

Alarms. An alarm is triggered for any of the following conditions:

- The chiller is enabled, the temperature sensor in the chilled water header is indicating a high temperature, and the chiller is OFF.
- The chilled water temperature sensor temperature rises above a preset setpoint.
- The chiller activates its trouble alarm.

Energy-Saving Measures. Chillers are sized to generate chilled water for design days, which are worst-case scenarios. In low-load conditions, the chilled water temperature can be raised, reducing chiller energy consumption. However, the chilled water setpoint can be raised only if none of the cooling coils using the chilled water are 100% open, which indicates that maximum cooling is required. In this case, the chilled water setpoint must be reduced. Chilled water temperature is modulated when an external controller generates an analog signal to the factory-installed chiller control panel indicating the desired chilled water temperature setpoint.

Chiller Room Air-Handling Units

Like boiler rooms, chiller rooms usually have a dedicated AHU. Under normal circumstances, this unit operates the same as a standard AHU serving most occupied building spaces. However, chiller rooms can be very hazardous due to the toxic refrigerants used inside the chiller. A leak would be a significant risk to the building staff. Therefore, a chiller room AHU includes extra safeties to control the airflow through the room in such an emergency.

A refrigerant detection unit provides a digital input to the controller indicating a refrigerant leak. A manually activated mushroom switch at each exit of the chiller room provides a digital input to the controller indicating that a leak is detected by building personnel. If a signal is received from either switch, the supply-air fan changes to a preset flow rate that is approximately 300 cfm below the rating of the exhaust fan and the return fan turns off. This clears the room while keeping the room at a negative relative pressure. These safety inputs are hardwired directly to the exhaust-air fan and return-air fan for action. The supply-air fan VFD is controlled by the building automation system.

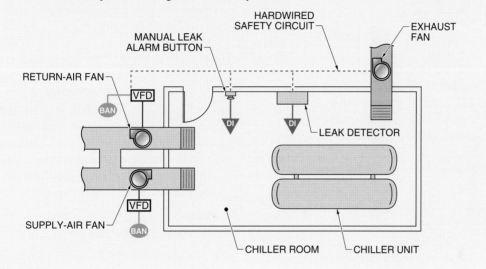

Cooling Tower Control

A cooling tower is a large heat exchanger discharging the heat absorbed by a chilled water system to the outside air. Cooling towers are more efficient than direct expansion units because the condenser fluid is exposed to the environment, thereby allowing it to take advantage of evaporative cooling. The condenser temperature can fall 10°F to 15°F below ambient dry bulb temperatures, thereby saving compressor energy.

Control of the cooling tower is managed by the controller based primarily on water outlet temperature. **See Figure 5-25.** The temperature is raised or lowered as necessary by changing the speed of the VFD operating the cooling tower fan.

Control Sequence. A typical sequence for controlling a cooling tower includes the following:

1. The controller monitors inputs from physical sensors.
   - A temperature sensor provides an analog input indicating the outside-air temperature.
   - A temperature sensor provides an analog input indicating the cooling tower water inlet temperature.
   - A temperature sensor provides an analog input indicating the cooling tower water outlet temperature.
   - The chiller unit control panel provides a digital input indicating its ON/OFF status.
2. The controller compares the measured water outlet temperature with a setpoint.
   - If the water outlet temperature is below the setpoint, the cooling tower fan's speed must be decreased.

- If the water outlet temperature is above the setpoint, the cooling tower fan's speed must be increased.

3. The controller generates an analog signal to the cooling tower fan VFD to modulate the fan motor speed.

Cooling Tower Control

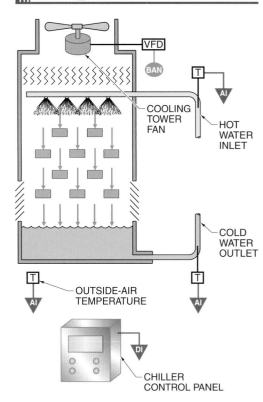

Figure 5-25. Cooling towers discharge the heat absorbed by a chilled water system by evaporating some of the water.

Heat absorbed from building spaces by chilled water is released to the outside air through a cooling tower.

Optional Controls. When there are multiple cooling towers, the header downstream of the last cooling tower should also have the supply and return temperatures monitored.

Alarms. An alarm is triggered for either of the following conditions:
- The individual cooling tower output rises above the setpoint.
- The supply header temperature rises above the setpoint.

Energy-Saving Measures. One technique to save energy is to set the cooling tower output setpoint based on chiller discharge temperature and outside-air temperature. The condenser setpoint resets from 85°F to 65°F as the outside-air temperature varies from 90°F to 50°F. This lower condenser temperature reduces the load on the chiller compressor. However, this type of control requires consultation with the chiller manufacturer, as chiller requirements may not allow the condenser water to fall below 7°F above the chilled water temperature, which could result in unstable operation and chiller damage.

Chilled Water Pumping Control

Chilled water pumping requires more control than hot water pumping because there must be constant flow through the chiller's evaporation coil or it will become unstable. A constant-volume chilled water system is a simple solution, but consumes a lot of energy. Instead, a primary/secondary system addresses both issues. **See Figure 5-26.** A primary pump ensures constant flow through the chiller, while a larger secondary pump only pumps what is needed by the cooling coils. Secondary pump motors are controlled with a variable-frequency drive (VFD). (The condenser side of the chiller, between the chiller and the cooling tower, is constant-volume pumping.)

When the system is fully loaded, the flow in the primary pump equals the flow in the secondary pump. In effect, the primary and secondary pumps are in series and the chilled water coming from the chiller goes out to the coils.

When the system is partially loaded, the flow in the primary pump is more than the flow in the secondary pump, so it needs to pull water from some other source than the coil chilled water return that is when the primary pump pulls from the water supply coming out of the chiller through a hydraulic bridge. This creates a parallel circuit where some water goes to the coils and some is routed back to the primary pump inlet.

Control Sequence. A typical sequence for controlling chilled water pumping includes the following:

1. The controller monitors inputs from physical sensors.
    - A digital signal from the chiller controller indicates the enabled/disabled status of the chiller. This signal also indicates the ON/OFF status of the primary evaporator and condenser pumps.
    - A differential static pressure sensor provides an analog input indicating the pressure differential between the chilled water supply pipe and the chilled water return pipe.
2. The controller compares the measured differential static pressure with a setpoint.
    - If the differential static pressure is below the setpoint, the secondary pump speed must be increased.
    - If the differential static pressure is above the setpoint, the secondary pump speed must be decreased.
3. The controller generates an analog signal to the secondary pump VFD to modulate the pump speed.

Optional Controls. Control of multiple chiller systems requires additional considerations. Configurations include both dedicated primary pumps and pumps combined into a header so that there is no single point of failure. In a multiple chiller installation, it is important to motorize the isolation dampers to the chillers and cooling towers. This shortens the amount of time it takes to bring a chiller on-line, making the chilled water temperature more stable, which helps stabilize the performance of the air-handling units and air terminal units.

Chilled Water Pumping Control

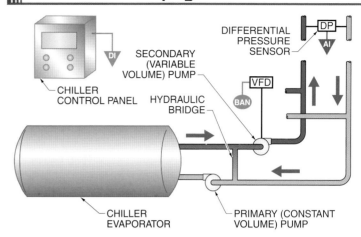

Figure 5-26. Chilled water pumping involves a pair of pumps that provide adequate flow without wasting energy.

Alarms. An alarm is triggered for the following condition:
- The differential static pressure sensor is not satisfied and the pump is already running at 100% rated capacity.

Chillers pump chilled water throughout a building to be used in cooling coils and then return it to be chilled again.

Heat absorbed by the water in the chiller condenser is rejected to the air in a cooling tower. A cooling tower uses evaporation to cool water. Air circulates through the tower by natural convection or is circulated by fans located in the tower. Cool water from the bottom of the tower is then circulated back to the condenser for reuse.

Chilled Water During the Winter

The industry standard terminology for producing chilled water during the winter is "free cooling." This refers to the fact that the chiller, a large energy consumer, does not need to run when the outside-air temperature is low.

When the outside-air temperature falls below the chilled water setpoint by approximately 5°F, diverting valves reroute the chilled water and cooling tower water through a "plate and frame" heat exchanger. Temperature sensors on the heat exchanger only monitor the heat exchanger for diagnostics. The cooling tower outlet water temperature is changed from a condenser water setpoint to a chilled water setpoint.

While chilled water may not be required for space cooling during the winter because the air-handling units can use the economizer mode, other equipment in the building may need cooling, such as computer room's air conditioning units, medical equipment, or food service refrigeration compressors.

In certain climates and building types, free cooling can actually use less energy than an AHU with an economizer. The additional pumping energy may be negligible in comparison to the energy required to keep a building properly humidified when a lot of outside air is used for economizer cooling.

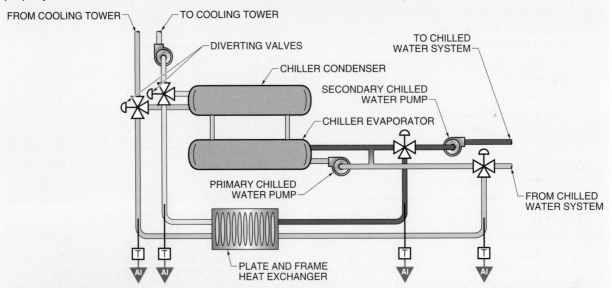

Summary

- Each HVAC system sequence specifically addresses a single application with a desired outcome.
- Many control sequences occur simultaneously to manage different aspects of comfort, such as temperature, humidity, and airflow.
- Sequences and controller programming require careful attention to these issues by the consulting-specifying engineer and controls contractor in order to appropriately manage the system as a whole.
- Control applications in HVAC systems may also include special considerations for safety functions.
- Some control applications allow variations in the sequence to reduce energy consumption.
- HVAC control applications can be divided by the primary unit or controller involved in executing the sequence.
- The supply-air fan moves conditioned air from the AHU to the building spaces via ductwork, enabling space conditioning to meet comfort requirements.
- The return-air fan assists the supply-air fan in moving conditioned air into the building spaces by drawing air back out of the spaces.
- The outside-air and return-air dampers control the relative proportions of the two components of supply air: outside air and return air from the building spaces.

- The economizer sequence modulates the outside-air and return-air dampers in order to maintain the mixed-air setpoint.

- Modulation of the exhaust-air damper is one of the more common methods to control building pressure.

- The cooling coil both chills (as measured in dry-bulb temperature) and dehumidifies the supply air and is controlled by modulating a cooling coil valve.

- The amount of moisture added to the air can be modulated by controlling the position of a valve in the humidifier.

- Differential pressure is monitored to determine filter condition.

- Smoke control sequences must be tightly regulated because failures can easily cause smoke to be spread throughout the building, endangering all occupants.

- The smoke evacuation mode creates a negative pressure in the smoke area by running only the return-air fan as an exhaust-air fan.

- The smoke purge mode creates a positive pressure in the smoke area by running only the supply-air fan.

- A combination evacuation/purge mode runs both the supply-air fan and the return-air fan at the same time, which completely ventilates the area with fresh outside air.

- A fan brings in outside air to pressurize the stairwell so that if someone exits a zone filled with smoke via the stairwell, the smoke is not introduced into the escape route.

- Air terminal unit controllers change the airflow rate into a room to change the temperature.

- The perimeter heating system is activated in relationship to when the heating device inside the air terminal unit is activated.

- An exhaust-air fan removes contaminated air from the building.

- Fine control of package boilers and chillers is managed by their respective factory-installed controllers, but can be enabled or disabled by the building automation system.

- The differential static pressure between the hot water supply pipe and the hot water return pipe indicates if there is enough energy to push water through the heating coils.

- External signals enable and modulate the cooling effect of direct expansion (DX) cooling coil units and receive feedback on their status.

Definitions

- An *air-handling unit* is a forced-air HVAC system device consisting of some combination of fans, ductwork, filters, dampers, heating coils, cooling coils, humidifiers, dehumidifiers, sensors, and controls to condition and distribute supply air.

- *Economizing* is a cooling strategy that adds cool outside air to the supply air. This is a very energy-efficient method of cooling but can only be used when the outside-air temperature falls below the supply-air temperature.

- A *terminal unit* is the end point in an HVAC distribution system where the conditioned medium (air, water, or steam) is added to or directly influences the environment of the conditioned building space.

- A *hydronic HVAC system* is a system that distributes water or steam throughout a building as the heat-transfer medium for heating and cooling systems.

- A *heat exchanger* is a device that transfers heat from one fluid to another fluid without allowing the fluids to mix.

Review Questions

1. How do some HVAC applications affect each other?

2. Explain the function of hardwired safety circuits in HVAC applications.

3. How can static air pressure setpoints be dynamically adjusted to save energy from supply- and return-air fan operation?

4. How does economizing save energy?

5. Explain the differences between the three different smoke purge sequences.

6. How does stair pressurization work?

7. Why is perimeter heating controlled separately from other heating functions?

8. What is the relationship of building automation systems to the control of package boiler and chiller units?

9. What is one method of sizing steam valves for heat exchangers?

10. Why does chilled water pumping require more control than hot water pumping, and how is this control achieved?

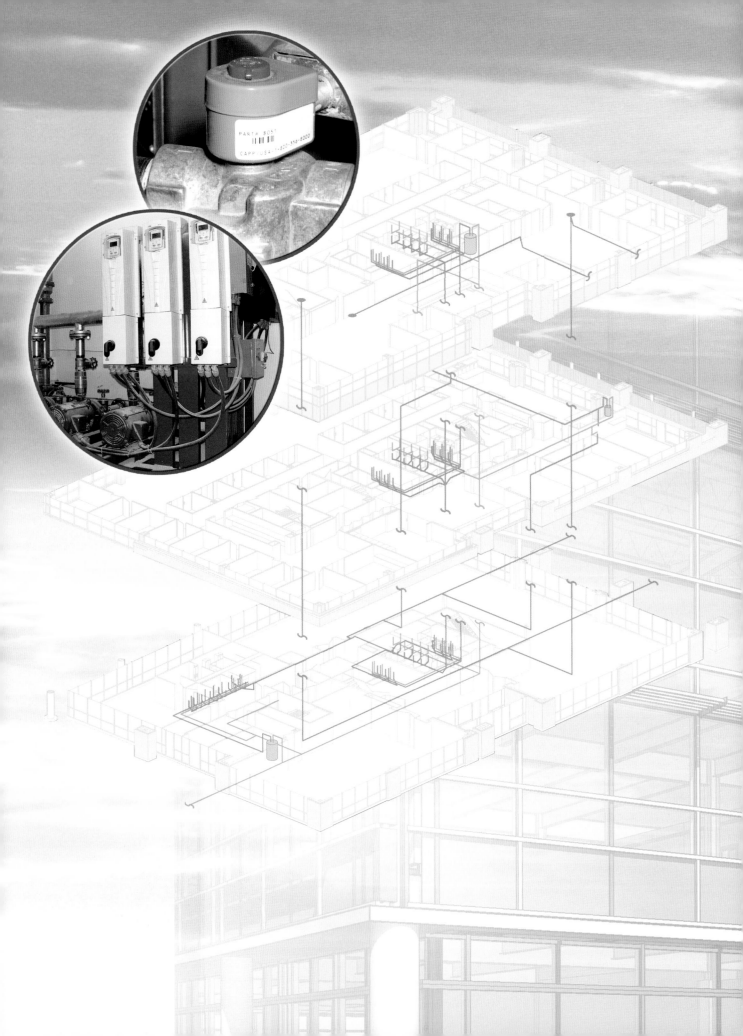

Chapter Six

Plumbing System
Control Devices and Applications

All plumbing systems require some degree of control to facilitate the operation of plumbing fixtures and carry away wastes. With growing interest in energy efficiency and water conservation, plumbing systems are becoming increasingly automated. Automated systems can optimally regulate water temperature and flow for various applications, even for different parts of the same system, greatly reducing wasted water and energy. With careful design and implementation, this control can also add extra benefits in convenience, hygiene, and reliability.

Chapter Objectives

- Describe the different parts of a building plumbing system, including the function and major components.
- Identify the four controlled characteristics of a plumbing water supply.
- Compare the different types of sensors for measuring the necessary characteristics of a water supply.
- Describe the different types of plumbing control devices and the effect each has on the water supply.
- Evaluate the control strategies and results of the most common plumbing control applications.

PLUMBING SYSTEMS

All buildings that are occupied by people, such as residences, office buildings, and manufacturing facilities, include a plumbing system. A *plumbing system* is a system of pipes, fittings, and fixtures within a building that conveys a water supply and removes wastewater and waterborne waste. These tasks are handled by different subsystems of the plumbing system.

Water Supply System

A *water supply system* is a plumbing system that supplies and distributes potable water to points of use within a building. *Potable water* is water that is free from impurities that could cause disease or other harmful health conditions. Potable water is drinkable.

The bacteriological and chemical quality of potable water must conform to state board of health requirements. Therefore, adequate plumbing is a major factor in public health and sanitation. The water in the water supply system must be kept separate from all other plumbing systems.

All plumbing systems include many of the same basic parts and subsystems that supply potable water and remove wastewater.

Potable water is supplied from the municipal water supply source through a water main. **See Figure 6-1.** A water service pipe extends from the water main to each building. Underground cocks (valves) on the service pipe at the water main and near the curb line allow the flow of potable water to a building to be turned on or off.

A water meter is installed on the water service as the water service pipe enters a building. The water meter measures the volume of water that passes through the water service. This allows the water utility to bill the building's owners for their water use. Valves on either side of the water meter allow the meter to be easily serviced or replaced.

The water supply system of a building consists of water distribution pipes, fittings, control valves, and fixtures. These components may be inside or outside of the building, but must be within the property lines to be considered part of the building's plumbing system. Landscape irrigation systems are considered part of a building's water supply system since they draw potable water from the same water mains.

Water Distribution Pipes. Water distribution pipes convey water from the water service pipe to the point of use. These pipes can be further classified according to their function within the building. A *building main* is a water distribution pipe that is the principal pipe artery supplying water to the entire building. Pipes are connected to the main to distribute the water supply to various areas of the building. A *riser* is a water distribution pipe that routes a water supply vertically one full story or more. A *branch* is a water distribution pipe that routes a water supply horizontally to fixtures or other pipes at the same approximate level. Branches may supply either individual fixtures or groups of fixtures.

A *fixture branch* is a water supply pipe that extends between a water distribution pipe and fixture supply pipe. **See Figure 6-2.** A *fixture supply pipe* is a water supply pipe connecting the fixture to the fixture branch. There is typically a valve between the fixture branch and fixture supply pipe to allow water to be shut off to individual fixtures without interrupting the rest of the system.

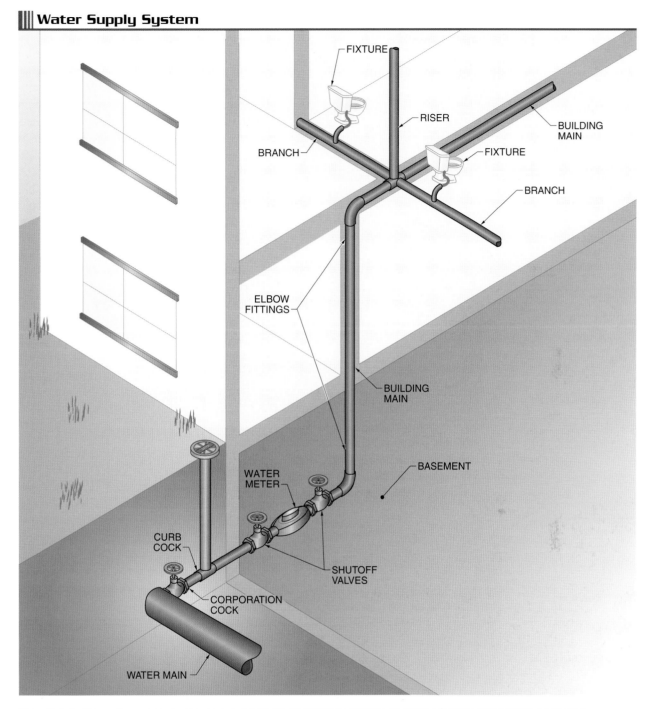

Figure 6-1. The water supply system conveys potable (drinkable) water to plumbing fixtures throughout a building.

Fittings and Valves. A *fitting* is a device used to connect two lengths of pipe. Fittings may be used to extend the length of a pipe only or may serve additional purposes, such as transitioning between different sizes or types of pipes or branching or changing the direction of a pipe.

Plumbing systems use various materials for pipes and fittings, depending on the use and code requirements. Common plumbing pipe materials include copper, cast iron, and polyvinyl chloride (PVC). Copper pipes and fittings are soldered together to distribute both hot and cold water supplies throughout a building.

Fixture Pipes

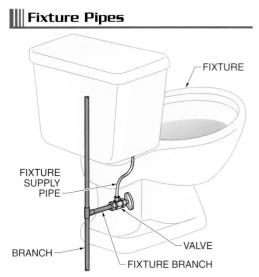

Figure 6-2. Fixture pipes connect fixtures to the main water distribution piping of the water supply system.

A *valve* is a fitting that regulates the flow of water within a piping system. There are many types of valves used in plumbing systems. Some are used only when work is being done on the plumbing system, such as during either installation or repair work, while others are used during the daily operation of the system.

Sioux Chief Manufacturing

Copper is the most common pipe material for water supply distribution piping. Copper pipes and fittings, such as elbows and tees, are soldered together to form waterproof piping.

Fixtures. A *fixture* is a receptacle or device that is connected to the water distribution system, demands a supply of potable water, and discharges the waste directly or indirectly into the sanitary drainage system. Common fixtures include lavatories (sinks), water closets (toilets), and bathtubs. While most fixtures are permanently connected to the plumbing system, some may be temporary.

An *appliance* is a plumbing fixture that performs a special function and is controlled and/or energized by motors, heating elements, or pressure- or temperature-sensing elements. Common appliances include water heaters, water softeners, and washing machines.

Controlling the flow of potable water and wastewater within fixtures is accomplished with fixture trim. *Fixture trim* is the set of water supply and drainage fittings installed on a fixture or appliance to control the water flowing into a fixture and the wastewater flowing from the fixture to the sanitary drainage system.

Sanitary Drainage System

A *sanitary drainage system* is a plumbing system that conveys wastewater and waterborne waste from the plumbing fixtures and appliances to a sanitary sewer. **See Figure 6-3.** *Sewage* is any liquid waste containing animal or vegetable matter in suspension or solution and/or chemicals in solution. A *sanitary sewer* is a sewer that carries sewage but does not convey rainwater, surface water, groundwater, or similar nonpolluting wastes. A sanitary sewer may be a public sewer, private sewer, individual building sewage-disposal system, or other point of disposal.

As wastewater and waterborne waste exits a fixture, such as a lavatory or water closet, gravity causes it to flow toward the sanitary sewer in a system of drainage pipes. Wastewater and waterborne waste first flow through a fixture trap, which provides a liquid seal to prevent the escape of sewer gases without affecting the flow of wastewater or waterborne waste through the trap. **See Figure 6-4.** The waste then enters a fixture drain. A *fixture drain* is a drainage pipe that extends from the trap of a fixture to the junction of the next drainage pipe. When multiple fixtures are on the same floor, this next pipe is a horizontal branch, which conveys the wastewater to the nearest stack. Individual fixture drains can also connect into the stack.

> Sanitary drainage piping must be equipped with cleanouts in several locations. A cleanout is an angled fitting with a removable cap or plug that allows access to the interior of the pipe for cleaning or removing stoppages.

Sanitary Drainage System

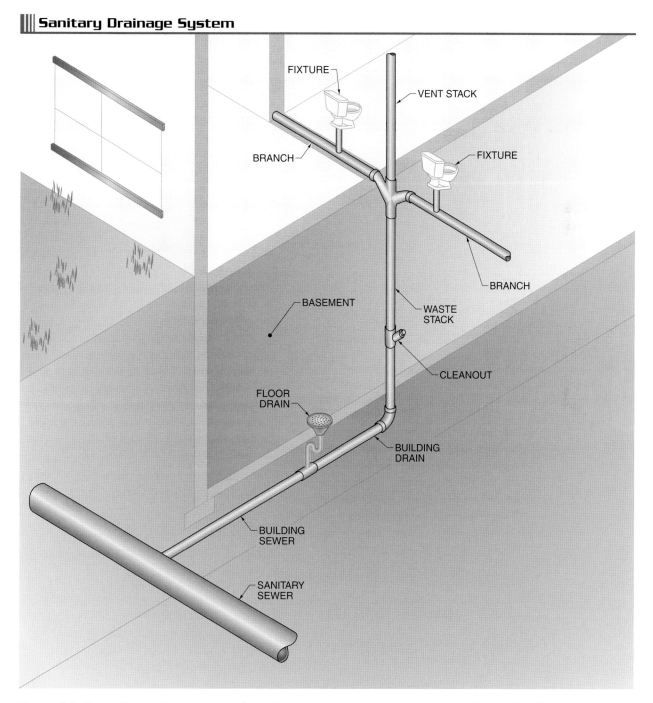

Figure 6-3. The sanitary drainage system collects all the wastewater and waterborne waste from various fixtures and conveys it to the sanitary sewer for disposal.

Fixture Traps

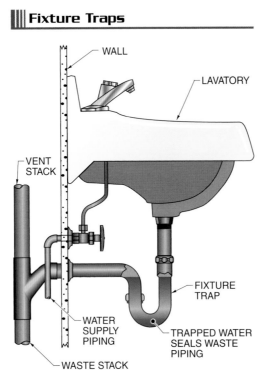

Figure 6-4. A fixture trap retains a small amount of wastewater in the bend of a pipe to seal against sewer gases entering the occupied area.

The building storm drainage system is sized based on the area that will collect rainwater, such as a building roof or parking lot, and the horizontal storm drain slope. A greater slope moves rainwater more quickly and efficiently through the system, allowing more rainwater to be conveyed.

A *stack* is a vertical drainage pipe that extends one or more floors. Stacks convey the wastewater or waterborne waste down to the building drain. A *building drain* is the lowest part of the drainage system, receives the discharge from all drainage pipes in the building, and conveys it to the building sewer. A *building sewer* is the part of the drainage system that connects the building drain to the sanitary sewer.

The entire system of drainage pipes must be arranged to allow all the wastewater and waterborne waste to flow unimpeded to a single, large sanitary drainage pipe. Therefore, horizontal branches must be installed with a slight slope to allow for natural flow caused by gravity.

Vent Piping System

A *vent piping system* is a plumbing system that provides for the circulation of air in a sanitary drainage system. The sanitary drainage system must be properly vented to ensure adequate removal of sewage and allow sewer gases to properly escape to the atmosphere. *Sewer gas* is the mixture of vapors, odors, and gases found in sewers. Also, ventilating the drainage system prevents trap siphonage and back pressure, which impedes the flow of wastewater to the sewer.

As wastewater and waterborne waste flow within the sanitary drainage system, air is displaced within the pipes. In order to keep the displaced air from impeding the flow of waste, the sanitary drainage pipes are vented to the outside air. The air in a drainage system exits the building through a vent pipe in the roof.

Vent pipes are extensions of the waste stacks in the sanitary drainage system. One or more vent pipes extend from above the highest connected horizontal drains to the roof. Fixtures on sanitary system branches can also be vented through individual vents that combine into vent branches and connect to the nearest stack vent.

Stormwater Drainage System

A *stormwater drainage system* is a plumbing system that conveys precipitation collecting on a surface to a storm sewer or other place of disposal. **See Figure 6-5.** These systems are common for draining rainwater and snowmelt from building roofs and parking lots. A *storm sewer* is a sewer used for conveying groundwater, rainwater, surface water, or similar nonpolluting wastes.

Stormwater drainage systems are similar to sanitary drainage systems in design. The difference is that stormwater drainage systems do not carry sewage. Rainwater enters a stormwater drainage system through roof or surface drains. Drains are covered by a tall strainer basket or flat slotted cover to prevent stones, leaves, and other debris from entering and clogging the system.

Stormwater Drainage System

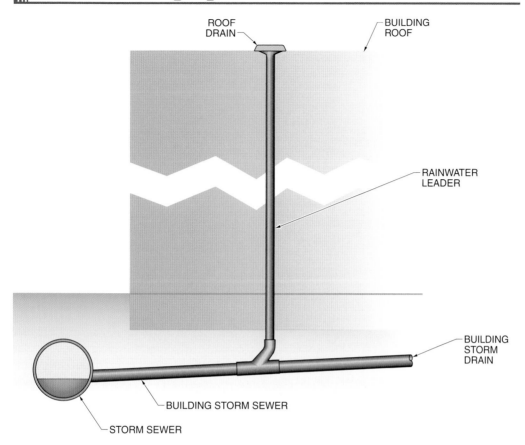

Figure 6-5. A stormwater drainage system collects precipitation from rooftops and other large surfaces and carries it away to storm sewers or retention ponds.

The drains convey the water into rainwater leaders. A *rainwater leader* is a vertical drainage pipe that conveys rainwater from a drain to the building storm drain or to another point of disposal. Rainwater leaders inside buildings usually run along columns or in vertical shafts constructed specifically for pipe. Leaders typically extend vertically from the base fitting, run through the floors of the building, and terminate below the roof where they connect to a roof drain.

Rainwater flows from the system of leaders into the building storm drain, which leads into the building storm sewer. If a municipal storm sewer is available, it receives the rainwater from the building storm sewer. If a storm sewer is not available, rainwater is piped to a drainage basin, such as a pond.

WATER SUPPLY

Controls in a plumbing system are typically located in the water supply system. Controls in wastewater systems are less common, though they can be important in certain situations. Vent piping systems are typically left completely open and unobstructed.

Four characteristics of the water supply that are commonly measured and controlled in a plumbing system are the temperature, pressure, flow, and level. This control allows for the most efficient operation of plumbing fixtures.

Water Temperature

Temperature is one of the most important characteristics of the water supply to building fixtures, but it is also relatively simple to measure and control.

Water is supplied to a building at a relatively cold temperature of about 40°F to 60°F, depending on the location and climate. This temperature is primarily influenced by the temperature of the ground, since the water supply is extracted from ground water (via wells) or from buried municipal water main pipes.

However, many building fixtures require water at a range of temperatures, so the water supply system is divided into hot and cold water systems. At the fixture, the desired temperature is achieved by mixing the hot and cold water. **See Figure 6-6.** The cold water supply is drawn directly from the building's main water supply, but the hot water supply is directed first through a water heater and then distributed in separate water distribution pipes to the necessary fixtures.

Hot and Cold Water Supplies

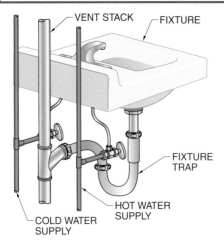

Figure 6-6. Some fixtures require both a cold water supply and a hot water supply, which are then mixed together at the fixture to dispense water at a desired temperature.

A *water heater* is a plumbing appliance used to heat water for the plumbing system's hot water supply. Water heaters create a steady supply of hot water, which is then distributed to plumbing fixtures throughout the building in a piping system similar to, and often parallel to, the cold water supply pipes. Much like boilers, water heaters produce heat either through the combustion of natural or liquefied petroleum gas (LPG) or

with electrical resistance elements. A thermostatic control activates the heating function as needed to maintain the set temperature. The water heater deals with this itself, without outside control signals.

Some fixtures require both a cold water supply and a hot water supply. Fixtures that do not require water at a specific temperature, such as water closets (toilets), receive only the cold water supply.

Water Pressure

Water pressure allows water to flow freely into pipes and fixtures when valves are opened. The force of flowing water may also make the water more effective for some applications, such as cleaning or rinsing, or the pressure may drive part of the fixture's operation, such as the flush of a water closet.

Water pressure also keeps contaminants from entering the potable water supply by preventing groundwater or wastewater from flowing into the water supply system pipes. The water supply system is a pressurized network of pipes. Water pressure is constantly pushing against all parts of the system. When a part of the system is opened, such as a faucet valve or a pipe crack, water is forced out. This keeps all other gases, liquids, and solids from entering the water supply system, unless forced into it intentionally with pumps.

Pressure is a force distributed over an area. Water pressure is typically measured in pounds per square inch (psi). Water in municipal mains is typically at a pressure of 45 psi to 60 psi. However, some water pressure is inevitably lost in the plumbing system between the main and the final point of use, a fixture usually being the final point of use. Head and friction are the primary sources of pressure loss. Pressure losses must be accounted for in a plumbing design so that adequate water pressure is available at each plumbing fixture for proper operation.

Pressure Loss Due to Head. Water has weight. Therefore, pressure in the water supply system must be high enough to push the water to the topmost plumbing fixture in the building, with enough pressure remaining to properly operate the fixture. The pressure required to overcome the weight of water to push it to a certain height is known as static pressure, or head. *Head* is the difference in water pressure between points at different elevations. Head is expressed as a height, commonly in feet.

A column of water 1′ high, with a cross-sectional area of 1 sq in., weighs 0.434 lb. Therefore, water exerts 0.434 psi of pressure for each foot of height. (Pipe diameter or shape has no effect on the pressure, since the force is exerted per cross-sectional area.) **See Figure 6-7.** Therefore, raising water requires 0.434 psi of pressure for every foot of height. Conversely, this can also be thought of as a pressure loss due to height (head). *Pressure loss due to head* is the loss of 0.434 psi of water pressure for every foot of height.

Pressure Loss Due to Head

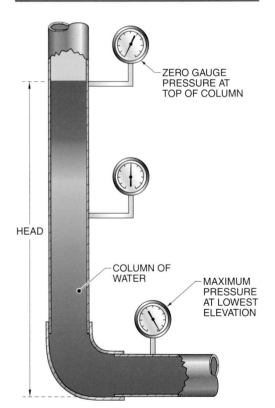

ZERO GAUGE PRESSURE AT TOP OF COLUMN

HEAD

COLUMN OF WATER

MAXIMUM PRESSURE AT LOWEST ELEVATION

Figure 6-7. *The weight of water exerts a pressure at the bottom of a water column in proportion to the height of the column, or head.*

For example, at a height of 76′, the pressure loss due to head is 33 psi (76′ × 0.434 psi/ft = 33 psi). If the water entering the building from the municipal water main was at a pressure of 45 psi, then there would be only 12 psi of pressure remaining at a height of 76′ (45 psi − 33 psi = 12 psi). This may not be adequate for the plumbing fixtures at that level. There would also be a significant difference in the operation of fixtures between the ground floor and the top floor.

Pressure Loss Due to Friction. *Pressure loss due to friction* is the loss of water pressure resulting from the resistance between water and the interior surface of a pipe or fitting. Different pipe materials have different resistances to water flow within the pipe. A coefficient of friction quantifies the pressure loss per length of pipe for a certain diameter, material, and flow rate. **See Figure 6-8.** Flow resistance also occurs as water changes direction as it passes through valves and fittings. The flow resistance effect of various pipe fittings is given as equivalent lengths of pipe. **See Figure 6-9.** For example, a 90° elbow for a 1″ pipe causes the same pressure loss due to friction as a 2.5′ length of 1″ pipe.

Pressure Loss Due to Friction*

Pipe Size†	Flow Rate‡								
	1	2	3	4	5	10	15	20	25
⅜	0.021	0.075	0.159	not recommended					
½	0.007	0.024	0.051	0.086	0.130				
¾	0.001	0.004	0.009	0.015	0.023	0.084			
1		0.001	0.003	0.004	0.006	0.023	0.049	0.084	
1¼			0.001	0.002	0.002	0.009	0.018	0.031	0.047
1½				0.001	0.001	0.004	0.008	0.014	0.021
2						0.001	0.002	0.004	0.005
2½		negligible					0.001	0.001	0.002
3								0.001	0.001

* psi/ft of pipe
† in in.
‡ in gpm

Figure 6-8. *Pressure loss can result from the friction of water flowing through pipes.*

Water expands when it freezes. If a pipe full of water freezes, the expansion of the water as it turns to ice exerts approximately 2000 psi of pressure against the inside of the pipe, which can cause pipes to burst.

Friction Allowance for Fittings and Valves*

Fitting Size[†]	Equivalent Tube Length[‡]								
	Standard Elbow		90° Tee		Coupling	Valve			
	90°	45°	Side Branch	Straight Run		Ball	Gate	Butterfly	Check
⅜	0.5	—	1.5	—	—	—	—	—	1.5
½	1.0	0.5	2.0	—	—	—	—	—	2.0
¾	2.0	0.5	3.0	—	—	—	—	—	3.0
1	2.5	1.0	4.5	—	—	0.5	—	—	4.5
1¼	3.0	1.0	5.5	0.5	0.5	0.5	—	—	5.5
1½	4.0	1.5	7.0	0.5	0.5	0.5	—	—	6.5
2	5.5	2.0	9.0	0.5	0.5	0.5	0.5	7.5	9.0
2½	7.0	2.5	12.0	0.5	0.5	—	1.0	10.0	11.5
3	9.0	3.5	15.0	1.0	1.0	—	1.5	15.5	14.5
3½	9.0	3.5	14.0	1.0	1.0	—	2.0	—	12.5
4	12.5	5.0	21.0	1.0	1.0	—	2.0	16.0	18.5
5	16.0	6.0	27.0	1.5	1.5	—	3.0	11.5	23.5
6	19.0	7.0	34.0	2.0	2.0	—	3.5	13.5	26.5
8	29.0	11.0	50.0	3.0	3.0	—	5.0	12.5	39.0

* Allowances are for streamlined soldered fittings and recessed threaded fittings. Double the allowances for standard threaded fittings.
† in in.
‡ in ft

Figure 6-9. Pipe fittings and valves have a significant effect on pressure loss due to friction. This effect is quantified in equivalent lengths of pipe.

The total pressure loss due to friction for a section of piping is the sum of the actual and equivalent pipe lengths (including those for every fitting) multiplied by the coefficient of friction. As more pipe, fittings, valves, and other devices are installed in a water supply system, the pressure loss increases.

Flow Pressure. *Flow pressure* is the water pressure in the water supply pipe near a fixture while it is wide open and flowing. A minimum of 8 psi of water pressure should be available to each plumbing fixture in order to function properly, though some fixtures require greater pressure. Hose bibbs require 10 psi and water closets and urinals require 15 psi to 25 psi (depending on the valve type). Insufficient flow pressure prevents water closets from flushing properly and results in inadequate flow rates from faucets.

Flow pressure is affected by the pressure losses due to head and friction. That is, the flow pressure at any point in a water supply system is determined by the available pressure at another point (such as the source) minus the pressure lost due to pipes, fittings, and elevation changes between them.

For example, if the available water pressure at the municipal water main is 50 psi, the pressure loss due to head is 33 psi, and the pressure loss due to friction is 2.5 psi, then the remaining flow pressure is 14.5 psi (50 psi – 33 psi – 2.5 psi = 14.5 psi). This pressure is adequate for many fixtures, but if this fixture were a water closet valve, which may require 25 psi of flow pressure, the fixture would not operate properly.

Water Flow

Water flow is the movement of water in pipes or channels. Water flow can be measured as a flow rate or a total flow. *Flow rate* is the volume of water passing a point at a particular moment. *Total flow* is the volume of water that passes a point during a specific time interval. For example, the flow rate of pumping water is typically given in gallons per hour (gph) or gallons per minute (gpm), while the total flow is the total number of gallons pumped.

Each plumbing fixture served by the water supply system has a specified flow rate. Multiple fixtures drawing water from the supply system increases the water demand. **See Figure 6-10.** The total of the minimum flow rates for every plumbing fixture within a building yields a total demand. However, plumbing fixtures in a building are rarely all used at the same time, so this estimate is unnecessarily high.

Minimum Flow Rates for Common Plumbing Fixtures

Type of Fixture	Flow Rate*
Lavatory faucet, standard	2.0
Lavatory faucet, self-closing	2.5
Kitchen sink faucet	3.0
Bathtub faucet	4.0
Laundry tub faucet	4.0
Shower head	4.0
Water closet flush valve	3.5
Drinking fountain	0.75
Sillcock or wall hydrant	5.0

* in gpm

Figure 6-10. Minimum flow rate requirements for different plumbing fixtures must be considered to ensure that each fixture receives an adequate water supply.

Based on the reasonable assumption that plumbing fixtures are not all used simultaneously, a better estimate of the total demand on a water supply system is determined from water supply fixture units. A *water supply fixture unit (wsfu)* is an estimate of a plumbing fixture's water demand based on its operation. A plumbing fixture, such as a water closet or lavatory, is assigned a wsfu value based on the:
- fixture flow rate when the fixture is used
- average time water is actually flowing when a fixture is being used
- frequency that the fixture is used
- type of building where the fixture is installed

For example, a domestic dishwasher for private use has a 2 wsfu demand and a commercial dishwasher for public use, such as in a restaurant, has a 4 wsfu demand. There is a difference because a domestic dishwasher uses less water and is not used as frequently as a restaurant dishwasher.

Water Level

There are only a few applications where water level is measured and/or controlled in a building plumbing system, but these can be extremely important for preventing undesirable water conditions. Level information is used to determine when to add or remove water from vessels by activating or deactivating pumps or valves. This prevents overflow of a vessel or supplies water to a vessel when it runs low. Level is normally measured in linear units of height, which may be translated into units of volume or weight. Some applications require continuous measurement, while others need to know only that the level is within the desired limits. **See Figure 6-11.**

Water Level

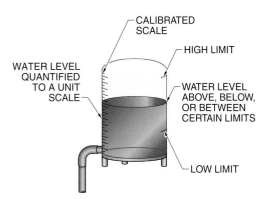

CALIBRATED SCALE

HIGH LIMIT

WATER LEVEL QUANTIFIED TO A UNIT SCALE

WATER LEVEL ABOVE, BELOW, OR BETWEEN CERTAIN LIMITS

LOW LIMIT

Figure 6-11. Like many other parameters, water level can be either quantified or only confirmed as being within certain limits.

PLUMBING SYSTEM CONTROL DEVICES

Plumbing controls are the devices used to control water in a plumbing system. Most plumbing control is related to water flow, but some may incorporate control of the water's temperature or level in certain vessels. Plumbing control equipment includes sensors, pumps, and control valves.

Water Condition Sensors

In order to control the various characteristics of water in a plumbing system, sensors are necessary to measure them. Not all plumbing systems include all of these devices, but they are increasingly important as a plumbing system is progressively more automated. These plumbing systems may include sensors to measure water temperature, pressure, flow rate, and level.

Immersion Temperature Sensors. Water temperature is measured to control water heating and temperature adjustment functions. Temperature sensors used with plumbing systems are very similar to those used with HVAC systems. Water-sensing temperature sensors may either measure an exact temperature or, like HVAC thermostats, provide a digital output based on a comparison of the water temperature to a setpoint. **See Figure 6-12.** Both types must be immersion instruments, rated for wet environments.

Immersion Temperature Sensors

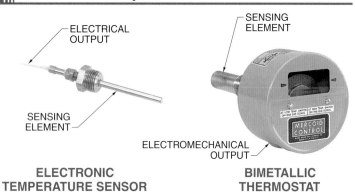

ELECTRONIC TEMPERATURE SENSOR

BIMETALLIC THERMOSTAT

Dwyer Instruments, Inc.

Figure 6-12. Several types of temperature sensors can be used with plumbing systems, though the appearance of different sensing elements can be similar.

The primary types of electronic temperature-sensing elements are thermocouples, thermistors, and resistance temperature detectors (RTD). These sensors change their electrical characteristics in proportion to temperature, either producing a small voltage or changing resistance. These changes may re read directly, or the sensors may include electronics to interpret these changes and convert them into standard analog signals or structured network messages

that correspond to the exact temperature in a standard scale, such as Fahrenheit.

Immersion thermostats may be either electronic or electromechanical and include normally open and/or normally closed contacts. Electronic thermostats are similar to electronic thermometers, except that they output only ON/OFF digital signals. Electromechanical temperature sensors incorporate a temperature-sensing element that moves slightly in response to temperature, such as a bimetallic strip. This movement is used to open or close physical electrical contacts, which provide a similar ON/OFF output signal. The thermostat is adjusted so that the signal change occurs at a specific temperature, which indicates that the water temperature is either above or below the setpoint temperature.

Regardless of the sensing method, immersion temperature sensors, which are used in wet environments, must have some special features. The sensing element is encased within a thermowell. A *thermowell* is a watertight and thermally conductive casing for immersion temperature sensors that mounts the sensing element inside the pipe, vessel, or fixture containing the water to be measured. **See Figure 6-13.** The thermowell allows a leak-proof conduit for signal conductors to pass outside the sensor. Special plumbing fittings may be necessary to mount thermowells.

Temperature Sensor Installation

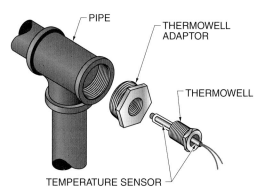

Figure 6-13. Special fittings allow the temperature sensor to be in contact with the water while the rest of the device remains dry and accessible.

Pressure Sensors. Since water pressure is so important for proper plumbing fixture operation, especially in tall buildings, the pressure is monitored by pressure gauges and pressure switches. A *pressure gauge* is a pressure-sensing device that indicates the pressure of a fluid on a numeric scale. Pressure switches are used in some applications where exact pressure is not needed. A *pressure switch* is a switch that activates at pressures either above or below a certain value. These pressure sensors are similar in operation to those used in HVAC systems, except that they are designed for wet environments.

Both types of pressure sensors include an elastic deformation pressure element, which is a piece of material that flexes, expands, or contracts in proportion to the pressure applied to it. Examples of elastic deformation pressure elements include diaphragms, bellows, and coiled tubes, which move back and forth in response to pressure changes. One side of the element is open to the pipe or vessel containing the water to be measured. The other side is open to the atmosphere, as water pressure is measured in reference to atmospheric pressure. The motion of the pressure element in response to the force exerted on it from the water is proportional to the water pressure. **See Figure 6-14.** A greater displacement indicates a higher pressure.

For pressure gauges, the motion is transferred to a transducer, where it is converted into an electrical signal. Additional electronics convert the transducer output into standard analog signals or structured network messages. For pressure switches, the motion is used to make or break an electrical contact when the applied pressure reaches a preset level. The switch can be adjusted to activate at a certain pressure. As the pressure rises or falls through the setpoint pressure, the contacts switch positions,

Flow Meters. A *flow meter* is a device used to measure the flow rate and/or total flow of fluid flowing through a pipe. Every plumbing system that draws water from a municipal water main has a flow meter installed on the water service pipe to measure and indicate water flow so that the building owner can be charged for water used. This particular flow meter is typically called a "water meter." However, additional flow meters may be installed at other points in the plumbing system to record water usage within sections of the system.

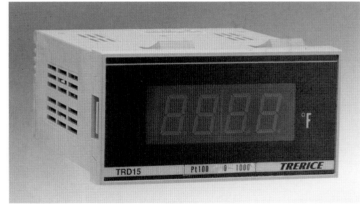

Trerice, H.O., Co.

The output of analog sensors must be interpreted and converted into a scaled value by additional devices, and can then be displayed on user interfaces. For example, the analog output from a temperature sensor is converted into degrees Fahrenheit and indicated on an LED display.

▓ Pressure Sensors

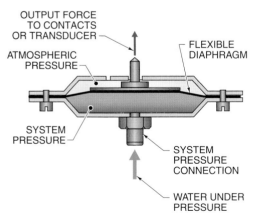

Figure 6-14. Pressure sensors include some type of elastic deformation pressure element together with the electronic means to interpret its movement.

Municipal water meters typically have the display registers mounted separately on the building's outside wall. This allows water utility workers to read the meter without entering the building.

Most common flow meters for plumbing systems are based on the rotation of mechanical spinners in a housing set in-line with the pipe. **See Figure 6-15.** The spinner is offset slightly so that the force of flowing fluid pushes against one side, causing the spinner to rotate. The spinner is often visible through a clear glass window in the flow meter. The speed of the resulting rotation is proportional to the flow rate of the water.

Water Meter

Dwyer Instruments

Figure 6-15. Water meters convert rotation induced by flowing water into a measure of its flow rate.

Boost-pump systems include automated controls that monitor water pressure and activate electric water pumps as necessary to maintain adequate system pressure.

For electronic monitoring, sensors inside the flow meter detect each revolution. Some flow meters output a pulse for each revolution of the spinner, which a separate device can then translate into a flow rate. For example, if a flow meter is designed to output one pulse for each 0.1 gal of water that flows through the meter, then a controller receiving 100 pulses within a minute can convert this information into a flow rate of 10.0 gpm. Other flow meters are already calibrated to convert the pulse rate into an equivalent flow rate. These then output a standard analog signal or structured network message based on the resulting flow rate.

Flow Switches. When the presence of flow, rather than the exact flow rate, must be monitored, flow switches are used. Flow switches include a vane mounted inside a pipe that is deflected from the force of flowing water. A certain amount of deflection, and therefore flow, activates the switch. It deactivates when the water flow falls below another setpoint. Switch connections include normally open and/or normally closed contacts. Flow switches in plumbing systems are used to indicate the operation of pumps or valves.

Level Switches. Level is not as commonly measured and controlled as other characteristics, even in many automation systems. However, there are some applications where monitoring water level is important, such as drainage sumps. A *level switch* is a switch that activates at liquid levels either above or below a certain setpoint.

In plumbing systems, the most common level-measuring devices are based on floats. **See Figure 6-16.** These devices are switches with normally open and/or normally closed contacts, which are used to either directly control another device, such as a pump, via a relay or input digital signals to controllers.

A float is a sealed chamber of air that floats at the top of the water, so its position is always a direct indication of the water level. The float is attached to the rest of the level-measuring device. Some use a rigid rod; when the float rises or falls with the water, the rod angle changes. Similar to vane-type flow switches, this movement activates a set of contacts within the switch housing.

Another type of float device utilizes an oblong chamber with a switch inside that is activated by orientation. The float is attached to the bottom of the vessel with a flexible cord that transmits the switch signal. When the level is low, the cord is slack and the float floats sideways. When the level rises, the float rises until the cord pulls it vertically, activating the switch.

Pumps

A *pump* is a device that moves water through a piping system. A pump moves water from a lower pressure to a higher pressure. It overcomes this difference by adding energy to the water. Pumps can operate either continuously or intermittently. Continuously operated pumps can be used to circulate water through a closed loop or boost pressure within a water system. Intermittently operated pumps can be used to empty sumps, draw water from wells, or operate plumbing appliances.

Water pumps are normally driven by electric motors. For smaller pumps, the motor is typically integrated into the pump housing. For pumps with a pumping rate above 100 gpm, the motor is normally connected by a coupling between the shaft and motor. The motors can be turned on or off with relays or motor starters. With a variable-frequency drive, the speed of the electric motor can be controlled to achieve the desired pumping rate.

Pumps can be categorized as either centrifugal pumps or positive-displacement pumps.

Centrifugal Pumps. Centrifugal pumps are the most widely used type of pumps. A *centrifugal pump* is a pump with a rotating impeller that uses centrifugal force to move water. An *impeller* is the bladed spinning hub of a fan or pump that forces fluid to its perimeter. **See Figure 6-17.** Water enters the pump through the center of the impeller. The impeller rotates at a relatively high speed, imparting a centrifugal force to the water, which moves it outward. The shape of the casing directs this water to the outlet.

Multiple impellers can be used in series to build higher pressures. In a multiple-stage centrifugal pump, the discharge from the first impeller enters the suction of the second impeller and increases the pressure even more. As many as four impellers may be on a single impeller shaft.

Another class of level gauges and switches sense the level of liquid by its electrical properties, such as conductivity, capacitance, and inductance. Some can only detect the presence or absence of water at sensor level, but others can measure level continuously on a scale.

Float Level Switches

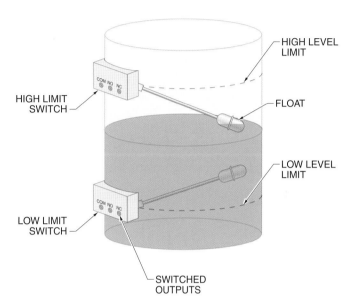

Figure 6-16. Mechanical level switches are actuated by the vertical position of a chamber that floats at the top surface of the water.

Centrifugal Pumps

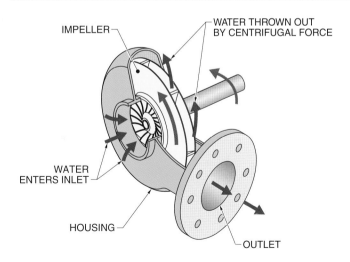

Figure 6-17. The rotating impeller of a centrifugal pump forces water to the perimeter of the pump housing, where it is then directed out through the outlet.

Positive-Displacement Pumps. A *positive-displacement pump* is a pump that creates flow by trapping a certain amount of water and then forcing that water through a discharge outlet. Positive-displacement pumps are further classified as either reciprocating-type or rotary-type, though both types operate by the same principle. Each type of pump includes a chamber for water that cycles between a small volume and a large volume. Water is drawn into the chamber from the pump inlet as the chamber volume is increasing. When the chamber is at maximum volume, the inlet is closed off and the outlet is opened. Then the volume decreases, pushing the water out.

Reciprocating-type pumps manipulate this chamber by moving some portion of it in a reciprocating (back-and-forth) motion. **See Figure 6-18.** This changes its volume. The movable portion is typically a piston or flexible diaphragm.

Rotary-type pumps create the displacement chamber by enclosing spaces within a circular housing. A rotor near the center includes vanes, lobes, or other projections that seal against the housing's inner wall. As the rotor rotates, the volume enclosed by these projections changes from small to large and back to small, moving water through the pump.

Positive-displacement pumps are typically self-priming and the resulting flow rate is independent of the water inlet pressure. However, some positive-displacement pumps create surges in flow and pressure with each cycle and are limited in capacity.

Valves

A valve is used to regulate water flow within a system. Valves are used to turn the water flow on and off, or to regulate the direction, pressure, and/or temperature of a fluid within the system.

Positive-Displacement Pumps

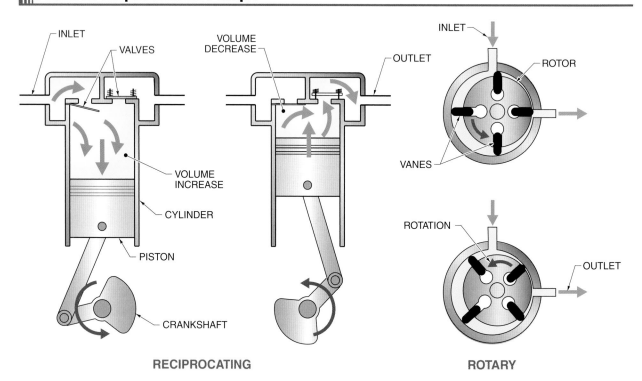

Figure 6-18. Positive-displacement pumps are classified by the method used to move water.

Most valves are operated by rotating the valve stem, which either raises or lowers an obstruction to water flow within the valve, or rotates a conduit for water flow into or out of position. When used in automation systems, mechanical valves are augmented with electromechanical actuators to allow for non-manual operation. Electromechanical actuation can be accomplished by mating a small electric motor to the rotating valve stem or by redesigning the valve to include electromechanical parts.

These motors require on-board controls to allow for rotation in both directions (for both opening and closing) and any position in between (for throttling). Complete electronic valves include the necessary electronics to operate the motor and accept input signals that indicate the desired valve position.

There are many designs for valves, which are classified by their intended use as either full-way or throttling valves. Either type can usually be used in the opposite function, but they will likely not perform optimally. There are also special types of valves for controlling water pressure and temperature. All of these types of valves are available with electronic actuators. In fact, some actuators can be purchased separately and fitted to existing (compatible) manually actuated valves to add automatic operation.

Full-Way Valves. A *full-way valve* is a valve designed to operate in only the fully open or fully closed positions. Full-way valves are also called shutoff valves. When open, the internal design permits a straight and unrestricted fluid flow through the valve, resulting in very little pressure loss due to friction. Full-way valves are installed in the building's water supply so the water can be shut off to individual rooms or fixtures without interrupting water service to other areas of the building.

Since they are intended for only ON/OFF operation, some mechanical full-way valves are designed to operate with only a 90° turn. A short rotation makes it easier and faster to change between the fully open and fully closed positions.

Examples of full-way valve designs include gate valves, ball valves, and solenoid valves. **See Figure 6-19.** Solenoid valves are examples of valves with completely integrated electromechanical components. A *solenoid valve* is a full-way valve that is actuated by an electromagnet. When supplied with an electric current, the integrated solenoid coil creates a magnetic force that pulls on a plunger. In a normally closed valve, this action opens the valve. This works against the force of a spring, which returns the plunger to the normal position when the coil is de-energized. Since the spring can slam the plunger back to its normal position with great force, solenoid valves may include a slow-closing feature to prevent damage.

Full-way valves may require only a digital input, which corresponds to the fully open and fully closed positions. This may be signaled with ON/OFF signals or with structured network messages.

Water Hammer

Water hammer can occur when a plumbing valve is closed quickly, forcing moving water to a sudden stop. This creates a momentary pressure surge in the water supply system, forcing water in the pipes to abruptly hit the ends of long, straight runs of pipe. The result is a loud banging noise and a jerking movement of some pipes. Over time, the repeated motion surges can damage pipes or fixtures, potentially causing leaks.

One way to avoid water hammer is to size water distribution pipes and fittings large enough to keep water flow velocities below 5 ft/s. Slower water has less energy to cause surges. Another option is to install water hammer arrestors in the distribution piping. Several may be placed in multiple locations near fixtures with quick-closing valves. Water hammer arrestors trap pockets of air in fittings or short pipe stubs. These air pockets absorb the surges by acting as a buffer. They are compressed when encountered by the waves of water pressure. These devices greatly reduce or eliminate the problems associated with water hammer.

▌▌▌Full-Way Valves

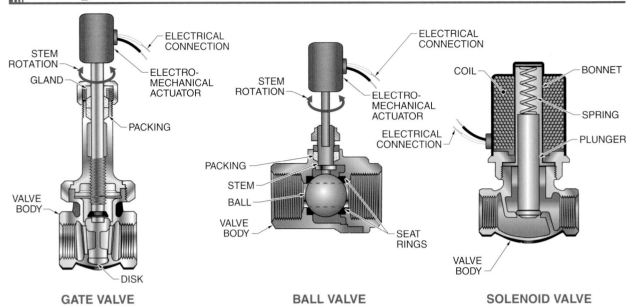

Figure 6-19. When fully open, full-way valves permit unrestricted flow through the valve.

Solenoid valves are compact electrically actuated valves, but are only capable of full-way operation.

Manual valves, including those converted to electronic actuation, often have some way to visually indicate the valve's approximate position at a glance. The stem and/or handwheel may rise when opening, or a long handle may move to one side.

Throttling Valves. A *throttling valve* is a valve designed to control water flow rate by partially opening or closing. Throttling valves are also called control valves. Due to the internal configuration, water flowing through the valve changes direction several times, resulting in flow resistance and a pressure drop. This makes them ideal for reducing water flow and pressure when required by the application. However, because they have some of this effect even when fully open, they are not recommended for use as full-way valves.

Throttling valves are installed on fixture supply pipes for individual fixtures. Examples of throttling valve designs include globe valves and butterfly valves. **See Figure 6-20.** Many throttling valves are required to be installed with the flow direction arrow pointing in the downstream direction.

Throttling valves require analog valves to determine intermediate positions. Many valve actuators accept standard analog signal types. Some are available for use with structured network message systems.

Three-Way Valves. A *three-way valve* is a valve with three ports that can control water flow between them. A *port* is an opening in a valve that allows a connection to a pipe. The use of the three ports as inlets or outlets may vary. Most often, water enters the valve from one port and the flow of that water can be directed into either of the two other outlet ports. **See Figure 6-21.** The valve may also allow a position that stops all flow through the valve. These types of valves typically use actuators that receive digital signals to rotate the valve stem into these fully actuated positions.

Three-way valves can also be designed to mix water from two inlet ports and discharge the result into an outlet port. A particular type of valve using this design is a thermostatic mixing valve. A *thermostatic mixing valve (TMV)* is a valve that mixes hot and cold water in proportion to achieve a desired temperature. **See Figure 6-22.** For many plumbing applications, the hot water supply is too hot and the cold water supply is too cold, but an optimal water temperature is achieved by mixing hot and cold water together. The thermostatic mixing valve can be set for all cold water (no hot water), all hot water (no cold water), or any position in between. For example, a position in the middle mixes 50% hot water with 50% cold water to produce warm water. Because of this range of possible positions, automated thermostatic mixing valves, or other types of three-way valves for mixing water supplies, require either analog signal inputs or structured network messages that include the desired valve position.

PLUMBING SYSTEM CONTROL APPLICATIONS

Control devices are integrated into a building's plumbing system to provide control and automation for certain aspects of its functionality. There are many possible control applications, depending on the specific system requirements. However, a few applications are relatively common.

The water pressure in some plumbing systems can be low or highly variable. This can be the case if the building draws water from a well or is at the end of a long municipal water main (caused by pressure loss due to friction in the mains). Fluctuating pressure may be a problem for overtaxed municipal water supply

systems or for buildings located near other facilities with significant water usage. For the plumbing system to operate adequately, some means is needed to boost and/or stabilize the water pressure.

||||| Throttling Valves

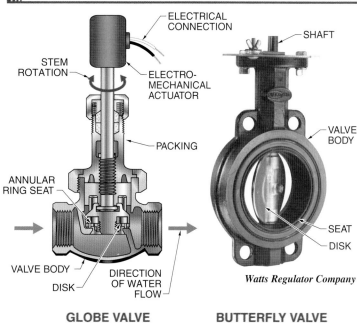

GLOBE VALVE   **BUTTERFLY VALVE**

Watts Regulator Company

Figure 6-20. Throttling valves are used in positions between fully open and fully closed in order to reduce water flow and pressure at the outlet.

||||| Three-Way Valves

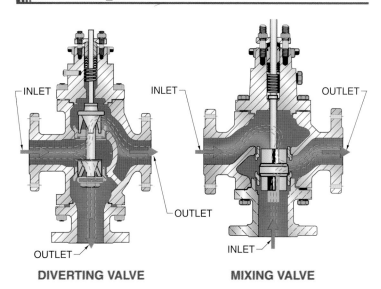

DIVERTING VALVE   **MIXING VALVE**

Figure 6-21. Three-way valves direct or mix water flow between three ports.

Thermostatic Mixing Valves

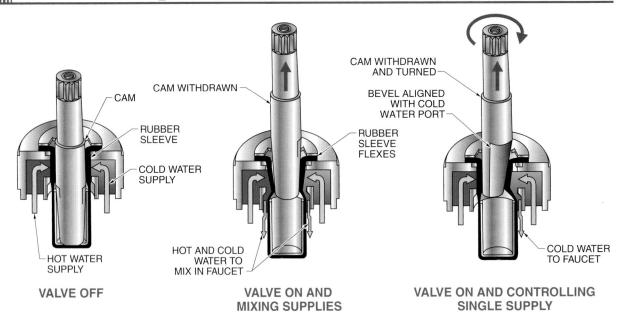

Figure 6-22. Thermostatic mixing valves mix water from the cold water supply and the hot water supply to dispense water at a desired temperature.

Boost Pumps

Boost pumps are the simplest method for increasing water pressure. A *boost pump* is a pump in a water supply system used to increase the pressure of the water while it is flowing to fixtures. **See Figure 6-23.** Boost pumps are fairly small and quiet and require no additional equipment. However, they cannot keep up with large water demands, so they are typically suitable only for residences or small office buildings.

Boost Pump

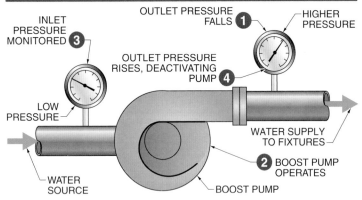

Figure 6-23. Boost pumps are installed in the water distribution system to increase the pressure of a low-pressure water supply.

Operating Conditions. Boost pumps are installed directly in the water supply system between the water supply (municipal main or well) and the rest of the water distribution system. Pressure switches monitor the water pressure on either side of the boost pump, controlling the pump to provide a certain water pressure as fixtures are opened and closed while protecting the water source from excessively low pressure.

Control Sequence. The following control sequence activates the boost pump as needed:

1. As fixtures draw water from a building's water supply system, the water pressure falls. A pressure switch on the outlet side of the boost pump activates if the water pressure falls to a low-pressure setpoint.

2. The activation of the pressure switch activates the boost pump via a relay or controller. The boost pump operates, increasing the pressure in the water supply system.

3. If the pressure of the water source falls to a low-pressure setpoint, the pressure switch on this side of the pump deactivates the pump. This protects sources such as wells from collapsing due to excessive water drawing.

4. As fixtures are closed, the water pressure in the water supply system increases. If the pressure rises to the high-pressure setpoint, the downstream pressure switch deactivates the pump.

Elevated Tanks

Water pressure changes according to height (head). In a column of water, the pressure at the bottom is greater than the pressure at the top. This principle is important for understanding why water pressure is lost when water rises in tall buildings, but it can also be used to solve this problem.

Water is typically supplied from underground mains. From this low reference point, all fixtures in the building are at a higher elevation, so the water pressure at those points is lower than the water pressure in the main. However, if water could be supplied from the top of the building, then all the plumbing fixtures would be below the supply and receive water at a pressure higher than that of the supply. The pressure would still vary between fixtures near the top and bottom of the building, but if the fixtures near the top receive a minimum adequate pressure, the higher pressures at lower elevations can be controlled with pressure-reducing valves. This scenario is accomplished with an elevated tank.

Operating Conditions. A large water tank is on or near the roof of the building and includes a pair of level switches to monitor the level of water within a specified range. **See Figure 6-24.** These tanks are not pressurized or sealed, other than measures needed to protect the potable water from outside contaminants. A pump moves water up to the tank from the water main and a check valve prevents the water from flowing back down through the pump. A control sequence is required to maintain an adequate supply of water in the tank.

Control Sequence. The following control sequence uses an elevated tank to control water supply system pressure:

1. Water usage in the building draws water from the tank, and the water level falls until the lower level switch is activated.

2. The activation of the low level switch activates the pump via a relay or controller. The pump operates, adding more water to the elevated tank.

3. The water level in the tank rises until the upper level switch is activated.

4. The activation of the upper level switch deactivates the pump via a relay or controller. This avoids overfilling the tank.

Elevated Tanks

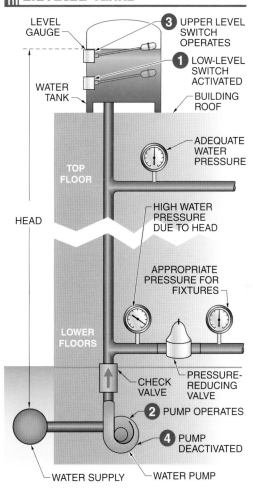

Figure 6-24. Elevated tanks provide a pressurized water supply to fixtures at lower elevations due to head (height).

The cycling time of a water pump for an elevated tank or hydropneumatic tank system is directly influenced by the size of the tank. A larger tank holds more water, allowing for more time between pump cycles.

Check Valves

A check valve is a valve that permits fluid flow in only one direction and closes automatically to prevent backflow (flow in a reverse direction). Check valves react automatically to changes in the pressure of the fluid flowing through the valve, and close when pressure changes occur.

Check valves are common in several parts of all plumbing systems, such as in the sanitary drainage line to prevent sewage from flowing back into a building. Check valves are also needed for several automation applications that utilize water pumps. The check valves prevent the water from back flowing and hold the pressure the pump built up on its outlet side.

Two common types of check valves are swing check valves and lift check valves. A swing check valve is a check valve with a hinged disk. In its normal operating condition, fluid flows straight through the valve and holds open the hinged disk. When backflow occurs, the hinged disk swings down into position, blocking flow. A lift check valve is a check valve with a disk that moves vertically. In its normal operating condition, fluid pressure forces the disk from its seat, allowing fluid to flow. When backflow occurs, the disk drops onto its seat, preventing backflow from occurring.

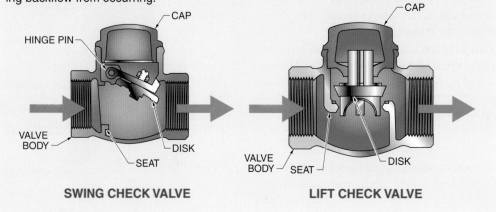

SWING CHECK VALVE   **LIFT CHECK VALVE**

Hydropneumatic Tanks

When a large roof tank is undesirable, pressure-boosting systems can employ a hydropneumatic tank. These tanks accomplish the same result as elevated tanks but they can be installed at any elevation, even in a basement. A *hydropneumatic tank* is a water tank incorporating an air-filled bladder that raises the pressure of water as it is pumped into the tank. **See Figure 6-25.**

As water is pumped into the tank, it displaces some of the space that was occupied by the flexible air bladder. Unlike water, air is compressible, so the air is forced to occupy less space. This increases the pressure exerted by the air, which increases the water pressure. When a fixture draws water from the system, the air pressure pushes water out of the tank to supply the fixture. As more water is drawn, the pressure gradually falls until it must be boosted again by the pump.

Systems with hydropneumatic tanks have similarities to both boost pump systems and elevated tank systems. Like boost pump systems, the water pressure is monitored to control the pump. Like elevated tank systems, a water supply is stored that can be used for some time after the pump shuts off, reducing the operating time of the pump.

Operating Conditions. Hydropneumatic tank systems require a pump to charge (fill) the tank. A pair of pressure sensors or switches are installed on either side of the pump. The downstream sensor controls the pump to maintain the water supply pressure within a specified range. The upstream sensor protects the water source, such as a well, from excessively low pressure.

IIII Hydropneumatic Tanks

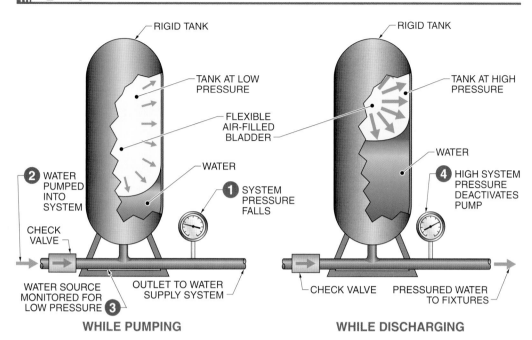

Figure 6-25. Hydropneumatic tanks can be installed at low elevations because they incorporate an air-filled bladder that is compressed by water, which exerts pressure on the water supply system.

Control Sequence. The following control sequence uses an elevated tank to control water supply system pressure:

1. As fixtures draw water from a building's water supply system, the water pressure falls. A pressure switch on the outlet side of the hydropneumatic tank activates if the water pressure falls to a low-pressure setpoint.

2. The activation of the pressure switch activates the pump via a relay or controller. The pump operates, adding water to the hydropneumatic tank and supplying water to fixtures.

3. If the pressure of the water source falls to a low-pressure setpoint, the pressure switch on this side of the pump deactivates the pump. This protects sources such as wells from collapsing due to excessive water drawing.

4. As fixtures are closed, the pressure within the hydropneumatic tank increases. If the pressure rises to the high-pressure setpoint, the pressure switch deactivates the pump.

Hot Water Loops

Hot water in a plumbing system's distribution pipes will cool to room temperature when it is not drawn regularly. This is why it usually takes several moments for water drawn from a faucet to get hot, since the cooled water must pass through the fixture first before recently heated water from the water heater is drawn. **See Figure 6-26.** This is not only inconvenient, but also wastes water.

One solution to this problem is to implement a hot water loop. A *hot water loop* is a closed circuit of hot water supply distribution pipes, including the water heater, through which hot water is circulated. A small pump circulates hot water through the loop from the water heater to the farthest fixture and back. **See Figure 6-27.** Therefore, the water in the pipes is always hot, and little water is wasted waiting for the water to become hot at the fixture.

Some hot water loops operate the circulation pump continuously, but if the heat loss through the pipes is low enough, it is possible to operate the pump only intermittently, conserving electricity. This requires that the system be automated.

||| Typical Hot Water Supply

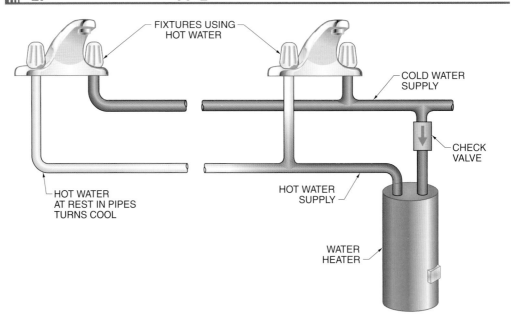

Figure 6-26. Without a hot water loop, there can be a delay in receiving hot water from the water heater at a fixture, which wastes water.

||| Hot Water Loop

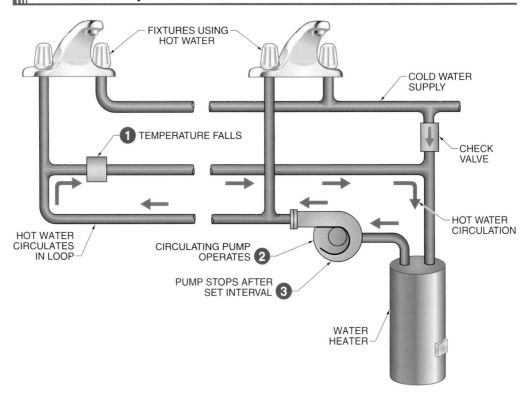

Figure 6-27. Hot water loops maintain a supply of hot water in a system of piping that is close to the fixtures requiring a hot water supply.

Operating Conditions. An immersion thermostat is installed at the farthest part of the loop from the outlet of the water heater, which is where the water is likely to be the coolest. A timer may be used to run the pump for a certain amount of time for each cycle. Schedules can also be used to limit the operation of the water heater and circulating pump only during the parts of the day when occupants are likely to use hot water, such as during business hours for office buildings.

Control Sequence. The following control sequence operates a hot water loop:
1. If the building is in occupied mode and hot water has not been drawn by a fixture for some time, the hot water in the loop begins to cool. The immersion thermostat activates when the temperature falls to the low-temperature setpoint.
2. The activation of the thermostat turns the pump ON via a relay or controller. The pump operates, circulating hot water through the loop. The activation of the pump may be used to also activate a cycle timer.
3. The pump is deactivated when either the thermostat registers a temperature above the high-temperature setpoint, or the pump has operated for a set amount of time, depending on the system design.

Lift Station

Wastewater systems are designed to move wastewater effectively with gravity, using as little active control as possible. However, due to facility design or surrounding terrain, sometimes wastewater must be actively pumped to a higher elevation. A *lift station* is a submersible pump in a sump pit that pushes wastewater to a higher elevation. Lift stations are very similar to sump pump systems. The sump pit is built below the lowest floor of a building. **See Figure 6-28.** Variations of lift stations include multiple sump pits for back-up use or emergency overflow.

If the sewer is located at a higher level than the lowest part of the building drain, a lift station is needed to pump the wastewater into the sewer. Lift stations can be used with sanitary sewers or storm water sewers.

Lift Stations

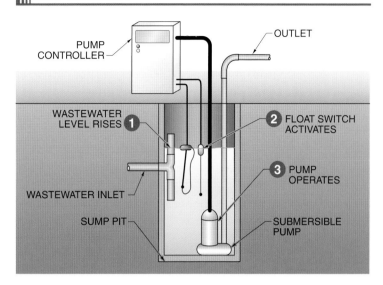

Figure 6-28. Lift stations pump wastewater to a higher elevation for disposal.

Operating Conditions. The pump is controlled by the wastewater level in the sump pit. The level controls include multiple level switches. These are often float-type switches, but other methods of measuring levels can be used. One switch operates at the highest permissible level and activates an alarm. This helps avoid overfilling the sump pit. One or two switches operate at the normal levels and operate the pump. One switch is effective, but using two reduces the cycling of the pump. A switch at a higher level turns the pump ON and a switch at a lower level turns the pump OFF.

Control Sequence. The following control sequence operates the lift station pump as needed:
1. As wastewater enters the sump pit, the water level rises.
2. The activation of a float-type level switch turns the pump ON via a controller.
3. The pump operates, drawing water out of the sump pit and lowering the wastewater level.
4. As the level falls, the switch deactivates, turning the pump OFF.
5. If the level rises too high, a level switch near the top of the pit activates an alarm.

Summary

- The different functions of a plumbing system are handled by separate systems.

- Controls in a plumbing system are typically located in the water supply system.

- The components of a plumbing may be inside or outside of the building, but must be within the property lines to be considered part of the building's plumbing system.

- Common fixtures include lavatories (sinks), water closets (toilets), and bathtubs.

- Four characteristics of the water supply that are commonly measured and controlled in a plumbing system are the temperature, pressure, flow, and level. This control allows for the most efficient operation of plumbing fixtures.

- Head and friction are the primary sources of pressure loss.

- In order to control the various characteristics of water in a plumbing system, water sensors are necessary to measure them.

- A pump creates a difference in pressure between the water at the pump's inlet and the water at the outlet. This causes the water to flow from the area of high pressure (inlet) to the area of low pressure (outlet).

- Pumps can operate either continuously or intermittently.

- There are many designs for valves, which are classified by their intended use as either full-way or throttling valves. Either type can be used in the opposite function, but will likely not perform optimally.

- Pressure-boosting systems are used to boost and/or stabilize the water pressure from the main water supply.

- Hot water loops keep the water in the pipes hot so that little water is wasted waiting for hot water at the fixture.

Definitions

- A *plumbing system* is a system of pipes, fittings, and fixtures within a building that conveys a water supply and removes wastewater and waterborne waste.

- A *water supply system* is a plumbing system that supplies and distributes potable water to points of use within a building.

- *Potable water* is water that is free from impurities that could cause disease or other harmful health conditions.

- A *building main* is a water distribution pipe that is the principal pipe artery supplying water to the entire building.

- A *riser* is a water distribution pipe that routes a water supply vertically one full story or more.

- A *branch* is a water distribution pipe that routes a water supply horizontally to fixtures or other pipes at the same approximate level.

- A *fixture branch* is a water supply pipe that extends between a water distribution pipe and fixture supply pipe.

- A *fixture supply pipe* is a water supply pipe connecting the fixture to the fixture branch.

- A *fitting* is a device used to connect two lengths of pipe.

- A *valve* is a fitting that regulates the flow of water within a piping system.

- A *fixture* is a receptacle or device that is connected to the water distribution system, demands a supply of potable water, and discharges the waste directly or indirectly into the sanitary drainage system.

- An *appliance* is a plumbing fixture that performs a special function and is controlled and/or energized by motors, heating elements, or pressure- or temperature-sensing elements.

- *Fixture trim* is the set of water supply and drainage fittings installed on a fixture or appliance to control the water flowing into a fixture and the wastewater flowing from the fixture to the sanitary drainage system.

- A *sanitary drainage system* is a plumbing system that conveys wastewater and waterborne waste from the plumbing fixtures and appliances to a sanitary sewer.

- *Sewage* is any liquid waste containing animal or vegetable matter in suspension or solution and/or chemicals in solution.

- A *sanitary sewer* is a sewer that carries sewage but does not convey rainwater, surface water, groundwater, or similar nonpolluting wastes.

- A *fixture drain* is a drainage pipe that extends from the trap of a fixture to the junction of the next drainage pipe.

- A *stack* is a vertical drainage pipe that extends one or more floors.

- A *building drain* is the lowest part of the drainage system, receives the discharge from all drainage pipes in the building, and conveys it to the building sewer.

- A *building sewer* is the part of the drainage system that connects the building drain to the sanitary sewer.

- A *vent piping system* is a plumbing system that provides for the circulation of air in a sanitary drainage system.

- *Sewer gas* is the mixture of vapors, odors, and gases found in sewers.

- A *stormwater drainage system* is a plumbing system that conveys precipitation collecting on a surface to a storm sewer or other place of disposal.

- A *storm sewer* is a sewer used for conveying groundwater, rainwater, surface water, or similar nonpolluting wastes.

- A *rainwater leader* is a vertical drainage pipe that conveys rainwater from a drain to the building storm drain or to another point of disposal.

- A *water heater* is a plumbing appliance used to heat water for the plumbing system's hot water supply.

- *Head* is the difference in water pressure between points at different elevations.

- *Pressure loss due to head* is the loss of 0.434 psi of water pressure for every foot of height.

- *Pressure loss due to friction* is the loss of water pressure resulting from the resistance between water and the interior surface of a pipe or fitting.

- *Flow pressure* is the water pressure in the water supply pipe near a fixture while it is wide open and flowing.

- *Flow rate* is the volume of water passing a point at a particular moment.

- *Total flow* is the volume of water that passes a point during a specific time interval.

- A *water supply fixture unit (wsfu)* is an estimate of a plumbing fixture's water demand based on its operation.

- A *thermowell* is a watertight and thermally conductive casing for immersion temperature sensors that mounts the sensing element inside the pipe, vessel, or fixture containing the water to be measured.

- A *pressure gauge* is a pressure-sensing device that indicates the pressure of a fluid on a numeric scale. Pressure switches are used in some applications where exact pressure is not needed.

- A *pressure switch* is a switch that activates at pressures either above or below a certain value.

- A *flow meter* is a device used to measure the flow rate and/or total flow of water flowing through a pipe.

- A *level switch* is a switch that activates at liquid levels either above or below a certain setpoint.

- A *pump* is a device that moves water through a piping system.

- A *centrifugal pump* is a pump with a rotating impeller that uses centrifugal force to move water.

- An *impeller* is the bladed, spinning hub of a fan or pump that forces fluid to its perimeter.

- A *positive-displacement pump* is a pump that creates flow by trapping a certain amount of water and then forcing that water through a discharge outlet.

- A *full-way valve* is a valve designed to operate in only the fully open or fully closed positions.

- A *solenoid valve* is a full-way valve that opens when supplied with an electric current and closes when the current stops.

- A *throttling valve* is a valve designed to control water flow rate by partially opening or closing.

- A *three-way valve* is a valve with three ports that can control water flow between them.

- A *port* is an opening in a valve that allows a connection to a pipe.

- A *thermostatic mixing valve (TMV)* is a valve that mixes hot and cold water in proportion to achieve a desired temperature.

- A *boost pump* is a pump in a water supply system used to increase the pressure of the water while it is flowing to fixtures.

- A *hydropneumatic tank* is a water tank incorporating an air-filled bladder that raises the pressure of water as it is pumped into the tank.

- A *hot water loop* is a closed circuit of distribution pipes, including the water heater, through which hot water is continuously circulated.

Review Questions

1. How is the water supply system different from all other portions of a building's plumbing system?

2. What are the four characteristics of a water supply that are controlled in a plumbing system?

3. Explain the two primary sources of pressure loss in a water supply system.

4. Why are some sensors used in plumbing systems divided into two parts?

5. How can the pumping rate be controlled with motor-driven pumps?

6. Why are throttling valves not recommended for use as full-way valves?

7. What is a common challenge when using automated fixtures?

8. What are the advantages of adding sensors to landscape irrigation controllers?

9. What are the advantages of including hot water loops in a plumbing system?

10. Explain the three primary techniques for increasing water pressure in an automated plumbing system.

Chapter Seven

Fire Protection System
Control Devices and Applications

Fire alarm systems have four basic functions: fire detection, alarm notification, fire protection system monitoring, and fire safety functions. Fire suppression systems, such as sprinklers, are separate systems that automatically activate to control fires. Fire protection system devices are typically only connected to and communicate with other fire protection devices. Only at the fire alarm control panel (FACP), the central controller of the fire protection system, does the system typically interface with other building systems to facilitate safe evacuation in the event of an emergency.

Chapter Objectives

- Differentiate between the roles and functions of fire alarm systems and fire suppression systems.
- Compare the different types of fire alarm signals and explain how each is triggered.
- Describe the wiring and signaling modes of the three types of fire protection device circuits.
- Identify common types of initiating devices and describe how each operates.
- Describe the operation of possible fire safety control functions.

FIRE PROTECTION SYSTEMS

The driving purpose behind fire protection systems is to save lives and protect property. Smoke and heat are deadly, so fire protection systems are also known as life safety systems. Smoke and heat can also damage or destroy buildings and their contents. For businesses, museums, libraries, historical structures, and other buildings with significant physical value, property protection is especially important. Since failures can lead to loss of life and property, fire protection systems must be extremely reliable.

A fire protection system consists of a fire alarm system and/or a fire suppression system. Some buildings have only a fire alarm system, but most commercial buildings—especially those with high occupancies—include both types of systems. The degree to which the two systems are integrated together varies.

Fire protection system equipment, including the control panels, supervisory devices, notification devices, and piping for fire suppression agents, is often painted red to distinguish it as a life safety building system.

Fire Alarm Systems

A *fire alarm system* is a system that detects hazardous conditions associated with fires (such as smoke or heat) and notifies building occupants. Fire alarm systems consist of initiating devices (inputs), a fire alarm control panel (FACP), and notification appliances (outputs). **See Figure 7-1.** The system monitors and displays the status of signals from the initiating devices. Fire alarm systems then respond by alerting building occupants with audible and visible signals.

Some fire alarm systems include additional features. The signals may be monitored off-premises by a fire department or supervising station. Alarms are automatically transmitted to the remote location, where personnel take the appropriate action and alert the necessary responders. This allows for quick response by emergency personnel to an alarm, even when the building is unoccupied.

Systems in automated buildings can be designed to take over the management of other building systems in order to facilitate safe evacuation and limit the spread of fire and smoke. Examples of these fire safety functions include smoke control, HVAC shutdown, elevator recall, door unlocking, and smoke door closure. A fire alarm system's specific features depend on the applicable codes and building requirements.

Fire Suppression Systems

Fire alarm systems alert building occupants to fire conditions, but do not directly address the fire hazard. Passive fire protection is the control of a fire through the resistance of the building itself to burn or to allow the fire to spread. Building codes require passive measures to control building fires. Typically, each large area and exit stairwell of a building is separated by walls or barriers with a 2 hr rating, which indicates the duration of its resistance to burning through. Penetrations through fire-rated walls or barriers must also be sealed in order to have the same rating. These fire protection measures allow time for occupants to evacuate during a fire.

Active fire protection, however, is a purposeful response to reduce or extinguish the fire. A *fire suppression system* is a system that releases fire suppression agents to control or extinguish a fire. **See Figure 7-2.** A fire suppression system does not replace the need for fire department response, but can allow occupants additional time to evacuate the building and can reduce the destructive capabilities of the fire.

Fire requires the following elements in order to ignite and continue burning: fuel, heat, oxygen, and combustion. If any one element is removed, a fire will be extinguished. A *fire suppression agent* is a substance that can extinguish a fire. Fire suppression agents are specifically designed to remove one or more of these four required elements of a fire. Suppression agents include water, foams, gases, or dry chemicals. The choice of suppression agent depends on the type of hazard. Suppression systems use piping, valves, and discharge heads to apply the agent, and may include sensors to monitor the system and detect actuation.

Fire Alarm Systems

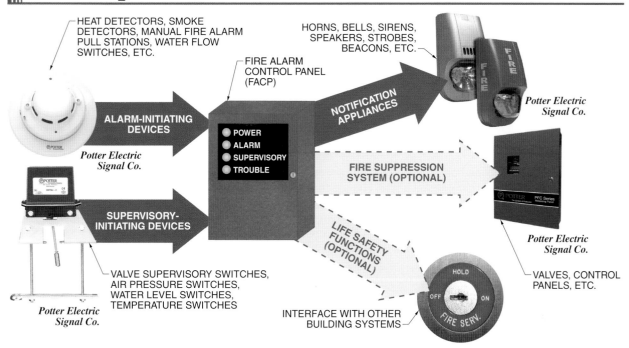

Figure 7-1. Fire alarm systems center around a fire alarm control panel (FACP), which receives input from initiating devices and sends signals to notification appliances and other devices as necessary.

When present, fire suppression systems are integrated with fire alarm systems, though in varying degrees. **See Figure 7-3.** Many fire suppression systems activate independently of the fire alarm system. The devices for releasing fire suppression agents are activated automatically from heat. In special applications, however, the fire alarm systems may be designed to receive alarm signals from the fire suppression system and/or directly control the activation of the fire suppression system. However, in either case, the fire alarm system always monitors the status of the fire suppression system for activation and faults and makes this information available to building personnel.

Code requirements for fire alarm systems and fire suppression systems are treated separately. If occupancy or fire hazards are low, buildings may not be required to have fire suppression systems, even if they do have a fire alarm system. However, high-rise, institutional, assembly, educational, and other types of structures are usually required to have fire suppression systems.

Fire Suppression Systems

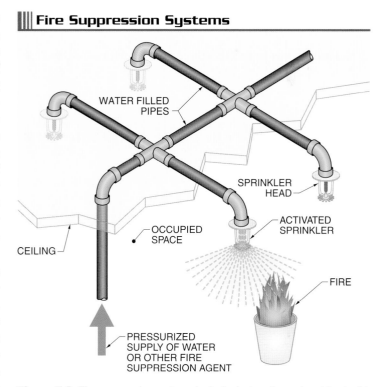

Figure 7-2. Fire suppression systems include devices throughout the building that release an extinguishing agent in the event of a fire.

Fire Suppression Agents

Fires require fuel, heat, oxygen, and the combustion reaction to continue burning. Removing any one or more of these elements with a fire suppression agent extinguishes the fire. The most common types of fire suppression agents are water, foams, gases, and dry chemicals.

Water

Water is by far the most common fire suppression agent and acts on both the oxygen and heat elements of a fire. When water contacts fire, it vaporizes. The rapid expansion of the water vapor fills a volume 1700 times greater than the volume of applied water. This displaces oxygen-containing air around the fire, effectively smothering the fire. Heat is required to change the state of water from liquid to vapor. Therefore, the vaporization of water also removes heat from the fire, cooling the burning materials. This also cools the surrounding areas that are not yet burning, reducing the chance that the fire will spread.

Water is most effective when widely dispersed over the fire area. This is usually accomplished with spray nozzles or sprinklers. Sprayed water maximizes the area of fire suppression, but the spray density must still be enough to adequately cool the burning materials. Recommended sprinkler types and specifications for various types of buildings and hazard risks are listed in National Fire Protection Association (NFPA) 13, Standard for the Installation of Sprinkler Systems.

Foams

One disadvantage of water is that it flows readily out of the fire area, and then is no longer effective. To address this issue, some systems add chemicals to the fire-fighting water supply to cause it to foam when dispersed from nozzles or sprinklers. Foams suppress fires in ways similar to plain water, displacing oxygen and cooling burning materials. However, since foams tend to stick to surfaces, they are effective for longer periods.

Gases

Some nonflammable gases can also be used as fire suppression agents. A gaseous suppressant must be heavier than air so that it sinks to the floor of a room, where most fires are located.

Carbon dioxide is a common gaseous fire suppression agent, though there are also many specialized gases. These specialized gases may be blends of inert gases or distinct compounds designed specifically for fire suppression. Proprietary fire suppression gases are known by their trade names, such as Inergen® and Halotron®. Gaseous suppression agents displace most or all of the oxygen surrounding the fire. Specially designed gas compounds are also designed to interfere with the combustion reactions.

Gaseous fire suppression systems are ideal for certain applications. This is because the gas, unlike water or foam, will not damage electronic equipment or delicate materials. Therefore, these systems are typically installed in telecommunications facilities, computer server rooms, laboratories, museums, and library vaults.

Unlike water and foam, however, gaseous fire suppression systems pose a unique hazard. Because these systems are so effective at displacing the oxygen from a large space, people trapped in the area can suffocate. To prevent such situations, these fire-suppression systems are pre-engineered systems with many precautions. Some require manual activation from within the protected area. Others activate automatically, but slightly delay agent release and include additional warning devices that advise immediate evacuation of the space. All include extensive warning placarding advising occupants of the hazard.

Dry Chemicals

Dry chemicals are powders of nonflammable materials that, when dispersed over burning materials, smother the fire and interfere with the combustion reaction. The most common dry chemicals are ammonium phosphate, sodium bicarbonate (baking soda), and potassium bicarbonate. These chemicals melt or decompose when heated by a fire, forming a crust over burning surfaces to suppress the fire. Dry chemical systems are used for kitchen hood systems, gasoline-dispensing stations, or other areas where flammable or combustible liquids are found.

Fire Protection System Integration

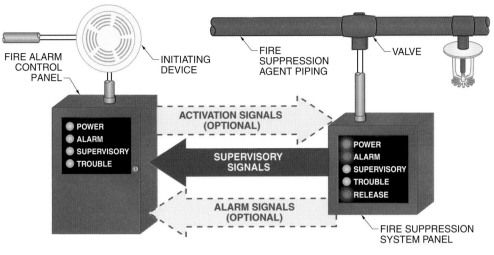

FIRE ALARM SYSTEM   **FIRE SUPPRESSION SYSTEM**

Figure 7-3. *Fire alarm systems monitor the integrity and actuation of fire suppression systems. Some systems are more closely integrated so that the fire alarm system activates the fire suppression system.*

FIRE ALARM SIGNALS

Fire alarm systems monitor the circuits of fire detection and fire suppression devices and wait for the electrical signals that indicate fire hazards or abnormal conditions. The three types of signals are alarms, supervisory signals, and trouble signals. **See Figure 7-4.** All signals are immediately communicated to the system operator, typically the building manager or on-duty facility personnel. Depending on the system and type of signal, the fire alarm system may then initiate further actions, such as activating the fire suppression system.

Alarms

An *alarm* is a signal that indicates a fire or other hazardous conditions. Alarm signals are initiated by a fire alarm-initiating device such as a manual fire alarm pull station, smoke detector, heat detector, automatic fire detector, water flow switch, or any other device of which activation is indicative of a fire. Alarm signals always have priority over all other signals.

The National Fire Protection Association (NFPA) reports that there were 1,602,000 fires in the United States in 2005, down 6% from the number of fires in 2000.

Fire Alarm Signals			
Signal	*Indicates*	*Sensing Devices*	*Examples*
Alarm	Fire condition	Alarm-initiating devices	Heat detectors; smoke detectors; manual fire alarm pull stations; water flow switches
Supervisory	Fire suppression system problems	Supervisory-initiating devices	Valve supervisory switches; air pressure switches; water level switches; temperature switches
Trouble	Wiring faults	Fire alarm control panel (FACP)	Open circuits; ground faults; loss of power

Figure 7-4. *A fire alarm system monitors control devices for alarm signals and then responds with the appropriate course of action.*

Conventional fire alarm control panels include enough terminals so that many fire alarm circuits can be connected.

Supervisory Signals

A *supervisory signal* is a signal that indicates an abnormal condition in a fire suppression system. This does not necessarily indicate a failure of a component or circuit, but it signals that something could still affect the proper operation of the system during a fire. An example of a condition causing a supervisory signal is a closed position of a valve that controls the water supply to a fire sprinkler system.

Trouble Signals

A *trouble signal* is a signal that indicates a wiring or power problem that could disable part or all of the system. Trouble signals indicate system problems that can prevent the fire protection system from functioning as intended in the event of a fire. Trouble signals are activated by disconnection of a device or circuit, loss of power supply, low power on the back-up battery, ground faults, or open circuits. Trouble signals must appear within 200 sec of the fault condition, and must reset within 200 sec after the fault is cleared.

FIRE ALARM CONTROL PANELS

All protected premises and supervising station systems utilize a fire alarm control panel (FACP). A *fire alarm control panel (FACP)* is an electrical panel that is connected to fire protection system circuits and interfaces with other building systems. The FACP monitors the circuits for signals and faults, monitors their power supply, and controls fire safety functions. The FACP is the "brain" of the fire protection system and is equipped with a user interface to provide operators with system information. Some FACPs include additional features, such as the capability of recording system events for troubleshooting, either with paper printouts or in electronic memory.

Fire Alarm Circuits

An *initiating device* is a fire protection system component that signals a change-of-state condition. Groups of fire alarm initiating devices are wired into circuits. **See Figure 7-5.** A circuit and its group of connected devices represent a single zone. Depending on the size of the building, there may be many zones, typically divided by floor and/or wing. The fire alarm zones may or may not match the building zones used by other building systems, such as the HVAC systems or security systems.

All of these zone circuits are connected to an FACP that monitors the devices on these circuits. In a very large building or campus-wide system, there may be multiple FACPs to monitor all the necessary circuits. These FACPs are then connected together to share alarm information. The FACP is also the point of integration for the fire alarm system into any other building systems.

There are two basic types of circuits: initiating-device circuits and notification appliance circuits. A third type, the signaling line circuit, is a variation of an initiating-device circuit, though the technology of its special capabilities can also be applied to notification appliance circuits. FACPs provide power to the fire alarm circuits, usually 24 VDC, which also facilitates the monitoring of circuit integrity. National Fire Protection Association (NFPA) 72, *National Fire Alarm Code*, requires all fire protection system circuits to be monitored for open circuits and ground faults that would severely impair the fire protection system in an emergency. When detected, these faults are interpreted as trouble signals by the FACP.

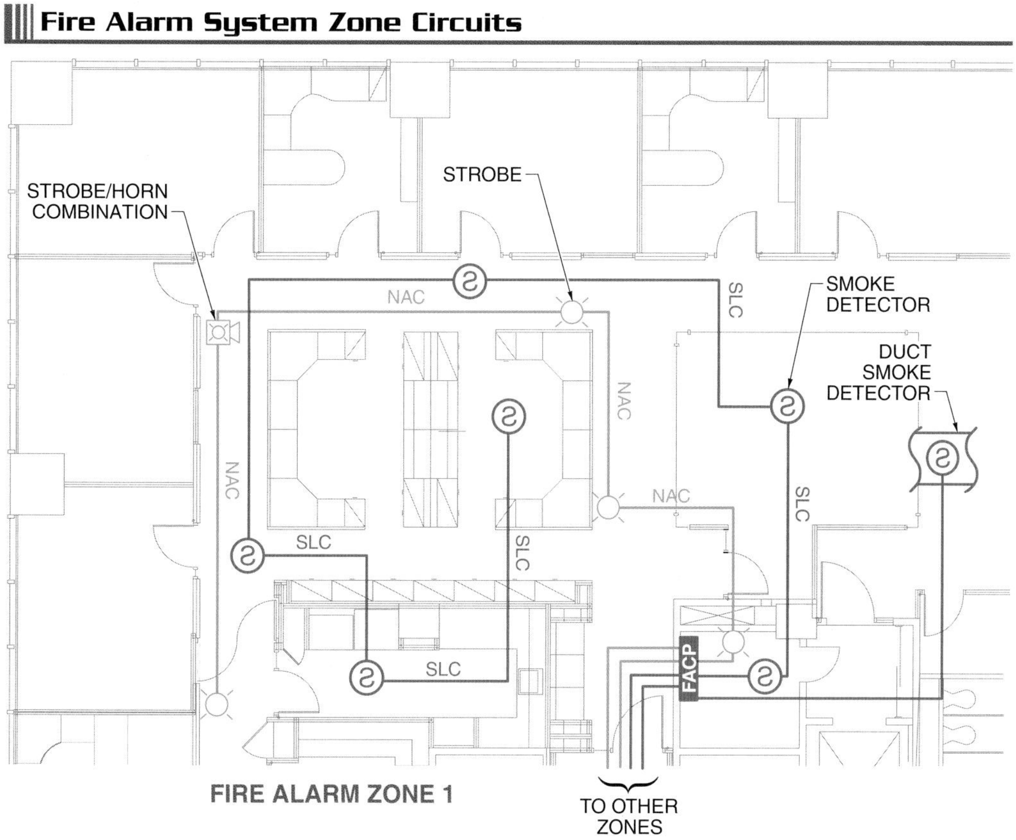

Fire Alarm System Zone Circuits

Figure 7-5. Fire alarm control panels (FACPs) interface with all other fire protection system devices through circuit loops grouped by device type and building zone.

Initiating-Device Circuits. An *initiating-device circuit (IDC)* is an electrical circuit consisting of fire alarm initiating devices, any of which can activate an alarm signal by closing its contacts. The initiating devices used with these circuits are of the conventional (non-addressable) type, using only the contact closure to initiate alarms.

The FACP applies a voltage to the circuit, usually 24 VDC. The circuit is closed at the far end with an end-of-line device such as a resistor or capacitor. An *end-of-line (EOL) resistor* is a resistor installed at the far end of an initiating-device circuit for the purpose of monitoring for circuit integrity. The FACP then monitors the circuit for any electrical changes, which indicate a signal. **See Figure 7-6.** A contact closure by any one of the devices shorts the circuit and reduces the resistance to nearly zero, which is interpreted as either a fire alarm or supervisory signal.

The alarm and supervisory signals appear the same to the FACP, but are differentiated by setting up the FACP to identify certain circuits with the appropriate signal, depending on the type of initiating devices on the circuit. For example, a circuit of alarm-initiating devices is associated with alarm signals by connecting the circuit to certain terminals in the FACP designated for alarm circuits. A contact closure on this circuit will then result in an alarm signal. Modules are available to add or to configure the different types of circuit terminals inside the FACP.

Some devices that require power to operate, such as smoke detectors, can receive power from the same two conductors that transmit the alarm signal. Others, however, require an extra pair of power supply conductors. These conductors must also be monitored.

▥ Initiating-Device Circuits

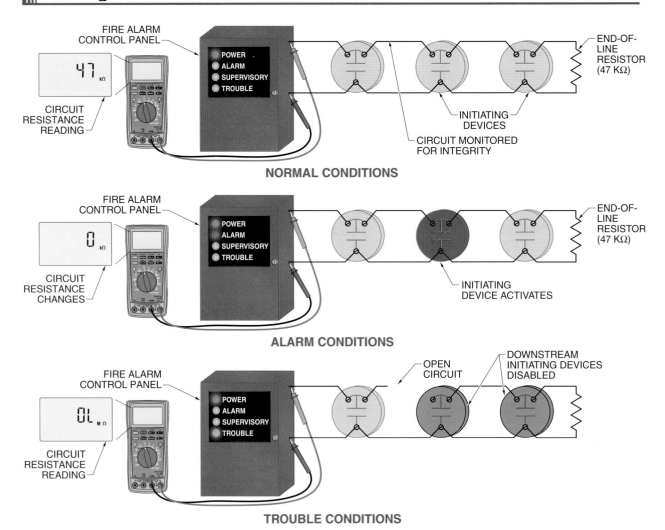

Figure 7-6. An initiating-device circuit relies on the electrical changes caused by contact closures and circuit faults to indicate signals from the group of initiating devices.

An open circuit condition is an infinite resistance that causes current flow to stop, while a ground fault is a low resistance with current leakage to ground. Both of these problems are indicated as trouble signals by the FACP, regardless of the type of devices on the circuit. Monitoring for open circuits and ground faults is particularly important because these conditions can disable some initiating devices. Under either condition, only the devices between the FACP and the fault will operate until the problem is remedied. All devices downstream of the problem could effectively be disabled.

This kind of initiating circuit is relatively reliable because of its simplicity, but has some drawbacks. First, alarm- and supervisory-initiating devices cannot normally be wired on the same initiating-device circuit zone. Most FACPs would interpret the resulting signals as either false alarms or false supervisory signals. Also, the FACP can only indicate signals as coming from a circuit group and cannot determine exactly which device has activated. If the circuit zone covers a relatively large area, such as the entire floor of a building, this can slow the response times to the exact location of a fire during alarm conditions.

Signaling Line Circuits. A *signaling line circuit (SLC)* is an electrical circuit of fire alarm initiating devices that communicate using addressable initiating devices. An *addressable device* is a device with a unique identifying number that can exchange messages with other addressable devices. Each addressable initiating device includes a microprocessor and software and can identify itself with a unique address. In this way, a signaling line circuit is similar to a computer network. The setting of the address, program, and other parameters varies by manufacturer. End-of-line devices may or may not be necessary, depending on the device manufacturer.

A fire alarm system with signaling line circuits operates differently than one with conventional initiating-device circuits. The FACP may either periodically poll each device to determine its status, or it may receive regular updates from each device with its status. Polling involves sending an electronic message to each device address and waiting for a response that includes information on either normal or signal conditions. **See Figure 7-7.** Update messages are received within specified time intervals. The signals can be designated as either alarms or supervisory signals. Failure of the device to respond to a polling request or failure to send an expected update, due to an open circuit or ground fault, results in a trouble signal. An FACP that is capable of signaling line technology is required for these circuits to operate.

Because the devices have unique addresses, the FACP can identify an individual alarming device, unlike conventional initiating-device circuit systems. By matching the device address with a database of device information, the FACP can indicate the type of detection and exact location of the device. For example, an FACP might indicate an alarm from device #058. The database reveals that device #058 is a smoke detector located in the electrical closet on the 12th floor. This information significantly shortens response times to a fire hazard or device malfunction.

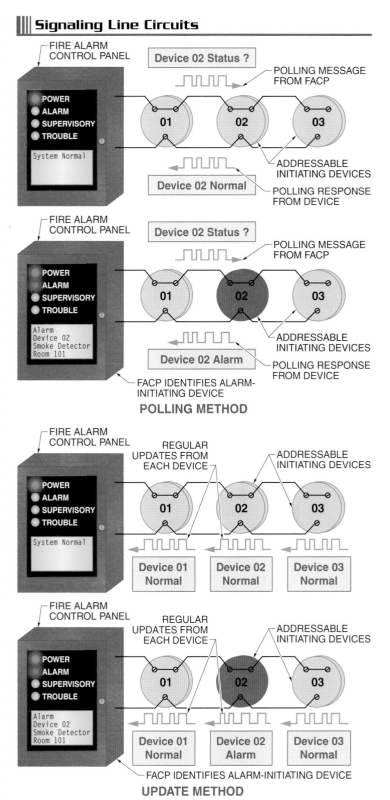

Signaling Line Circuits

POLLING METHOD

UPDATE METHOD

Figure 7-7. Devices on a signaling line circuit communicate much like a computer network. Each device is capable of sending and receiving digital messages containing commands or information.

Notification Appliance Circuits. A *notification appliance circuit (NAC)* is an electrical circuit of fire alarm notification appliances that are activated by the fire alarm control panel (FACP) under certain signal conditions. A *notification appliance* is a fire alarm system component that provides an audible and/or visible indication of alarm signals. A conventional notification appliance circuit is wired similarly to a conventional initiating-device circuit. This allows entire groups of notification appliances to be activated at once. The appliances in a circuit are either all ON or all OFF.

Notification appliance circuits are also monitored with a small current, but a blocking diode inside the appliance prevents the monitoring current from activating the appliance circuitry under normal conditions. **See Figure 7-8.** Similar to initiating-device circuits, an open circuit or ground fault will interrupt the current flow and cause a trouble signal. In an alarm condition, the FACP reverses the circuit polarity, allowing current to flow in the opposite direction. The diode is then forward-biased, which allows current to activate each appliance.

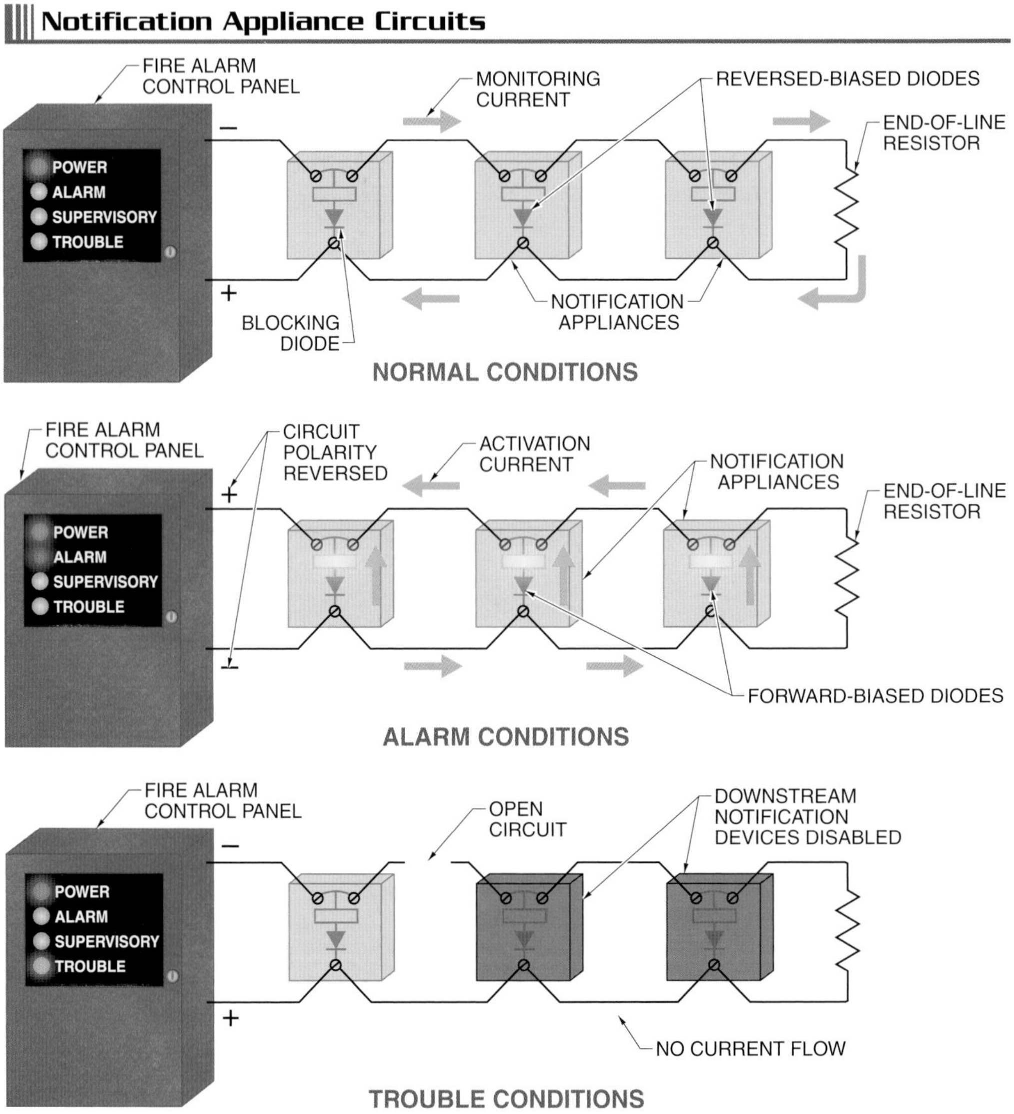

Notification Appliance Circuits

Figure 7-8. Notification appliances are wired together into a circuit in such a way that they can all be activated simultaneously by the FACP.

Circuit Classes

Circuit class indicates the ability of the circuit to withstand a single fault condition, such as a ground fault or open circuit. NFPA 72, *National Fire Alarm Code,* defines the operational requirements of two different classes under specific fault conditions.

Class A circuits require a pair of return conductors from the last device to the FACP. This allows the entire circuit to continue to operate with an open circuit or ground fault condition, so they are very reliable. The outgoing and return conductors must be routed separately to prevent a fire from disabling the entire circuit at once. The FACP must also contain special controls that sense the fault conditions in the circuit and provide power and communications in two directions. However, Class A circuits can only tolerate a single fault. With two faults, the devices between the two faults cannot communicate with the FACP.

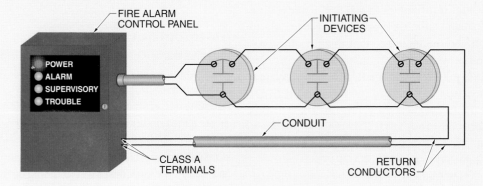

CLASS A CIRCUIT

Class B circuits do not have these requirements and, therefore, are inherently less reliable than Class A circuits. Class B circuits have a single pathway from the FACP to the devices. In a conventional device circuit, an end-of-line (EOL) device is installed as the last device on the circuit to provide a path for monitoring current so that an open circuit or ground fault will be detected and reported.

Signaling line circuit technologies may or may not require the use of an EOL device since they monitor by exchanging messages with the devices instead. However, an open circuit or ground fault on an addressable Class B circuit will still cause the loss of all devices downstream from the fault.

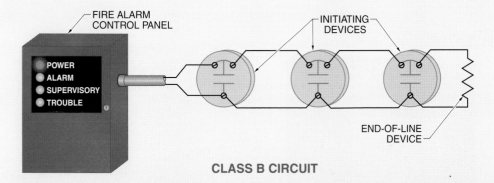

CLASS B CIRCUIT

The system designer is responsible for choosing the circuit class based on fire protection goals and the hazards involved in the premises. Higher-risk applications generally demand the use of more reliable Class A circuits. For example, most hospital fire alarm systems are designed with Class A circuits. Lower-risk applications, such as office buildings, often use Class B circuits.

Like initiating-device circuits, signaling line circuit technology can be applied to notification appliances, though this is a relatively new innovation. With compatible equipment, this allows for addressable control of individual appliances. The advantages include being able to quickly identify the location of circuit faults and selectively choose which notification appliances activate for various alarm situations.

Annunciators

System information is usually displayed on an annunciator. An *annunciator* is a fire alarm notification appliance that displays fire protection system modes, status, and alarms. **See Figure 7-9.** Most annunciators are physical panels on the front of the FACP with LEDs or LCD displays to represent the various pieces of information. The same information can also be displayed on remote annunciators connected to the main FACP.

Many addressable FACPs can be customized to display system information within a drawing of the building floor plan. For annunciator panels, the floor plan is physically drawn on the front panel and the system information is shown with indicators such as LEDs. Graphic annunciators can also be shown on computer monitors, which offers more options for displaying floor plans and system information. Graphic annunciator information allows responders to quickly determine the location of the fire, particularly in unfamiliar buildings.

FIRE PROTECTION SYSTEM CONTROL DEVICES

Signals to the FACP are provided by initiating devices installed throughout the building. Initiating devices are designed to cause either alarm or supervisory signals. Outputs from an FACP can vary. All FACPs can activate alarm notification devices during fire alarm conditions. Additional output capabilities depend on the FACP's integration with a fire suppression system and other building systems.

Alarm-Initiating Devices

Initiating devices produce either alarm or supervisory signals, depending on the type of device and how it is programmed or wired. Alarm-initiating devices cause alarm signals, which indicate a fire condition. The most common types of alarm-initiating devices are heat detectors, smoke detectors, manual fire alarm pull stations, and water flow switches. Most initiating devices are available in either conventional versions, which only close a set of contacts to cause an alarm, or in addressable versions, which include electronics that allow communication with alarms using digital messages.

Heat Detectors. A *heat detector* is an initiating device that is activated by the high temperatures of a fire. Activation of a heat detector results from the ambient air temperature either reaching a fixed, high temperature or rising at a high rate.

||||| Annunciators

Gamewell-FCI
CONVENTIONAL

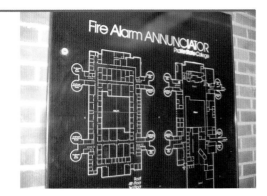

GRAPHIC

Figure 7-9. Annunciators display fire protection system information to building managers and system operators in an easy-to-read format.

Fixed-temperature heat detectors are designed in one of several configurations: fusible links, bimetallic strips, rate-compensated detectors, and heat-sensitive cable. **See Figure 7-10.** These heat detectors close a set of electrical contacts when the temperature reaches a certain setpoint, initiating an alarm.

A fusible-link heat detector includes a small plunger and spring arrangement that is held compressed by a small piece of material that melts at the setpoint temperature, such as 135°F. When the material melts during a fire, the plunger is released and the spring causes it to close the contacts. This type of heat detector is effective, but nonrestorable, meaning that once activated, it must be replaced.

Bimetallic-strip heat detectors also respond to setpoint temperatures. Strips of two different metals are bonded together, each with a different coefficient of thermal expansion. When heated, one metal expands more than the other metal, causing the strip to bend in one direction. This movement closes the set of contacts. The advantage of bimetallic-strip heat detectors is that they return to their position when cooled, reopening the contacts and restoring the function of the device.

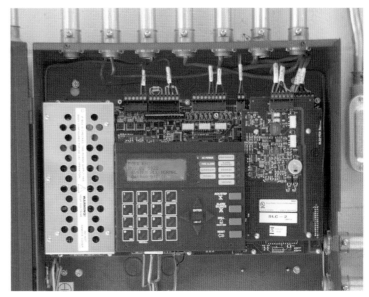

Initiating devices are connected in circuits and the wiring is run through conduit to the fire alarm control panel, where it is connected to specific terminals.

A radiant energy fire detector is an initiating device used in specialized applications. The device is activated by the specific wavelengths of light produced by the fuels expected to burn in the protected space. Most operate in the ultraviolet or infrared light ranges. Some detect light in both ranges, which helps prevent nuisance alarms.

Fixed-Temperature Heat Detectors

PLUNGER HELD DEPRESSED
CONTACTS
COMPRESSED SPRING
HEAT COLLECTOR
FUSIBLE LINK

FUSIBLE-LINK

CONTACTS
BIMETALLIC STRIP

BIMETALLIC-STRIP

CONTACTS
METAL ENCLOSURE

RATE-COMPENSATED

OUTER JACKET
CONDUCTORS
SECTION EXPOSED TO HEAT
HEAT-SENSITIVE POLYMER
CONDUCTORS
SHORT

HEAT-SENSITIVE CABLE

Figure 7-10. Fixed-temperature heat detectors include various methods of closing a set of contacts when the ambient temperature reaches a certain point.

A disadvantage of fusible-link and bimetallic-strip detectors is the significant time lag between when the ambient temperature reaches the setpoint and when the device absorbs enough heat to activate the sensor. Rate-compensated heat detectors are designed to activate much faster. These detectors encase a set of electrical contacts in a metal enclosure. The contacts are held apart in such a way that when the metal enclosure expands from heat, the contacts will close. The enclosure takes advantage of its large thermal exposure and mechanical design to quickly absorb heat and activate the contacts.

Heat-sensitive cable is also known as a line-type heat detector. This detector activates in a way similar to fusible-link heat detectors, except that it monitors a large, long area corresponding to the length of the cable. The cable includes two conductors separated by a heat-sensitive plastic. The two conductors act like a set of contacts. Under normal conditions, the plastic insulates them electrically and the circuit is open. Under high temperatures, however, the heat melts the plastic and the conductors contact and short the circuit, just like closing a set of contacts. A short at any part of the cable can cause an alarm. This type of detector is nonrestorable. However, if the shorted portions of the cable can be found, these sections can be patched with new cable.

Rate-of-rise heat detectors activate only if the temperature rises rapidly in a short period, such as a rise in temperature ranging between 12°F per minute and 15°F per minute. These devices react to changes in air pressure caused by extreme heat. A diaphragm inside the device flexes under sudden high pressure, closing contacts. **See Figure 7-11.** As a back-up measure, many rate-of-rise heat detectors also include a fusible-link fixed-temperature detector. When this link melts, it releases a spring that pushes a plunger up on the diaphragm to close the contacts. The fusible-link backup ensures that the device can detect a slowly burning fire that does not raise the temperature fast enough to activate the rate-of-rise portion of the heat detector.

Electronic heat detectors are the most versatile type of heat detector. Since they are controlled by software, they can be programmed to activate at a fixed temperature for a certain rate-of-rise. If not damaged by the fire, these

devices are also restorable. The sensing element in electronic heat detectors is typically a thermocouple or thermistor. When the resulting output voltage (thermocouple) or resistance (thermistor) reaches the setpoint, the on-board electronics initiate an alarm signal.

Rate-of-Rise Heat Detectors

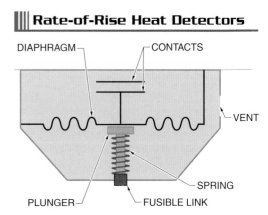

Figure 7-11. *Rate-of-rise heat detectors respond to rapidly increasing temperatures by detecting the corresponding increase in air pressure.*

Smoke Detectors. A *smoke detector* is an initiating device that is activated by the presence of smoke particles. In addition to gases, smoke is composed of solid particles and liquid droplets suspended in the air, often invisible to the unaided eye. Smoke detectors sense these smoke particles as the particles enter a small sensing chamber within the device. The primary difference among smoke detectors is the technology used to detect the particles. The two basic types are photoelectric and ionization smoke detectors.

Photoelectric smoke detectors are the most common type of smoke detector. This type senses light scattered by smoke particles within the sensing chamber. An infrared LED light source and sensing photocell are both aimed into the sensing chamber, but not at each other. **See Figure 7-12.** When no smoke is present, the light from the source does not reach the sensing photocell. However, smoke particles will scatter some light onto the photocell, which results in an alarm condition. The device resets once smoke concentrations fall below threshold levels.

Photoelectric Smoke Detectors

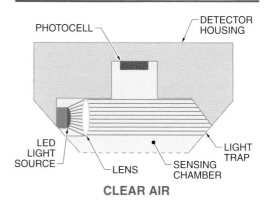

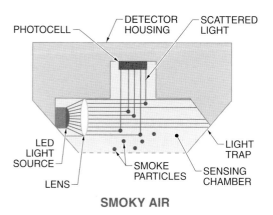

Figure 7-12. Photoelectric smoke detectors sense the presence of smoke particles by watching for the scattering of light they cause.

Ionization Smoke Detectors

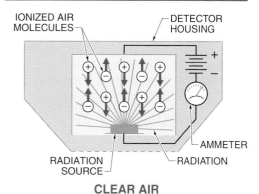

Figure 7-13. Ionization smoke detectors sense the presence of smoke particles by the change they induce in the current flow across a chamber of ionized air.

Ionization smoke detectors use a small radioactive source to ionize (electrically charge) the air in the sensing chamber. **See Figure 7-13.** This causes the air to conduct a small current across the chamber. However, smoke particles entering the sensing chamber will interfere with the normal current flow. The reduction in current is detected by the device, which initiates an alarm.

Manual Fire Alarm Pull Stations. A *manual fire alarm pull station* is an initiating device that is manually operated by a person to cause a fire alarm signal. This allows building occupants to initiate a fire alarm even if no other detectors have yet sensed the fire. These devices are placed at an accessible height at exits or in high-occupancy locations.

Smoke detectors are installed on or near ceilings, even in unfinished rooms, because smoke rises. This allows for early detection of fire hazards.

The actuation of a manual fire alarm pull station may be single-action or double-action. **See Figure 7-14.** Single-action stations require only one action on the part of the user, such as a pulling action. Double-action stations require two sequential actions, such as lifting and pulling or breaking glass and pulling. Inside, a manual fire alarm pull station is essentially a simple switch that closes a set of contacts, causing an alarm signal. Manual fire alarm pull stations may be either conventional or addressable devices. Resetting a manual fire alarm pull station may require a special key or replacement of broken glass.

Manual Fire Alarm Pull Stations

Potter Electric Signal Co.   *Fire-Lite Alarms*

SINGLE-ACTION   **DOUBLE-ACTION**

Figure 7-14. Building occupants can directly initiate a fire alarm signal by using a manual fire alarm pull station.

Water Flow Switches. A *water flow switch* is an initiating device that detects the flow of water in an automatic sprinkler fire suppression system. A waterflow switch is required for each sprinkler zone. This is a unique type of alarm-initiating device in that it does not directly sense the fire, as do smoke and heat detectors, but rather senses the automatic response of a fire suppression system to the fire.

The two basic types of water flow switches in fire protection systems are vane-type and pressure-type switches. **See Figure 7-15.** Vane-type water flow switches are the most common and are

used in sprinkler system pipes that already contain water. When an automatic sprinkler system is activated, the water in the pipe will suddenly flow, pushing the vane inside the pipe to one side, which closes a set of contacts in the switch housing. To avoid accidental actuation from normal surges in water system pressure, the switch response can be delayed up to a maximum of 90 sec. This delay is called "retard" and is adjusted by turning a screw or switch located inside the housing. After actuation, vane-type switches reset once water is no longer flowing.

Water Flow Switches

Potter Electric Signal Co.

Figure 7-15. The activation of a water-based fire suppression system can be detected with water flow switches installed in the sprinkler piping.

Vane-type water flow switches cannot be used in dry sprinkler system pipes because the sudden flow of water can damage the vane, possibly causing a blockage that interrupts water flow. Instead, these systems must use a pressure switch that senses a sudden drop in air pressure. Pressure switches can also be used in water-filled sprinkler pipes, where they sense the increase in water pressure upon sprinkler activation.

Notification Appliances

The FACP responds to alarms by activating notification appliances installed throughout the building. **See Figure 7-16.** There is a wide variety of available notification appliances. Audible notification appliances include bells, horns, mini-horns, chimes, speakers, and sirens. Visible notification appliances include strobes, LCDs, LEDs, printers, computer monitors, and electronic signs. The types of installed notification appliances are determined by applicable codes and the building system specifications. Codes also include requirements for appliance locations, spacing, and sound or light intensity.

When the FACP receives a fire alarm signal, its primary response is to alert all occupants of the building to the fire hazard. This involves energizing a large number of notification appliances on the notification appliance circuits. In many systems, the FACP will also automatically alert a supervising station or the local fire department via phone or data network.

FACPs also annunciate supervisory and trouble signals. Notification appliances for supervisory and trouble signals are typically located at the FACP or remote annunciators. The annunciator may distinguish between signal types by the indicator color or by an audible tone. For example, supervisory or trouble signals can be easily identified by LEDs of different colors.

Non-alarm notification appliances can also be wired as separate notification appliance circuits. This allows them to be activated separately for trouble or supervisory signals. Even though the building occupants are in no immediate danger, the system problem must be addressed promptly by building managers and system operators. Annunciation in other parts of the building beside the FACP allows personnel to respond quickly to the problem.

> The circuits of notification appliances often include devices that provide visual signals, as well as other devices that provide audible signals. Including both types of signals allows both hearing-impaired and sight-impaired occupants to be notified of a fire alarm.

Notification Appliances

Potter Electric Signal Co.
HORNS

BELLS

Potter Electric Signal Co.
STROBES

Potter Electric Signal Co.
STROBES AND SPEAKERS

Potter Electric Signal Co.
SPEAKERS

Figure 7-16. A wide variety of notification appliances is available to suit the particular requirements of local codes, building owners, and occupants.

The color of the liquid inside a water sprinkler's bulb corresponds to the temperature range at which the bulb will break. 135°F to 170°F is the most common activation temperature range and is represented by the color red. Green represents an activation temperature range of 175°F to 225°F.

Fire Suppression Initiation Devices

Many fire suppression systems are automatic, requiring no input from the fire alarm system. Being a separate system, fire suppression systems can operate even when a fire alarm system fails. Some types of fire protection systems, however, are directly activated by signals from the fire alarm system. This is particularly common for the release of fire suppression agents, such as gases, that are potentially harmful for building occupants. This situation requires greater control over the fire suppression system by the FACP in order to manage occupant evacuation prior to agent release.

In either case, the activation of the fire suppression system must be detected by the fire alarm system. Initiating devices connected to the fire suppression system cause alarm signals upon system activation.

Automatic Fire Suppression Devices. Automatic water sprinklers are the most common type of automatic fire suppression system. Water flows through the sprinklers when the sprinkler heads self-activate after being heated to their setpoint. A common type of sprinkler head includes a small frangible (easily breakable) glass bulb that plugs the water pipe opening above the water deflector. **See Figure 7-17.** The bulb is filled with a liquid that expands with high temperatures. When this temperature

is reached, the bulb breaks, unblocking the water pipe opening and allowing water to flow out of the sprinkler head. Heat activates these sprinkler heads individually. A small fire may activate only one or two sprinkler heads in the entire sprinkler system.

Automatic Sprinkler Heads

Figure 7-17. A common type of automatic sprinkler head self-activates by use of a heat-sensitive glass bulb that blocks the opening of the water piping.

Automatic water sprinklers are classified as either wet-pipe sprinklers or dry-pipe sprinklers. **See Figure 7-18.** Wet-pipe sprinklers always have water in the piping up to the sprinkler head. Therefore, wet-pipe sprinkler systems should only be used in environments not subject to freezing temperatures. Dry-pipe sprinklers, however, can be used in potentially freezing environments because that part of the system contains only compressed air. When activated, the dry sprinkler allows the air in the pipe to escape, which trips a water valve located in the upstream piping. All water-holding piping must be located in a heated environment. With the valve open, water then flows through the dry piping to the actuated sprinkler.

A preaction system is a type of dry-pipe sprinkler, but requires the preceding activation of an automatic detector, such as a smoke detector, before the preaction water valve will allow water to flow into the pipes. This preaction valve converts the dry-pipe system into a wet-pipe system. Only with the preceding fire detection will actuation of a sprinkler release water. This extra fire detection requirement is important to avoid accidental water release in areas that hold valuable collections, such as museums and libraries.

Automatic Sprinklers

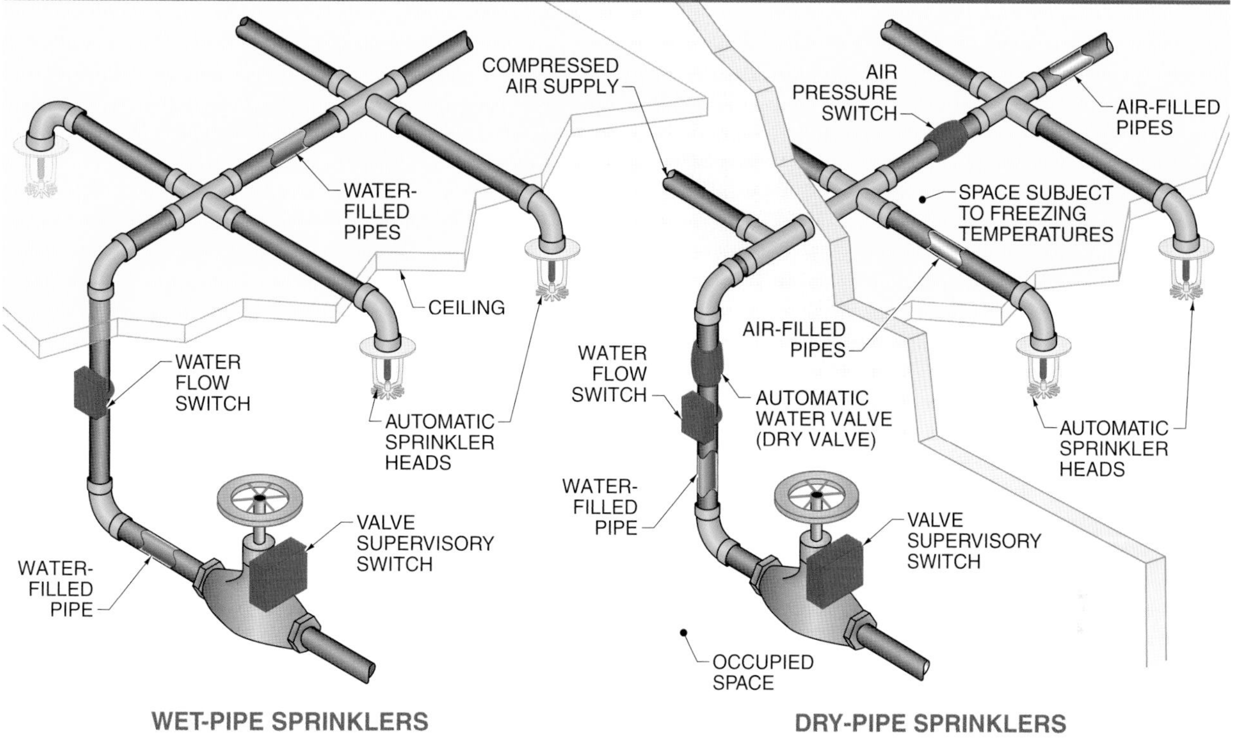

WET-PIPE SPRINKLERS   **DRY-PIPE SPRINKLERS**

Figure 7-18. Both wet-pipe and dry-pipe automatic water sprinklers include initiating devices to monitor for supervisory signals.

While automatic water sprinklers do not need to be activated by the fire alarm system, they are monitored for flow conditions by water flow switches. This causes a fire alarm signal, which ensures that the FACP is in alarm mode even if it had not yet received an alarm signal from another alarm initiating device.

Signal-Activated Fire Suppression Devices. Unlike automatic fire suppression systems that are activated by a mechanical action, signal-activated fire suppression systems require an electrical control signal to release a suppression agent. This is often the case for chemical and gaseous fire suppression systems, and some types of water-based fire suppression systems.

There are two possible sources of the activation signals. **See Figure 7-19.** First, the FACP can directly activate the fire suppression system, but only if it is designed and listed for releasing service. In this case, the alarm signal can come from any initiating device in the fire alarm system. However, many signal-activated systems are engineered packages that include their own releasing panels and initiating devices, such as smoke detectors and heat detectors. Alarm signals from the fire suppression system's initiating devices go to the separate releasing control panel, which then sends a signal to activate the fire suppression system. These control panels must be listed for releasing service and are connected to the FACP to share fire suppression system monitoring and activation information.

Activation signals are commonly only needed to open valves in the fire suppression system, since most fire suppression agents are stored under pressure and the systems are designed to activate with the simplest action. Opening a key valve in the system quickly releases the agent throughout the system. Some water-based fire suppression systems also utilize fire pumps to move large volumes of water into the system. These usually have their own separate controls that manage the pump startup and operation as soon as the activation signal is received.

Signal-Activated Fire Suppression

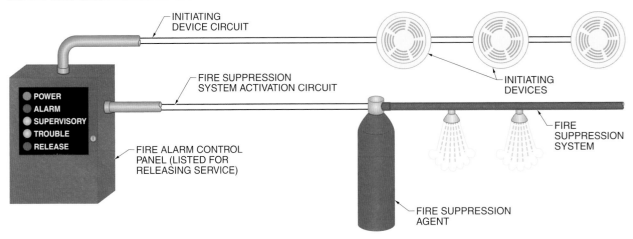

INTEGRATED FIRE SUPPRESSION SYSTEM ACTIVATION

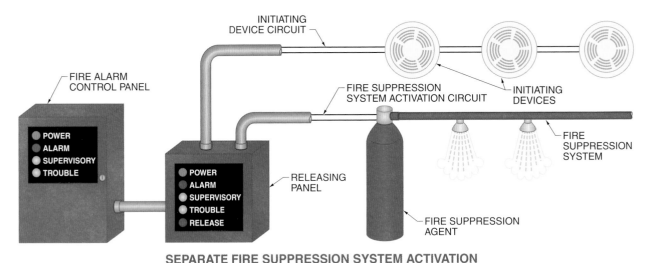

SEPARATE FIRE SUPPRESSION SYSTEM ACTIVATION

Figure 7-19. Signals for activating fire suppression systems can come directly from the FACP or from a separate control panel that interfaces with a separate set of initiating devices.

Supervisory-Initiating Devices

Closed supply valves and low system pressure are two examples of problems that can prevent adequate fire suppression system operation. Therefore, fire alarm systems are required to monitor all conditions essential to the proper operation of the fire suppression systems and to report impairments as supervisory signals. Supervisory-initiating signal devices are used to cause supervisory signals for fire suppression system faults. When using conventional initiating devices, those for supervisory roles are wired on separate

circuits than those for alarm roles because the two signals must not be confused with each other. Supervisory-initiating devices include valve supervisory (tamper) switches, air pressure switches, water level switches, and temperature switches.

Valve Supervisory (Tamper) Switches. A closed valve in suppression agent piping is a serious impairment to the fire suppression system. Therefore, valve positions are monitored by a valve supervisory switch to ensure that they remain open at all times. A *valve supervisory switch* is an initiating device that indicates when a valve is not fully open. Most valve switches use a microswitch and cam arrangement that closes a set of contacts, causing an off-normal signal within two turns of the handwheel, or upon one-fifth of the total handwheel rotation between fully open and fully closed. The supervisory device must not interfere with the operation of the valve.

A common type of valve supervisory switch mounts to an existing valve with an outside stem. A switch module clamps to the side of the valve and extends a trip rod toward the valve stem. **See Figure 7-20.** The trip rod sits in a groove in the stem when the valve is fully open. As the valve is closed, the groove moves toward the valve body and the rod slips out of the groove. This moves the rod sideways enough to trip the switch in the attached module.

A post indicator valve (PIV) is another type of valve supervisory device with a built-in indicator for valve position (OPEN or SHUT). PIVs control building water supply and are mounted so that the fire department can easily see the valve position upon arrival. Since it is usually located outside, a PIV may include a tamper switch and lock.

Buildings with sprinkler systems but no fire alarm system to monitor water flow must have a water flow-activated mechanical alarm, bell, horn, or siren to annunciate sprinkler activation. Water gongs are water operated and require no electricity. Sometimes these devices are even required for fire alarm systems.

Air Pressure Switches. In dry-pipe sprinkler systems, maintaining adequate air pressure prevents tripping of the dry valve under normal conditions. Pressure switches monitor the air pressure in dry pipe systems as insufficient air pressure may cause the system to trip. Pressure switches cause supervisory signals for air pressure deviations greater than ±10 psi.

Water Level Switches. Some water-based fire suppression systems include water reservoirs because of uncertain reliability of the municipal water supply. These reservoirs must be monitored with water level switches to ensure an adequate supply of water at all times. Most are float-type water level switches that close a set of contacts when the level falls to a certain level below nominal. This sends a supervisory signal to the FACP. Code requirements specify allowable deviations of 3″ or 12″ in the water level, depending on the tank type.

▥ Valve Supervisory Switches

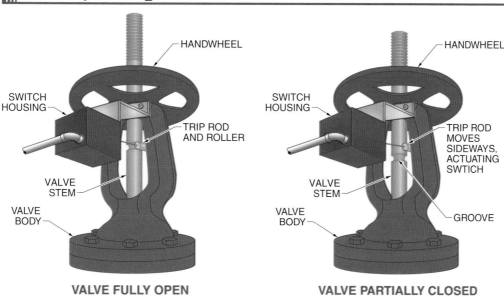

VALVE FULLY OPEN

VALVE PARTIALLY CLOSED

Figure 7-20. Valve supervisory switches ensure that valves controlling the supply of water to a fire-suppression system are fully open.

The installation of electric motor fire pumps ensures that a fire suppression water supply has adequate flow and pressure. These pumps can be used to pump water from sources such as retention ponds for fire suppression use.

Temperature Switches. Wet-pipe sprinkler systems could still be installed in areas that are occasionally subject to freezing temperatures, so these occurrences should be monitored by the fire protection system. Notification of falling temperatures allows operators time to address the problem before pipes and sprinklers are permanently damaged by ice forming and expanding in the pipes. Room and water temperature switches are used to supervise temperatures in these areas. Many temperature switches use a bimetallic strip that bends in one direction when cool and the opposite direction when heated, similar to restorable heat detectors. These switches close a set of contacts when the temperature drops to 40°F, causing a supervisory signal.

FIRE PROTECTION SYSTEM CONTROL APPLICATIONS

Most fire protection systems, both conventional and addressable, have the same main control functions and communicate primarily with their own devices. However, FACPs are sometimes used to interface with other building systems, particularly in automated buildings. This allows the FACP to take over control of other systems in an emergency. A *fire safety control function* is an integration of a fire protection system with other building systems and is intended to make the building safer for evacuating occupants and/or control the spread of fire hazards.

The FACP typically uses programmable outputs to communicate with other systems. The FACP may directly activate or deactivate devices, or instruct other controllers to initiate their own preprogrammed fire alarm mode. The FACP may be programmed with one or more fire safety control function sequences. During fire conditions, fire protection system signals have priority over all other building system communications.

With integrated building automation systems, there is practically no limit to the variety of scenarios possible, but there are several control functions that are very common. Systems typically involved in fire safety control functions include HVAC, elevator, and door systems.

HVAC System Shutdown

Smoke can often be just as dangerous as the fire. HVAC systems designed for efficient air circulation and ventilation can exacerbate the problem by quickly spreading toxic smoke throughout the building if they continue to operate normally during a fire. Smoke detectors are required by code to test the air inside the HVAC ductwork. Some are mounted within the duct itself and some are mounted in an exterior housing with sampling tubes to test air within the duct. **See Figure 7-21.** Either type quickly detects smoke in the system and initiates the smoke control response by the FACP.

Shutting down the HVAC system is the most common fire safety control response. This is a simple but effective way to help contain smoke and allow occupants time to escape the building through relatively smoke-free areas.

Operating Conditions. The FACP signals a control relay or module that causes the HVAC controller to de-energize. This stops all of the fans and prevents smoke from being recirculated throughout the area.

Control Sequence. The control sequence for shutting down the HVAC system during a fire is as follows:

1. The FACP receives an alarm signal from an alarm initiating device circuit.

2. In addition to fire alarm and fire suppression functions, the FACP initiates the HVAC shutdown sequence.
3. The FACP outputs a digital signal or structured network message to a special relay to completely de-energize the HVAC system.

Duct Smoke Detectors

Figure 7-21. Duct smoke detectors can be mounted outside the ductwork, but sampling tubes or long sensors should be used to continuously test the air inside the duct for smoke particles.

Elevator Recall and Shunt Trip

Elevators can become dangerous during a fire. Both the fire and the water from the fire suppression system can damage the elevator motors, doors, cables, controls, and emergency brakes. Elevator occupants can become trapped in an inoperable and unsafe elevator. Wet brakes or wet electrical equipment can cause erratic elevator operation or dangerous conditions.

Per ANSI Standard A17.1, *Safety Code for Elevators and Escalators,* elevators are required to be immediately recalled to a designated level in the event of a fire alarm. This safety feature prevents elevators from being called to floors with fire hazards, which could trap elevator occupants. Also, the main power line to the elevator must be disconnected upon, or prior to, the discharge of water from a sprinkler system in the elevator shaft or machine room. This shutdown is known as shunt trip and occurs regardless of the elevator's location or occupied status.

Active Smoke Control

Instead of a complete shutdown, a more sophisticated program can make use of the HVAC system to actively control the spread of smoke or even draw smoke out of the building. Smoke control activates fans and dampers to create positive and negative pressures in certain building areas. Negative pressure is created in fire areas to exhaust the smoke, while nonfire areas are pressurized to minimize smoke infiltration. A common application is to pressurize stairways, which helps occupants to quickly exit the building.

Since the location of the fire may affect the best response for fire control, multiple sequences may be programmed into the FACP to address different scenarios. The FACP program then utilizes the appropriate sequence in accordance with information from initiating devices about the fire hazards' locations. This type of sophisticated control may require the use of signaling-line circuits and addressable initiating devices, which allows the fire hazard location to be identified.

However, smoke control with HVAC systems is a highly regulated function since control or device problems in this sequence can worsen the fire hazards by spreading smoke through the building. Therefore, it requires engineered coordination between the fire alarm system and the HVAC system.

Operating Conditions. Shunt trip can be initiated by quick-acting heat detectors located near the elevator sprinklers or water flow switches with no time delay. Like all other fire alarm-initiating devices, they must be connected to the fire alarm control panel for monitoring the integrity of the wiring circuits. Some systems, upon detection of heat or smoke, hold the shutdown and suppression system activation until the elevator car stops at a landing.

Control Sequence. The control sequence for elevator recall and shunt trip is as follows:

1. The FACP receives an alarm signal from a fire-detecting initiating device, such as a smoke detector.

2. In addition to fire alarm functions, the FACP initiates the elevator recall sequence.

3. The FACP outputs a digital signal to a special input connection on the elevator controller to put the elevator system into recall mode.

4. The FACP receives an alarm signal from a fire suppression-initiating device, such as a water flow switch.

5. The FACP outputs a digital signal to a special relay to completely de-energize the elevator system.

Rated fire doors are tested to withstand a fire for a specified period. There are 20-, 30-, 45-, 60- and 90-minute rated fire doors. These doors must be certified by an approved testing laboratory.

Fire Doors

Doors can be both a help and a hindrance during fire emergencies. When closed, doors help control the spread of fire and smoke by containing building areas. However, doors must also allow people to move freely during an evacuation. Some doors include their own fire-related control devices and are not integrated with the fire alarm system. However, many are automatically closed or unlocked by the building fire alarm system.

Fire doors within building hallways are held open by electromagnets. **See Figure 7-22.** The magnetic force holds the door open against the opposing force of a spring and damper that tries to slowly close the door. When the magnetic holders are de-energized, the doors are allowed to close.

Doors that are normally locked must be unlocked to allow occupants to leave the building and responders to enter. Electrically held locks are unlocked automatically by the fire alarm system. Security concerns do not change this requirement, although some building codes permit manual unlocking of doors.

Fire alarm systems can also be used to activate other evacuation aids, such as pre-recorded instructions or flashing lights.

Refer to Quick Quiz® on CD-ROM

Operating Conditions. Electromagnetic door holders and electrically held locks are designed to default to a safe condition (from a fire hazard point of view) is case of power loss. For example, the electromagnetic door holders require a constant electrical current to hold the door. Interruption of the current causes the door to close. This also simplifies integration with the fire alarm system because relays can be used to easily control the electrical power to the doors with a programmable digital output from the FACP.

Electromagnetic Door Holders

Figure 7-22. *Magnetic door holders de-energize in the event of a fire alarm, releasing the doors and preventing the migration of toxic smoke into nonfire areas.*

Control Sequence. The control sequence for closing and unlocking fire doors is as follows:

1. The FACP receives an alarm signal from an alarm-initiating device circuit.

2. In addition to fire alarm and fire suppression functions, the FACP initiates the fire door safety control sequence.

3. The FACP outputs a digital signal or structured network message to the relay(s) controlling the electromagnetic door holders, de-energizing the holders.

4. The FACP outputs a digital signal to the relay(s) controlling the electrical door locks, which unlocks the normally locked doors.

Summary

- Fire protection systems consist of fire alarm systems and fire suppression systems.

- Fire alarm systems consist of initiating devices (inputs), a fire alarm control panel (FACP), and notification appliances (outputs).

- Many fire suppression systems activate independently of the fire alarm system.

- Some fire suppression systems are directly controlled by the fire alarm system.

- The fire alarm system always monitors the status of the fire suppression system for activation, power loss, and ground faults.

- Supervisory signals and trouble signals both indicate problems that can prevent the fire protection system from functioning as intended in the event of a fire.

- The fire alarm control panel (FACP) monitors the circuits for signals and faults, monitors its power supply, and controls fire safety functions.

- Conventional initiating devices use only contact closure to initiate alarms.

- A signaling line circuit is a variation of an initiating-device circuit that uses addressable device technology.

- An addressable initiating device can send and receive information via encoded digital signals and can identify itself with a unique address.

- Addressable FACP and initiating-device fire alarm systems can indicate the type of detection and exact location of the device.

- Initiating devices produce either alarm or supervisory signals, depending on the type of device and how it is programmed or wired.

- Smoke detectors sense tiny smoke particles as the particles enter a small sensing chamber within the device.

- A water flow switch does not directly sense the fire, as do smoke and heat detectors, but instead senses the automatic response of a fire suppression system to the fire.

- Audible notification appliances include bells, horns, mini-horns, chimes, speakers, and sirens. Visible notification appliances include strobes, rotating beacons, LCDs, LEDs, printers, computer monitors, and electronic signs.

- System mode, status, and signal information is displayed on one or more annunciators in the fire alarm system.

- Automatic water sprinklers are the most common type of automatic fire suppression system.

- Signal-activated fire suppression systems require a control signal from a separate unit to release a suppression agent.

- Many signal-activated systems are engineered packages that include their own control panels and initiating devices, such as smoke detectors and heat detectors.

- Activation signals are commonly only needed to open valves in the fire suppression system.

- Fire alarm systems monitor fire suppression systems and report impairments as supervisory signals.

- When using conventional initiating devices, those for supervisory roles are wired on circuits than different from those for alarm roles because the two signals must not be confused with each other.

- An optional feature of FACPs involves interfacing with other building systems for fire safety functions.

- Systems typically involved in fire safety functions include HVAC, elevator, and door systems.

Definitions

- A *fire alarm system* is a system that detects hazardous conditions associated with fires (such as smoke or heat) and notifies building occupants.

- A *fire suppression system* is a system that releases fire suppression agents to control or extinguish a fire.

- A *fire suppression agent* is a substance that can extinguish a fire.

- An *alarm* is a signal that indicates a fire or other hazardous conditions.

- A *supervisory signal* is a signal that indicates an abnormal condition in a fire suppression system.

- A *trouble signal* is a signal that indicates a wiring or power problem that could disable all or part of the system.

- A *fire alarm control panel (FACP)* is an electrical panel that is connected to fire protection system circuits and interfaces with other building systems.

- An *initiating device* is a fire protection system component that signals a change-of-state condition.

- An *initiating-device circuit (IDC)* is an electrical circuit consisting of fire alarm initiating devices, any of which can activate an alarm signal by closing its contacts.

- An *end-of-line (EOL) resistor* is a resistor installed at the far end of an initiating-device circuit for the purpose of monitoring circuit integrity.

- A *signaling line circuit (SLC)* is an electrical circuit of fire alarm initiating devices that communicate using digital signals.

- An *addressable device* is a device with a unique identifying number that can exchange messages with other addressable devices.

- A *notification appliance circuit (NAC)* is an electrical circuit of fire alarm notification appliances that are activated by the fire alarm control panel (FACP) under certain signal conditions.

- A *notification appliance* is a device with a unique identifying number that can send and receive messages with other addressable devices.

- An *annunciator* is a fire alarm notification appliance that displays fire protection system modes, status, and alarms.

- A *heat detector* is an initiating device that is activated by the high temperatures of a fire.

- A *smoke detector* is an initiating device that is activated by the presence of smoke particles.

- A *manual fire alarm pull station* is an initiating device that is manually operated by a person to cause a fire alarm signal.

- A *water flow switch* is an initiating device that detects the flow of water in an automatic sprinkler fire suppression system.

- A *valve supervisory switch* is an initiating device that indicates when a valve is not fully open.

- A *fire safety control function* is an integration of a fire protection system with other building systems that is intended to make the building safer for evacuating occupants and/or control the spread of fire hazards.

Review Questions

1. What are ways in which fire alarm systems and fire protection systems are typically integrated together?

2. Explain the difference between a supervisory signal and a trouble signal.

3. What is the primary advantage of using addressable initiating devices?

4. What is the difference between fixed-temperature heat detectors and rate-of-rise heat detectors?

5. How does a fire alarm system detect the activation of an automatic fire suppression system (such as water sprinklers)?

6. How are graphic annunciators particularly useful for emergency responders?

7. Why should conventional alarm-initiating devices and conventional supervisory-initiating devices not be wired together on the same circuit?

8. Why must fire suppression systems be monitored for supervisory signals?

9. How can the HVAC system in a building be used to facilitate the evacuation of building occupants during a fire?

10. Why are the control sequences for signal monitoring different between conventional-type systems and addressable-type systems?

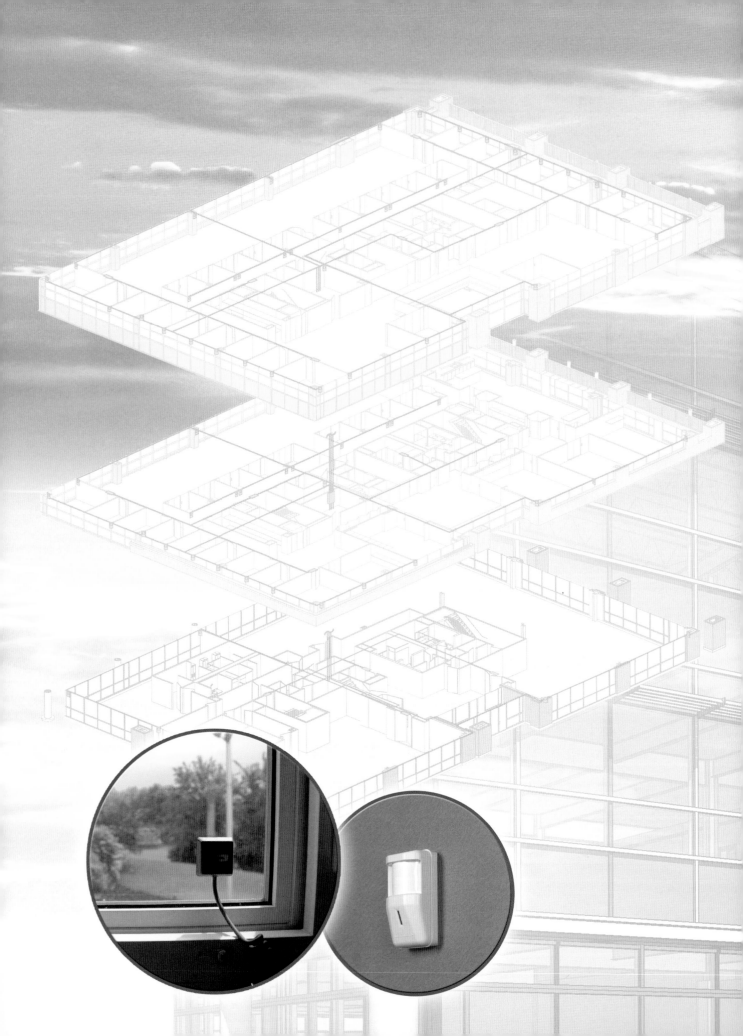

Chapter Eight

Security System
Control Devices and Applications

The security system industry has grown significantly in the past 20 years, mostly due to innovations in electronic technology and decreasing equipment costs. This growth has put security systems in many more industrial, commercial, and residential buildings, undoubtedly contributing to a falling burglary rate in the United States. Security systems provide automated, around-the-clock security for a building and its contents, even without on-site security personnel. Security systems also offer several opportunities for integration with other automated building systems, particularly fire alarm and access control systems, thus creating more sophisticated features.

Chapter Objectives

- Compare the different types of security systems with respect to protection type and supervision.
- Contrast the behaviors and arming options of different types of security zones and modes.
- Describe the features of control panel operation that help reduce false alarms.
- Identify the common types of sensing, notification, and interface devices used with security systems.
- Describe how security systems can integrate with other building systems for additional functionality.

SECURITY SYSTEMS

A *security system* is a building system that protects against intruders, theft, and vandalism. The configuration and operation of security systems are somewhat similar to those of fire alarm systems. The system is managed by a central control panel that interfaces with all other security devices and, as necessary, other building systems.

The control panel receives input information from sensing devices that detect intrusion into the secured area. **See Figure 8-1.** Depending on the on-board programming and system capabilities, the control panel can then take certain actions to alert the authorities and/or the building occupants to the security breach. The purpose of these actions may be to activate certain on-site output devices or to communicate directly with off-site security personnel.

Besides intruder detection for the notification of authorities, one of the main functions of a security system may be to deter intruders. Some security devices are installed to be purposely noticeable, discouraging some would-be intruders from attempting to break in. Also, making an intruder's detection immediately known, such as with sirens or flashing lights, may drive an intruder away before any damage is done.

The individual devices and actions that make up a security system are each relatively simple, but can be arranged and programmed in ways to provide feature-rich security protection options. These variations determine the area or objects being protected, the method of alarm notification, the defined functions of the devices in certain areas, and the active status of the device for different alarm modes.

Security System Protection

Security system protection includes three different types: point-of-entry protection, specific-area protection, and spot protection. Security systems can include any combination of these types of protection, depending on the application, size, and type of the system.

Point-of-Entry Protection. *Point-of-entry protection* is a type of security system protection that monitors the specific points that might allow entry or exit from a secure area. Point-of-entry sensors provide a signal whenever someone attempts to enter through a point-of-entry. A *point-of-entry* is a potential opening in the envelope or perimeter of a building or area through which a person may enter. **See Figure 8-2.** This includes doors and windows, but may also include other access points such as ventilation shafts, service-access openings, and fence gates.

Security Systems

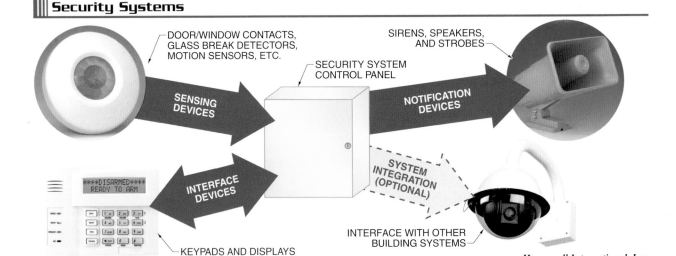

DOOR/WINDOW CONTACTS, GLASS BREAK DETECTORS, MOTION SENSORS, ETC.

SIRENS, SPEAKERS, AND STROBES

SECURITY SYSTEM CONTROL PANEL

SENSING DEVICES

NOTIFICATION DEVICES

INTERFACE DEVICES

SYSTEM INTEGRATION (OPTIONAL)

****DISARMED**** READY TO ARM

KEYPADS AND DISPLAYS

INTERFACE WITH OTHER BUILDING SYSTEMS

Honeywell International, Inc.

Figure 8-1. Security systems are managed by a security system control panel, which receives input signals from sensing devices and sends output signals to notification devices and/or other building system devices.

Point-of-Entry Protection

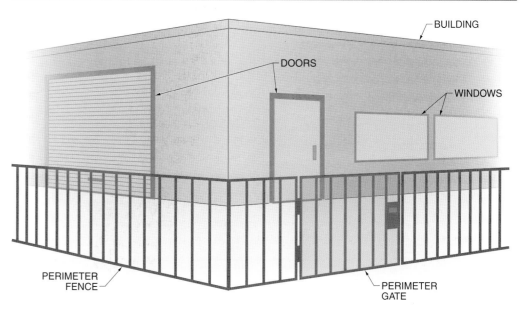

Figure 8-2. Point-of-entry protection secures a facility by monitoring all the possible places that an intruder might penetrate.

Specific-Area Protection. *Specific-area protection* is a type of security system protection that monitors only one defined location. Specific-area protection is designed to protect areas that might require security protection when nearby areas are in use. If the area is enclosed, such as a room, then point-of-entry protection is used to secure the area. However, the area to be protected may not be surrounded by walls or barriers, or those barriers may not provide an adequate perimeter so that points-of-entry are defined. In this case, specific-area protection is implemented to define and secure this area. **See Figure 8-3.**

Spot Protection. *Spot protection* is a type of security system protection that monitors one particular object. **See Figure 8-4.** A variety of objects may require spot protection, such as safes, cases, doors, or drawers. Spot protection may involve alarming for the object being approached, touched, moved, or tampered with. This means spot protection systems can be implemented in a large number of ways and with many different types of sensing devices.

Specific-Area Protection

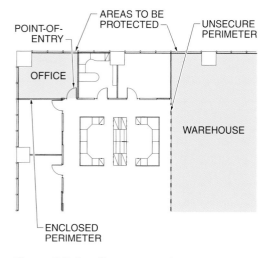

Figure 8-3. Specific-area protection secures an area even if it does not have defined points-of-entry.

Security System Supervision

Security systems may include additional features, but all involve monitoring and notification for an abnormal condition. The method of notification is used by security system professionals to classify different system types.

A security system can be classified as a local system, supervisory system, or central station system. The primary difference between these systems is how the information or notification of an abnormal condition is processed.

▓ Spot Protection

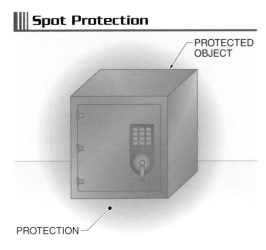

Figure 8-4. Specific objects can be secured with spot protection, which uses sensing devices and security modes focused on a small area.

Local Security System. A *local security system* is a self-contained security system that does not require full-time surveillance or off-premises wiring. **See Figure 8-5.** Many security systems installed in residential and small commercial buildings fall under this category. They are common when cost must be minimized or where other types of systems are not practical. Most local alarm systems are designed specifically for intrusion detection, though some provide options to include fire protection functions.

Local security systems are self-contained systems. These systems do not require constant human supervision and are enclosed within one specific building or area. A local alarm does not transmit alarms to responding authorities. It only triggers an audible signal, such as a siren, at the premises. Even though they are relatively inexpensive, the controls allow a variety of user features, such as battery back-up systems and custom programming keypads. These two features make these units easy to maintain and operate.

▓ Local Security Systems

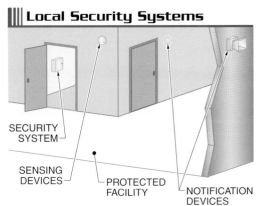

Figure 8-5. Local security systems are contained entirely within the building they protect.

Supervisory Security System. A *supervisory security system* is a security system that is monitored continuously by on-site security personnel. **See Figure 8-6.** The security personnel can help investigate minor alarm problems and cancel known false alarms before they are transmitted to the responding authorities. The supervisory station may also provide monitoring for fire protection systems and process systems.

▓ Supervisory Security Systems

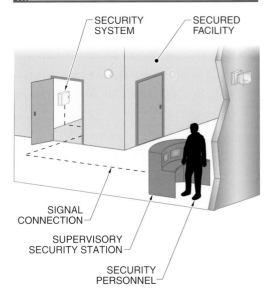

Figure 8-6. A supervisory security system requires that security personnel be on site continuously to monitor alarms and trouble conditions.

Some supervisory monitoring stations provide access to the status of each alarm point, often with a graphic display to easily locate any problems. Other supervisory stations simply provide a summary of signals from one or more zones, each of which has its own protection area.

Any problems that might prevent the alarm system from operating properly are also indicated at the supervisory station. Since supervisory security systems are typically more complex than local security systems, they are often computer based or computer controlled. These systems are not self-contained and often require additional peripheral equipment.

In many cases, a supervisory security system is integrated with an access control system, which controls the entry of certain individuals into secured areas. The supervisory security system can monitor both security devices and access control devices, such as keypads or magnetic strip cards.

The disadvantage of supervisory security systems is that they require 24 hr monitoring. These systems are generally limited to high-security buildings and continuously occupied facilities, such as hospitals, universities, airports, and industrial plants.

Central Station Security System. A *central station security system* is a security system that is monitored by an off-site station. **See Figure 8-7.** If an abnormal condition occurs, the system transmits a signal, typically over a phone line, to the central station. The monitoring station responds by sending the appropriate security or other personnel to the alarm location to investigate the problem. In some systems, remote audio or video surveillance can be used to help rule out false alarms before personnel are dispatched.

The off-site monitoring station is often a commercial security company. Commercial security companies offer this service to industrial, commercial, and residential customers for a fee. A contract between the security company and building owner establishes the services, fees, and terms of service. The advantage of central station security systems is that 24 hr protection is available to businesses or facilities that cannot employ their own 24 hr security personnel.

Central Station Security Systems

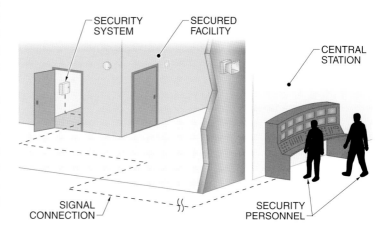

Figure 8-7. *Central station security systems rely on off-site security personnel to monitor a facility security system remotely.*

Security Zones

A *zone* is a defined area protected by a group of security system sensing devices. A building typically includes at least several zones defined by location, size, and security requirements. For example, the main entry of a building may be assigned to Zone 1 of the system, while a rear entry of the building is assigned to Zone 2.

Zones can also be programmed to behave in certain ways under certain conditions. **See Figure 8-8.** A *zone type* is a definition of the type of alarm a zone initiates and the corresponding reaction of the control panel. It may be important to have different zone types in a system, depending on the type of protection. For example, in areas prone to triggering sensing devices unintentionally, the appropriate zone type may be one that delays the initiation of an alarm. A motion detector near an entry door can be programmed as an interior zone with a delay follower. This means that the motion detector has the same trigger delay as the entry door if the entry door is opened first.

There are many common zone types, including delay zones, instant zones, interior zones, stay/away zones, 24-hour zones, and null zones. The naming terminology for zone types may vary between manufacturers, but the definitions are primarily the same.

Security Zone Types

Zone Type	Alarm Type	Active	Delay
Delay zone	Security	When armed	Programmable
Instant zone	Security	When armed	None
Interior zone	Security	When armed	Programmable
Stay/away zone	Security	Only in armed (away) mode	None
24-hour zone	Security, trouble, or fire	Always	Programmable
Null zone	None	Never	None

Figure 8-8. Security zones can be defined as a certain type, which determine many of their characteristics.

Delay Zones. A *delay zone* is the zone type that allows a certain period of time to elapse before a triggered sensor initiates an alarm. Delayed zones are used primarily for doors, allowing a person exiting or entering the building time to arm or disarm the system before the control panel triggers an alarm. The delay period can typically be configured during initial system programming.

A building may define more than one delay-zone type if there are exit/entry points that may require different delay periods. For example, primary entry points such as a front door or employee entrance are commonly assigned as a delay 1 zone type with a short time delay. A delay 2 zone with a longer programmed delay is commonly used for secondary entry points that may require more time, such as a door that is located farther away from a security keypad.

Honeywell International, Inc.
Central station security systems provide off-site security personnel to monitor and respond to system alarms and trouble signals.

Instant Zones. An *instant zone* is a zone type that indicates an immediate alarm condition when a sensing device in the zone is activated. An instant zone has no time delay before the control panel signals an alarm, but the zone can be armed and disarmed as needed. This zone type is commonly used for possible points-of-entry that are not normally used, such as service or emergency exits.

Interior Zones. An *interior zone* is a zone type used to differentiate interior sensing devices. Interior zones are areas that must be protected while the security system is armed. This zone type may be used for other manufacturer-specific functionalities.

Stay/Away Zones. A *stay/away zone* is a zone type that is programmed to ignore signals from its sensing devices when the system is armed, but when there is no exit from the premises. This zone type is commonly used in residential applications when the occupants arm the security system at night but must be able to walk past an interior motion sensor without causing an alarm.

24-Hour Zones. A *24-hour zone* is a zone type that is always active (never disarmed). Most 24-hour zones initiate security alarms and are used for initiating devices that must always be available, such as a panic button for triggering a silent alarm.

Variations of 24-hour zones can also be defined by the type of alarm signal they initiate. A *24-hour supervisory zone* is a zone type that is always active and initiates a trouble signal. These signals are triggered by sensing devices that detect wiring and equipment problems in the security system. A *24-hour fire zone* is a zone type that is always active and initiates a distinct fire alarm tone at the speakers or sirens. This zone type is used exclusively for fire protection devices that are connected to the security system.

Null Zones. A *null zone* is an unused zone. This type of zone is used to make unused zones inactive. This may be set as a temporary condition for zones during installation, maintenance, or reconfiguration, or for building areas that have become inactive.

Security System Modes

Unlike fire alarm systems, which must always be in an active and ready condition, security systems can be placed into a variety of states of readiness. These modes correspond to different building occupation scenarios or different levels of required security. The various modes are common between different manufacturers, though the terminology may vary slightly.

Only certain users should be allowed to change between different modes, so this action requires authentication. *Authentication* is the process of identifying a person and verifying their credentials. A variety of methods can be used for authentication, but the most common is the use of alphanumeric codes known only to authorized individuals. The code is entered into the system at an interface device, and then the desired system change is accepted. Some codes, for lower security levels, allow a user to change between certain security modes. Other codes, for higher security levels, allow a user to access other functions, such as partitioning, changing user codes, changing times and dates, and changing a security level.

Disarmed Mode. A *disarmed mode* is a security system mode in which the control panel ignores all inputs from sensing devices. When the security system is in the disarmed mode, it is on standby. This mode is commonly used when a building is occupied during the day. When a door opens and closes, the change of state in the door-sensing device will still be detected by the control panel and possibly even displayed with system status, but the control panel will not initiate an alarm. The exception to this is that any device in a 24-hour zone will trigger an alarm.

Honeywell International, Inc.

Arming a security system requires the user to enter an authorized access code and armed mode, which is typically done at a keypad.

Armed (Away) Mode. An *armed (away) mode* is a security system mode that makes all security-sensing devices active. This is the most common mode for activating the security system when there will be no occupants within the protected building or area. In this mode, any door, window, motion sensor, or other sensing device will trigger an alarm signal.

Armed (Stay) Mode. An *armed (stay) mode* is a security system mode that makes all perimeter-sensing devices active while signals from the sensing devices (that were defined as being stay/away zone types) are ignored. This allows the building or area to be protected from outside intruders while people can occupy the building without triggering false alarms. For example, door and window protection is activated while interior motion sensors are inactive.

Partitions

A *partition* is a portion of a security system that operates under a different program than the rest of the system. Partitioning allows the system to be subdivided into two or more separate systems using the same control panel and control devices. **See Figure 8-9.** This is used when there is a need to have one part of a building armed while another part of the building remains disarmed.

▥ Partitions

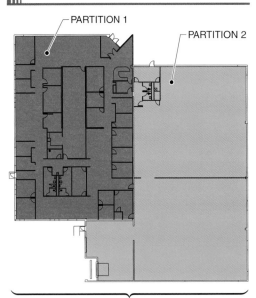

Figure 8-9. Buildings can be partitioned into areas according to different security needs when using a simple security system.

PARTITION 1

PARTITION 2

ENTIRE BUILDING PROTECTED
BY ONE SECURITY SYSTEM

Honeywell International, Inc.

Security system control panels are typically plain enclosures housing the system electronics and circuit connections for the entire security system.

The security system is powered via the control panel. Most panels operate on 16 VAC to 18 VAC (from a Class 2 plug-in transformer) and include a 12 VDC back-up battery. The back-up battery must be sized to power all system devices in the event of an outage.

For example, in a mixed-use building with both office and warehouse space, the office staff leaves at 5 PM and the office portion of the building requires security protection. The office partition is armed while the warehouse partition remains disarmed because the warehouse staff continues to work after 5 PM. When the warehouse staff leaves, the warehouse partition is armed. Partitions are armed and disarmed individually with partition codes. However, a global code can be used to arm or disarm both partitions simultaneously.

SECURITY SYSTEM SIGNALS

Security systems monitor the circuits of sensing devices and wait for the electrical signals that indicate abnormal conditions. Similar to fire alarm systems, security systems indicate alarm and trouble signals. These signals have essentially the same meaning as the related fire alarm signals, except that they pertain to building security and system operation.

Alarms

An *alarm* is a security system signal that indicates an intrusion or other alarm condition. An alarm signal is initiated by various sensing devices that detect intrusion into the secured space. Upon receiving an alarm, the control panel system activates notification devices. If the system includes a supervisory station, either on site or off site, the control panel sends the alarm signal to the security personnel.

Trouble Signals

A *trouble signal* is a signal that indicates a wiring or power problem that could disable part or all of the system. These signals are primarily initiated by the control panel itself, which monitors device circuits for electrical changes that indicate wiring or equipment faults. For example, remodeling work in an office space may cause security system circuits to inadvertently be cut. This open-circuit condition is identified by the control panel as a trouble condition. Many security systems include a built-in diagnostic system that also monitors AC power, battery charge, telephone connections, and wireless devices for trouble conditions.

Trouble conditions do not require an alarm response by municipal authorities but do require a prompt response by building personnel to correct the problem quickly. Trouble signals are typically annunciated at security stations (on site or off site) and at system keypads. More information about the trouble condition can typically be accessed at keypads by entering a special diagnostic code.

SECURITY SYSTEM CONTROL PANELS

The control panel is commonly referred to as the "brain" of the security system because it makes all decisions for the security system. All field devices for the security system—input and output devices—are connected to the control panel. **See Figure 8-10.**

The control panel is programmed with instructions for various scenarios, and it makes decisions based on the status of the system and field devices. Control panels are either preprogrammed by the manufacturer or custom programmed by the system installer. Control panels are programmed at a keypad or with software running on a computer connected to the control panel. The computer can be connected locally through a USB or serial port, or from any other location through a network connection.

Security System Circuits

The control panel relies on numerous field devices throughout the building or protected area to detect alarm conditions. Large buildings are typically divided into zones, and devices within the same zone are grouped together into circuits.

Security system devices can be grouped into either open-loop or closed-loop circuits. **See Figure 8-11.** An *open-loop circuit* is a security system circuit with devices that are all normally open. The devices are connected in a parallel circuit configuration, and an alarm is triggered when any device closes. All fire protection-initiating device circuits

are open-loop circuits. This type of circuit is commonly used for 24 hr devices such as smoke detectors.

However, circuits for security systems are primarily closed-loop circuits. A *closed-loop circuit* is a security system circuit with devices that are all normally closed. The devices are connected in series and an alarm is triggered when any device opens. This type of circuit is used for door/window contacts, motion sensors, and glass break detectors. Variations of closed-loop circuits can be used to detect wiring faults.

Control Panel Connections

Figure 8-10. All sensing device, notification device, and system integration connections are made at the control panel.

Sensing-Device Circuits

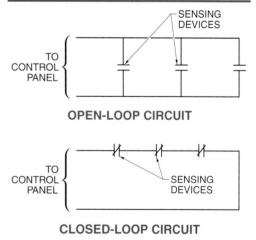

Figure 8-11. Open-loop circuits are common for 24-hour zones, such as fire alarm circuits, while closed-loop circuits are typically used for security zones.

Wireless

Due to the many field devices, security systems are good candidates for wireless infrastructure, particularly for retrofit or upgrade applications. Wireless systems communicate between devices with radio frequency (RF) signals, which requires little physical wiring. This makes them faster to install and highly flexible with sensor placement. The wireless devices are usually powered with a battery that must be replaced periodically. Most common security system sensing devices are available in wireless versions. A disadvantage associated with wireless systems is that some building materials, such as concrete or metal, affect the range of the transmission. In this case, signal repeaters may be necessary.

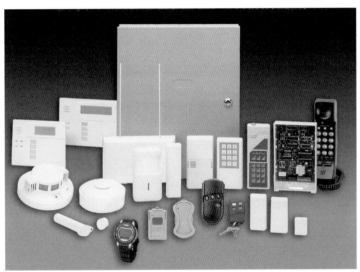

Honeywell International, Inc.

Security systems can include a large variety of devices that interface with the security system control panel.

Nonsupervised Circuits. Security system circuits without monitoring devices are nonsupervised circuits. A *nonsupervised circuit* is a security system circuit of field devices that cannot monitor the integrity of the wiring. A short-circuit wiring fault, for example, cannot be detected with a nonsupervised circuit because it will be indistinguishable from normal conditions. **See Figure 8-12.** Also, an open-circuit wiring fault is detected as an alarm, since it is indistinguishable from a device activation. Therefore, a non-supervised circuit cannot be an open-loop circuit.

Control panels with nonsupervised circuits may require special programming to eliminate circuit supervision software features because they will not work with this type of circuit. Therefore, these systems cannot take advantage of the supervision capabilities on any other circuit because nonsupervised circuits cannot be mixed with supervised circuits in the same system.

Supervised Circuits. The configuration of closed-loop circuits can be used to supervise (monitor) the wiring for trouble faults when resistors are added to the circuit. A *supervised circuit* is a security system circuit of field devices that can monitor the integrity of the wiring. This type of circuit requires end-of-line resistors. An *end-of-line (EOL) resistor* is a resistor installed at the far end of an initiating-device circuit for the purpose of monitoring the circuit. A security system monitors the integrity of each circuit by measuring its resistance. There are two common types of supervised circuits for closed-loop security system circuits: a single end-of-line resistor configuration and a double end-of-line resistor configuration.

The single end-of-line resistor configuration is the most common configuration for security systems. It includes an end-of-line resistor in series with the sensing devices. This configuration monitors the circuit for wiring faults but cannot distinguish between an open due to wiring trouble and an open due to sensing-device activation. **See Figure 8-13.** For example, under normal conditions, the control panel measures the resistance of the circuit as approximately equal to the known resistance of the end-of-line resistor. If a sensing device is activated, its contacts open, the circuit resistance increases to infinity, and the control panel indicates an open condition. However, if a wiring problem causes an open circuit, the control panel indicates the same condition. Both conditions result in an alarm activation. Short-circuit conditions, however, can be detected because the circuit resistance drops to zero.

Nonsupervised Circuits

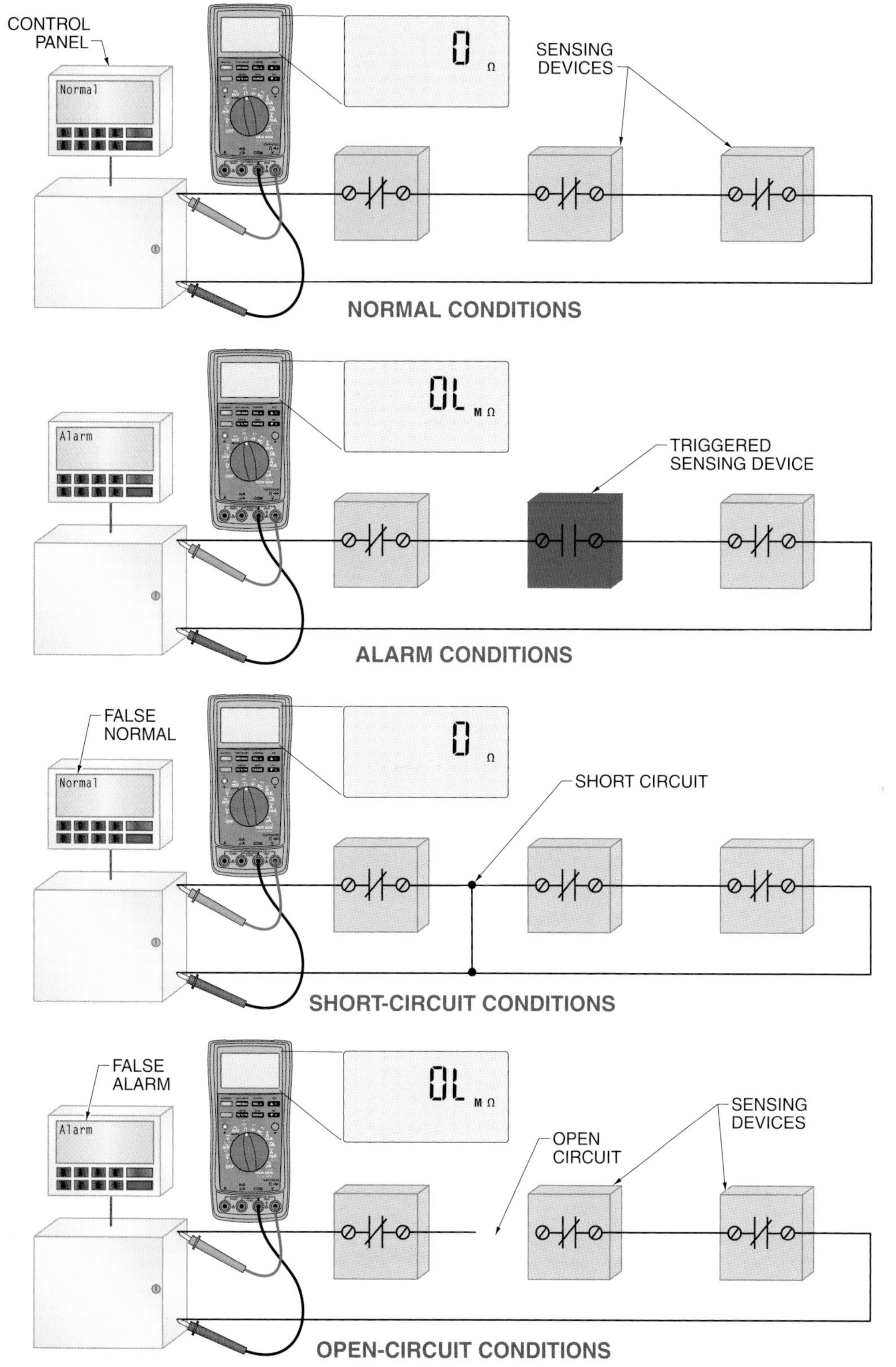

Figure 8-12. Nonsupervised circuits are effective at detecting alarms but cannot detect trouble conditions due to wiring faults.

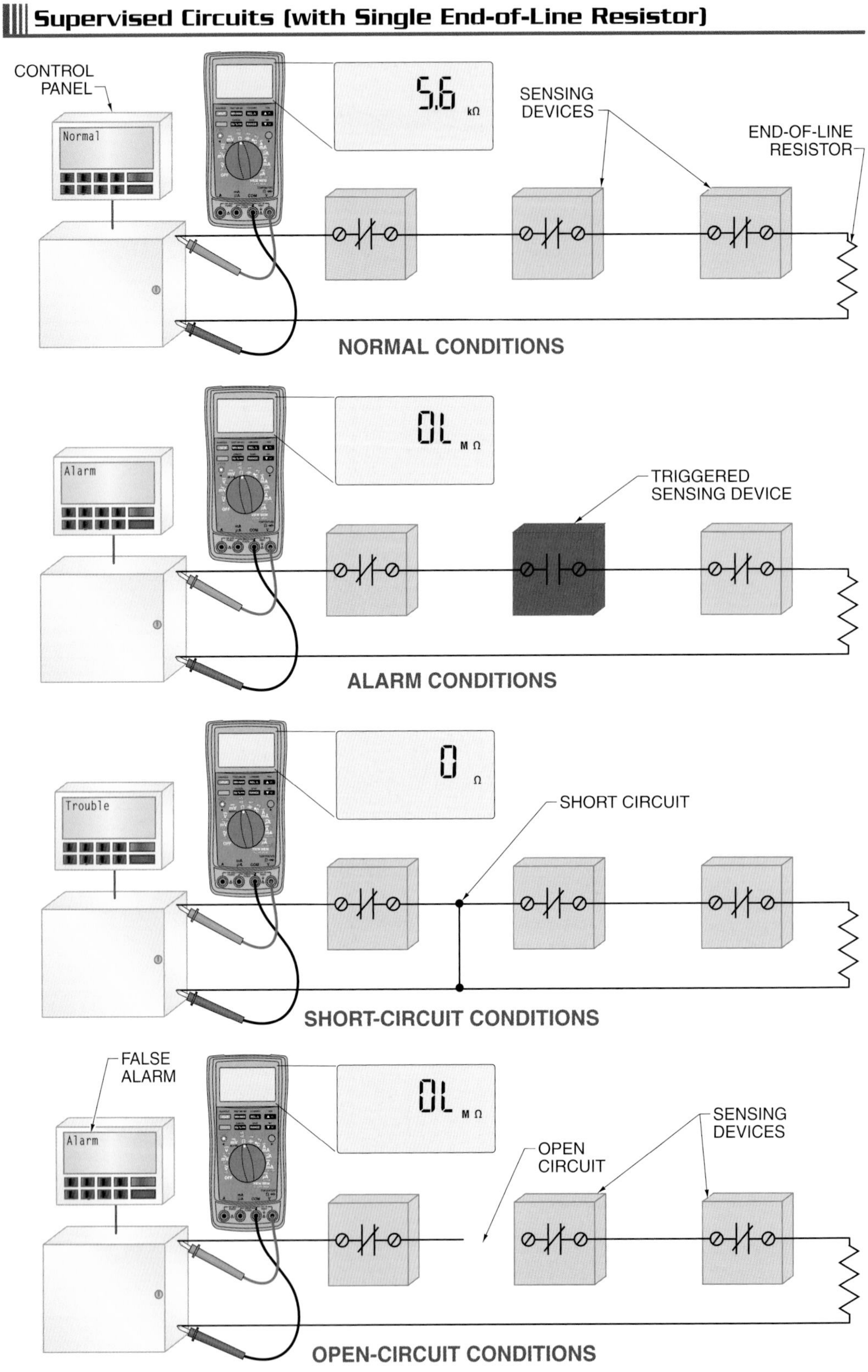

Supervised Circuits (with Single End-of-Line Resistor)

Figure 8-13. Supervised circuits with a single end-of-line resistor can accurately detect short-circuit faults but cannot distinguish between triggered sensing devices and open-circuit wiring faults.

A double end-of-line resistor configuration, however, is able to distinguish an open-circuit condition from an alarm condition. **See Figure 8-14.** This type of circuit includes two end-of-line resistors: one connected in series and one connected in parallel with the sensing device. A circuit wired in this configuration measures only the resistance of the series resistor value under normal conditions. If the loop is shorted, the control panel measures almost no resistance. If the loop is opened, the control panel measures infinite resistance. If a sensing device opens upon activation, the control panel measures the combined resistance of both resistors.

Double end-of-line resistor circuits are most common for high-security applications, which typically limit the circuits to only one sensing device each.

Addressable Circuits. An *addressable circuit* is a circuit of addressable devices that communicate using digital signals. An *addressable device* is a device with a unique identifying number that can exchange messages with other addressable devices. An addressable circuit operates much like a simple computer network, with detailed structured network messages passing between specific devices and the control panel without interfering with the other devices in the circuit. **See Figure 8-15.** This type of circuit does not use an end-of-line resistor.

These addressable devices communicate information about status and alarms directly to the control panel. The control panel may either periodically poll each device to determine its status, or it may receive regular updates from each device with its status. The failure of a device to respond to a polling request or the failure to send an expected update due to an open circuit or other fault results in a trouble signal.

Because the devices have unique addresses, the control panel can identify an individual alarming device instead of only the alarming zone. Information about the device can be quickly accessed and provided to security personnel to identify the precise location of the detection. Addressable circuits are also commonly used with fire alarm systems where they are called signaling-line circuits.

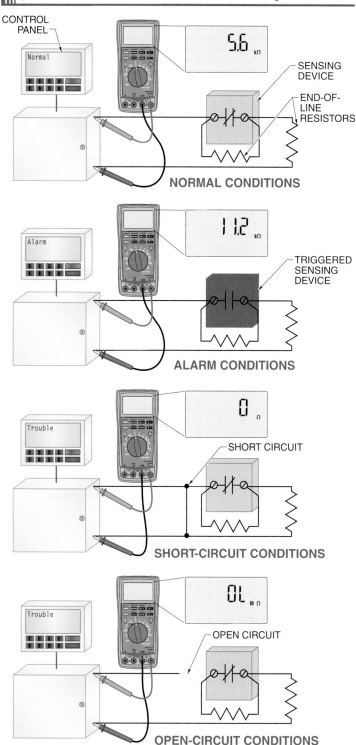

Supervised Circuits (with Double End-of-Line Resistors)

Figure 8-14. A supervised circuit with double end-of-line resistors can accurately distinguish between normal conditions, alarms, short circuits, and open circuits by measuring the resistance of the circuit.

Addressable Circuits

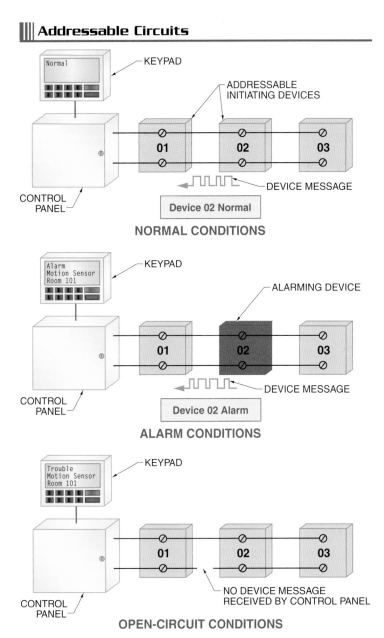

NORMAL CONDITIONS

ALARM CONDITIONS

OPEN-CIRCUIT CONDITIONS

Figure 8-15. *Addressable circuits use intelligent devices, each with a unique identification, and a rich exchange of information to accurately monitor the circuit and identify individual triggered devices.*

According to a Security Industry Alarm Coalition (SIAC) survey, the following were the most common causes of false alarms (respondents could choose more than one): improper arming/entry (68%), improper arming/exit (56%), pets (22%), equipment malfunction (15%), window/door left ajar (15%), other user error (11%), and installation problems (5%).

False-Alarm Mitigation Features

A *false alarm* is an activation of an alarm without any evidence of an actual or attempted intrusion. False alarms can result in the needless dispatch of police, fire, or other authorities to the facility, wasting their time and effort. Authorities may also assess fines for excessive false alarms, which can be costly.

Improving installation and training practices helps mitigate false alarms due to improper installations or user errors. System features can also be incorporated into the security system control panel to provide procedural solutions to prevent accidental false alarms. Per ANSI/SIA CP-01, *Control Panel Standard—Features for False Alarm Reduction,* several of these measures are required. Also, NFPA 731, *Standard for the Installation of Electronic Premises Security Systems,* covers the proper application, location, installation, performance, testing, and maintenance of security systems and their components.

Voice Verification. *Voice verification* is a security system control panel feature that allows two-way communication between the central station and the facility where the security system is installed. Two methods can be used for voice verification. After the control panel transmits an alarm signal to the central station, the telephone line can be latched to allow a building occupant to communicate verbally with the central station. Alternatively, a two-way voice communication module can be installed at the facility. When an alarm signal is received at the central station, the central station initiates a callback that allows the station operator to listen in and communicate through a speaker in the secured building.

Fire Alarm Verification. *Fire alarm verification* is a security system control panel feature that requires two trips of a fire alarm-initiating device within a specified amount of time before it transmits a fire alarm signal. When one device trips, the security panel immediately resets by removing power to the tripping device. After power is restored, the control panel monitors the device for a second trip within the specified amount of time (typically 2 min). If the device trips again, the panel will transmit the fire alarm signal.

False Alarms

Alarms caused by natural phenomena, disruptions in telephone or power service, or failure of equipment at a central station are not considered false alarms, since they cannot be controlled by the building personnel. False alarms are caused by user error, installation error, or faulty on-site equipment.

User-related alarms are the largest single cause of false alarms but are largely preventable. After an installation has been completed, it is the contractor's responsibility to provide adequate training on the proper operation of the security system, including how to cancel a false alarm. In many cases, the recipients of the contractor's training are key building personnel who will then train additional users of the system. The quality of the initial and subsequent trainings greatly affects the number of false alarms caused by user errors.

Installation error is another significant cause of false alarms. Installation errors include incorrectly installed equipment, installations that exceed equipment limits, incorrectly configured equipment, and inappropriately applied equipment (equipment used in an unintended manner). Many times, the issue is either a lack of quality installer training or a lack of familiarity with new system innovations.

Unlike comprehensive training for customers and installers, defective or faulty equipment is difficult to control. As with any electronic equipment, components eventually fail as they age. The most effective ways to reduce false alarms caused by equipment are to install listed components from reputable manufacturers, test the system regularly, and address any problems immediately.

Programmable Abort. *Programmable abort* is a security system control panel feature that delays the transmission of an alarm signal for a short period to allow a manual cancellation. This feature can eliminate many accidental false alarms by allowing time for an authorized user to cancel the alarm before emergency personnel are automatically alerted. The delay is typically programmed for 30 sec or 60 sec. All other functions, such as the siren and annunciation at the keypad, function normally. Other systems can approximate this feature by transmitting a second, authenticated "cancel" signal to the supervising or central station.

Audible/Visual Exit Delay Countdown Warning. An *audible/visual exit delay countdown warning* is a security system control panel feature that audibly and visually indicates the amount of exit delay time remaining before the system is armed. An *exit delay* is the programmed amount of time between when a security system is commanded to be armed and when the control panel begins registering alarm signals from sensing devices. The exit delay allows the person arming the security system time to leave the detection area before their presence causes an alarm.

The keypad beeps every 5 sec and the exit light is solidly lit until the last 10 sec of the exit delay period. During the last 10 sec, the keypad will beep every second and the exit light will flash, indicating to the user that this is his or her last chance to exit the building before the system is armed.

Cross Zoning. *Cross zoning* is a security system control panel feature that requires two predetermined zones to trip before an alarm signal is transmitted. This feature is very useful in systems with a motion sensor with frequent nuisance tripping. If a single motion sensor trips, this detector starts a countdown. Only if a second zone is tripped within the specified amount of time is an alarm signal transmitted. Cross zoning can be used with any zone type except fire, making it very flexible.

Swinger Shutdown. *Swinger shutdown* is a security system control panel feature that disables a zone after repeated triggering during an armed period. A control panel with this feature will ignore a specific zone after it triggers a certain number of alarms. This can prevent numerous alarm signals due to malfunctioning equipment or other problems, such as a door blown open in the wind. Though that particular zone is disabled, all other zones remain active.

SECURITY SYSTEM CONTROL DEVICES

Most security systems use many of the same basic control devices, though different systems may implement them in varying combinations or numbers, depending on the desired type and level of protections and the specifics of the application. High-security applications will also install advanced devices in addition to common detectors.

There are three types of security system control devices. Sensing devices are the inputs into the system that detect conditions that cause security actions. Notification devices are the outputs that alert building occupants, off-site monitoring stations, and/or local authorities to an alarm. Security systems also include interface devices, since security systems require regular interaction with building users to arm and disarm the protection features. All security system control devices are connected directly to the control panel.

Sensing Devices

Input devices used with security systems are sensing devices. A *sensing device* is a security system device that sends an input signal to the control panel when triggered by some security condition. The control panel uses this input information to take security actions. There are many different sensing devices made for security systems. Most security system sensing devices open a normally closed-loop circuit when triggered. This changes the electrical resistance of the circuit, causing the control panel to register an alarm.

Door/Window Contacts. A *door/window contact* is a sensing device that indicates whether a door or window is fully closed. Door/window contacts consist of two parts: a magnet that is mounted on the door or window and a reed switch that is mounted on the frame. **See Figure 8-16.** The reed switch includes two small metal strips separated by a small gap and enclosed in a gas-filled, hermetically sealed glass tube. The stiffness of the metal strips keeps them in a normal state, typically open, unless actuated by the magnet. When the door or window is fully closed, the two halves of the device are very

close to each other and the magnet pulls the contacts closed. When the two parts separate as the door or window is opened, the contacts revert to their normal state.

Door/Window Contact Operation

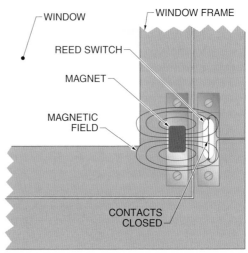

CLOSED WINDOW

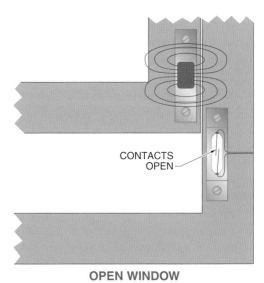

OPEN WINDOW

Figure 8-16. Door/window contacts use a small magnet to move the tiny contacts of a reed switch open or closed.

The magnet actuates or releases the reed switch contacts at a certain distance, known as the make/break distance. The make/break distance will vary between switches, with typical values being between ½″ and 2″, depending on the size and strength of the magnet and the reed switch. Door/window contacts are also available in many different sizes, colors, and mounting configurations. **See Figure 8-17.**

Glass Break Detectors. Many commercial buildings have either windows that cannot be opened or so many windows that wiring each one with door/window contacts is impractical. Glass break detectors are ideal for these circumstances. Glass break detectors include acoustic glass break detectors, shock sensors, and devices that include a combination of the two technologies.

An *acoustic glass break detector* is a sensing device that detects the noise (sound frequencies) of breaking glass. **See Figure 8-18.** Older versions of these detectors were prone to false alarms because the frequency range that they detected was also produced by scraping metal, squealing tires, and jingling keys. Newer technologies have improved the reliability of these devices by fine-tuning the frequency range, as well as incorporating technologies to recognize the acoustic patterns of breaking glass. Other technologies look at low, medium, and high frequencies in addition to specifics such as peak amplitude, signal duration, and average ambient sound in the room before triggering an alarm.

One of the biggest advantages to using an acoustic glass break detector is that one detector can cover multiple windows within a single room. Acoustic glass break detectors can be mounted on a wall or ceiling. They are typically line-of-sight devices, meaning they cannot detect through doors or around corners. They also have a maximum range, with 25′ being the most common limit.

A *shock sensor* is a sensing device that detects the physical vibration of breaking glass. A shock sensor is mounted directly to the glass or the frame of the window to detect the specific 5 kHz (kilohertz) shock frequency of breaking glass. **See Figure 8-19.** Like older-technology acoustic sensors, older shock sensors were once also prone to false alarms. Also, many of the first shock sensors had manual sensitivity adjustments that were difficult to tune. Newer shock sensors use modern electronics and piezoelectric transducers and are much more reliable.

Door/Window Contacts

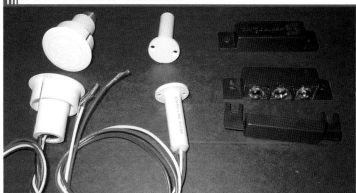

Figure 8-17. Door/window contacts are available in many sizes, colors, mounting types, and switch configurations.

Acoustic Glass Break Detectors

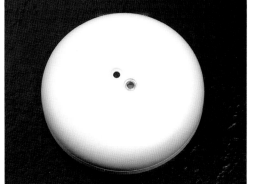

GE Security

Figure 8-18. Acoustic glass break detectors are commonly used in commercial applications to protect many windows in a single room.

The reed switch of a set of door/window contacts is a tiny glass vial containing a movable metal strip that touches another metal contact. Some reed switches include three metal contacts, providing both normally open (NO) and normally closed (NC) switch connections.

Shock Sensor

Figure 8-19. *Shock sensors are commonly used for single-window applications when it is cost prohibitive to use an acoustic sensor.*

Still, neither acoustic-type nor shock-type glass break detectors are 100% immune from false alarms. To further improve the accuracy of glass break detectors, a dual-technology glass break detector combines both types of detection technologies in a single device. For the device to trigger an alarm, it must first sense the acoustic signals, followed by the shock waves traveling through the walls and ceiling. If both of these signals are not present within a specified amount of time, the detector will not initiate an alarm signal.

Motion Sensors. Many large areas are difficult to protect with door or window devices, such as large open rooms, atriums, corridors, or file rooms. Additionally, an intruder may enter the building through a roof, vent, wall, or other point-of-entry not monitored by a door or window device. In these instances, protection can be accomplished by using motion sensors. A *motion sensor* is a sensing device that detects movement within the coverage area of the detector. **See Figure 8-20.** Motion sensors are common inputs for other building systems, where they are also known as occupancy sensors. The two most common motion sensor technologies used for security systems are passive infrared (PIR) and microwave.

A *passive infrared (PIR) sensor* is a sensor that activates when it senses the heat energy of people within its field-of-view. The PIR sensor detects the rapid change in temperature caused by a person within or passing through the detection area. The infrared energy from the detection area is focused on the pyroelectric sensor with either a mirror or lens.

A *microwave sensor* is a sensor that activates when it senses changes in the reflected microwave energy caused by people moving within its field-of-view. This technology requires movement to detect a person. Microwave sensors have difficulty detecting slow-moving objects, including people, and they are prone to false activations from physical vibrations and certain moving objects. Microwave sensors are not affected by temperature changes, sunlight, air turbulence, or humidity, and the microwaves even penetrate most building materials, providing a larger potential detection area.

Motion Sensors

Honeywell International, Inc.

Figure 8-20. *Motion sensors use either passive infrared (PIR) or microwave technology to detect the presence of people within a protected area.*

Both types of motion sensors have advantages and limitations. Dual-technology motion sensors take advantage of this by combining both sensing technologies into a single device. This reduces false activations by requiring that both sensors register a detection in order for the device to signal an alarm.

An ultrasonic sensor is another type of motion sensor that is commonly used as an occupancy sensor for controlling lighting, HVAC, and other noncritical systems. Ultrasonic sensors are often more sensitive than PIR or microwave sensors. Therefore, they are more prone to false triggers. In applications such as lighting control, this is a minor issue. However, for this same reason, they are generally considered unsuitable for life/safety applications, such as security systems.

Panic/Duress Buttons. Panic/duress buttons are used to manually cause an alarm. A *panic/duress button* is a sensing device that causes an alarm signal when manually activated. **See Figure 8-21.** These are often used to summon emergency personnel during burglaries without alerting the intruder to the alarm. Therefore, the system is often designed to initiate a silent alarm when a panic/duress button is activated.

Panic/Duress Buttons

GE Security

Figure 8-21. Panic/duress buttons are hidden alarm devices for manual activation during in-progress burglaries.

These devices are commonly located under a desk or counter, out of public view, and connected to a 24-hour zone. A panic/duress button can be a single momentary pushbutton, latching pushbutton, wireless pendant, or a dual-action device that requires two buttons to be pushed. A special type of panic/duress button is a bill trap, which triggers an alarm when the last bill in a cash register is removed.

Pressure Sensors. Pressure sensors detect the presence of a person in a certain floor area by detecting the pressure of their steps. Two of the more common types of pressure sensors are floor mats and stress sensors. **See Figure 8-22.**

Pressure Sensors

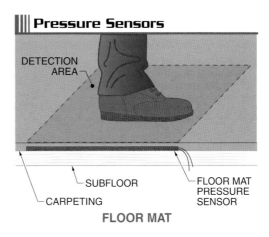

FLOOR MAT

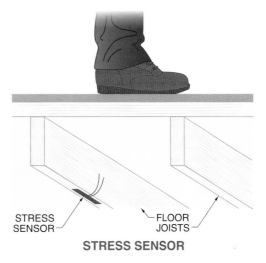

STRESS SENSOR

Figure 8-22. Floor mats and stress sensors are hidden sensing devices that detect the weight of a person crossing a secured threshold.

A *floor mat* is a sensing device that detects the steps of a person walking on it. Floor mats are typically installed under the carpet in locations where an intruder is likely to travel, such as in a doorway or at the bottom or top of a set of stairs. The mats sense the weight of an intruder and trigger an alarm when the measured weight exceeds a setpoint, such as 60 lb. A disadvantage of these devices is they gradually wear out and, if they are installed under a carpet, they can be difficult to replace.

A *stress sensor* is a sensing device that measures a weight load over a specific area. The sensor is glued with epoxy under a floor joist or floor truss. An electronic processing circuit analyzes the measured stress signals and initiates an alarm when the stress exceeds a setpoint. The stress sensor, like the floor mat, is installed in common areas or high-traffic areas, but can also be installed in outdoor areas to protect an area such as a deck or a boat dock. However, stress sensor installation requires access to the floor from below.

Environmental Sensors. Environmental sensors detect potentially serious problems in the surrounding space so that they can be remedied quickly. An *environmental sensor* is a sensing device that detects abnormal conditions in the environment. These conditions could be abnormal values of temperature, moisture, carbon monoxide, toxic gases, or other damaging or harmful conditions. **See Figure 8-23.** Environmental sensors can detect problems that could damage equipment, ruin inventory, shut down the building, or even pose a safety hazard.

Environmental Sensors

TEMPERATURE SENSOR   **WATER SENSOR**

Honeywell International, Inc.

Figure 8-23. Environmental sensors detect abnormal ambient conditions such as temperature extremes, or moisture that can cause safety hazards or damage to equipment.

The most commonly used environmental sensor is a temperature sensor. These sensors monitor the temperature within a specific area and trigger a trouble signal or an alarm signal if the temperature exceeds a predetermined level. The sensor is similar to temperature sensors used in other building systems, except that those used in security systems output only a contact closure signal that indicates that a temperature limit has been reached.

The two common temperature sensors used with security systems are fixed-temperature and adjustable-temperature sensors. A fixed-temperature sensor triggers an alarm when the temperature exceeds a factory-defined limit. For example, the most common type is a cold temperature sensor that activates when the temperature drops below 40°F. An adjustable-temperature sensor triggers an alarm when a user-defined temperature limit, either high or low, has been exceeded. Excessively cold temperatures can freeze water pipes, while excessively high temperatures can cause electrical equipment to overheat.

Other types of environmental sensors are used to detect water in areas such as a basement. There are also detectors that can monitor the moisture in a specific area, and there are sensors available that can detect gases that are toxic to humans.

Notification Devices

A *notification device* is a device that alerts building occupants of an alarm condition. The notification device also serves as a deterrent to an intruder because it indicates that an alarm has occurred and, in most cases, that police have been dispatched.

Most alarm notification devices are either audible, such as sirens or speakers, or visual, such as strobes. **See Figure 8-24.** Including both types of notification devices in a security system ensures that both visually impaired and hearing-impaired persons are alerted. A third type of notification device is an automatic phone dialer.

Speakers. A *speaker* is a device that produces sound from electrical audio signals. Speakers can be configured to produce simple siren tones or play a voice recording, depending on the equipment features and programming. There are many types of speakers available, including interior and exterior models and models with various volume levels and power requirements.

Control panels initiate the audio alert, and the sound is produced by a speaker. In between, though, there must be circuitry to generate the audio signals compatible with the speaker. A *siren driver* is a circuit that produces the frequency, volume, and pattern of a desired sound. For example, most siren drivers produce a warble or steady tone for loud warning alerts. The selection of the type of tone depends on which terminals on the siren driver are connected to the speaker.

Some drivers include voice annunciation capability. A voice driver generates voice sound signals. Recordable voice drivers include a built-in microphone to record voice audio, allowing any alert message to be recorded (within the recording time limitations of the device). The voice driver may even store multiple messages. The appropriate one is then selected and played back depending on the alarm conditions. For example, one message may give evacuation instructions in the event of a fire alarm, while another message, in the event of a burglary alarm, may alert an intruder that the alarm has been activated and the police have been dispatched. Some drivers with pre-recorded messages include versions in more than one language.

Siren drivers are integrated into the security system's audible notification circuits in three different ways. **See Figure 8-25.** Some security system control panels include built-in siren driver circuitry as a feature of the panel. The speakers are connected directly to the siren driver terminals on the control panel for the audio signal.

Siren Device Integration

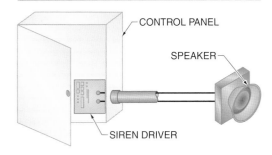

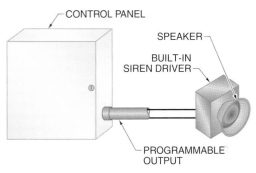

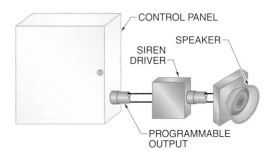

Figure 8-25. Control panels initiate audio alerts through siren drivers, which can be connected to the speakers in three different ways.

Notification Devices

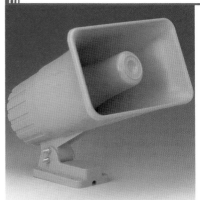

Honeywell International, Inc.
SIRENS

GE Security
SPEAKERS

STROBES

Figure 8-24. The most common notification devices are sirens, speakers, and strobes.

For control panels that do not include a built-in siren driver feature, separate siren driver units are available. These are connected between the control panel and the speaker. They are activated by signals from programmable outputs or the bell output on the control panel.

Alternatively, siren driver circuitry may be included with a speaker, though only siren-producing drivers are commonly integrated in this way. In fact, this type of speaker unit is commonly known as a siren, since it is a self-contained speaker and siren driver unit capable of producing a siren tone when activated.

Strobes. A *strobe* is a visual notification device that flashes a bright light at regular intervals. These devices are used in both fire alarm systems and security systems. A security system strobe is commonly round in shape and available with many different lens colors, such as red, amber, blue, and clear. The strobe color is purely a design choice.

Strobe lights are commonly used with security systems as a back-up notification device. Sometimes a strobe is installed on the exterior of the building to assist law enforcement officials in determining which building has triggered an alarm condition. This is very helpful for multibuilding campuses or multiunit buildings.

Automatic Phone Dialers. An *automatic phone dialer* is a notification device that dials telephone numbers and communicates a trouble or alarm condition to the recipients. Telephone numbers are programmed into the unit to be dialed for different alarm conditions. When an alarm is triggered, the dialer dials each telephone number, one at a time, until it connects with a receiver. The dialer plays prerecorded or computer-synthesized voice messages that announce to the receiver the origin of the call and the nature of the problem. A touch-tone code may be required from the recipient to acknowledge the message and prevent the dialer from attempting to connect with other recipients.

Dialers can be used to notify personnel of any type of event or trouble that requires immediate attention, including intruder alarms and environmental conditions. Integrating a dialer in this manner can be a low-cost alternative to central station monitoring for facilities that do not require high security.

Dialers

A security system "dialer" is an often-misunderstood term in the security industry. It is sometimes used to describe the internal electronics that some security systems use to communicate with the off-site monitoring station. Rather, this circuit is more accurately called a digital alarm communicator. The function of the digital alarm communicator is to send digital signals over a phone line, cellular transmission system, or another communication means when the alarm system is triggered.

Interface Devices

An *interface device* is a device that facilitates communication between a person and the security system for sharing information or controlling the system. The security system must be able to indicate the alarm status, operating mode, and programmed behavior of each zone to security personnel. There must also be a means with which to change the operating modes and programs. Security systems require authentication before interfacing with users to ensure that only authorized personnel are allowed access to system configurations.

The most common security system interface devices are keypads and displays. However, since security systems may be integrated with access control systems, many access control devices can also be used as security system interface devices.

Keypads. The keypad is the most common interface device connected to the security system. A *keypad* is an interface device used to display the status and control the operation of the security system. Keypads include an array of pushbutton keys, LED status indicators, and sometimes a small liquid crystal display (LCD). **See Figure 8-26.** The keys are used to enter authentication codes and select desired system modes. Keypads may include only the numbers 0 through 9, sometimes with letters also assigned to the number keys, similar to a telephone layout. Special keys may be provided as on-board panic or emergency buttons.

Keypads

Honeywell International, Inc.

Figure 8-26. Keypads allow authorized users access to the status and mode of the security system.

LED status indicators provide status information, such as armed or disarmed, at a glance, but detailed information is shown on the LCD display. This may include zone types, alarm types, and which zone initiated an alarm.

Keypads may also include audible annunciation of the system state, such as arming delay in progress. For some security systems, the initial programming can be completed at the keypad.

Displays. Small displays are included with some keypads for basic display features, but many security systems support stand-alone displays for detailed information monitoring. Common display types include LED displays, fixed LCDs, custom LCDs, and touchscreen displays.

LED displays light certain LEDs that correspond to zones and their status. LED displays are inexpensive and easily readable at a distance, but are limited in display capabilities. Fixed LCDs include standard information titles and modifier words that are activated as needed to indicate status. The information is somewhat limited because the display information cannot be customized. For example, a fixed LCD may display "ZONE 1," with an LED indicating the armed state. Zone 1 may be an entry area, but the display cannot be changed to reflect an entry area identifier.

A custom LCD allows nearly any type of character or information to be shown and is limited only by the size of the screen. These displays can be customized to the customer's specifications. A touchscreen display is a display with keypad or other selectable features available by touching certain areas of the screen. **See Figure 8-27.**

Touchscreen Displays

Honeywell International, Inc.

Figure 8-27. Touchscreen displays can be used to show customized information about the status of a security system and allows any type of selectable interface features.

Access Control Devices. It is common to integrate the security and access control systems in a building together. Access control devices may improve security by contributing additional devices for authenticating system users. There are many types of access control devices. All use some personal characteristic to allow only certain individuals to arm or disarm a security system or to cancel an alarm.

A *biometric device* is an interface device that evaluates specific human physical characteristics for authentication. These characteristics are unique to each individual and are not easily falsified. Biometric authentication includes fingerprint scanning, iris scanning, facial recognition, and voice recognition.

A *card reader* is an interface device that evaluates the magnetic or electrical characteristics of a special handheld keycard for authentication. These devices are very useful when an organization has many employees who need to access the building and the company chooses not to assign code numbers to all employees.

A *key switch* is an interface device that arms or disarms a security system with a special key. Key switches are an older technology but are still common in new security systems.

A *wireless key fob* is an interface device that arms or disarms a security system from a handheld wireless transmitter. This can be used from nearly anywhere in or near the protected area.

SECURITY SYSTEM CONTROL APPLICATIONS

A security system can be used to control aspects of other building systems, such as HVAC functions, lighting, and locks, according to the state and status of the security system. Outputs from the security system can activate relays to control specific devices, use auxiliary modules to interface with other controllers, or utilize network-based devices to share information across building systems.

In the past, many of these applications were limited to residential or light commercial applications such as a small office or small retail establishments. Advances in technologies and systems integration have facilitated large-scale opportunities. However, due to the wiring and programming involved, many of these integration features must still be planned for at the time of initial system installation.

Some manufacturers offer equipment specifically designed to integrate multiple systems. For example, given the configuration and operational similarities between fire alarm systems and security systems, it is becoming more common for the functions provided by these two systems to be combined into one control panel. Access control is also often integrated with security systems.

Controlling Lighting or HVAC Systems

Security systems have the ability to control lighting and/or HVAC settings after a specific event, such as arming or disarming the security system. Since the arming of a security system implies that the authorized occupants are no longer in the building, this type of status change can be used to control other systems that are based, at least in part, on building occupancy. **See Figure 8-28.** For example, with no occupants, the light fixtures can be switched off and the HVAC system adjusted to an energy-saving mode.

These operations may normally have used clock switches that are very effective at controlling the system based on time, or even sunrise and sunset, but cannot account for unusual circumstances, such as employees working late. They also require reprogramming or reconfiguration if normal schedules change. However, using the security system armed/disarmed status is considerably more accurate at determining the occupancy status of the building. Even the activation of the armed (stay) mode can be used to selectively control other building systems in certain building areas.

Operating Conditions. The security system must be configured to interface with other systems based on its state and status. This may involve using a programmable output connection at the security system control panel as a way to send ON/OFF signals to another device or controller. Alternatively, the control panel may offer special connections in order to communicate with other controllers.

On the side of the lighting or HVAC controllers, the controllers must be able to accept an input that either directly controls its connected devices, or signals the controller to initiate an internal program. For example, a lighting controller may accept a digital input that controls whether the unit is in a building-occupied mode or an after-hours mode.

Control Sequence. The control sequence for a security system to control lighting or HVAC functions is as follows:

1. When the security system is armed, the control panel initiates any system integration functions.
2. The control panel activates a signal on a certain output that is connected with other building system controllers.
3. A lighting and/or HVAC controller receives the signal that the building is now unoccupied and initiates any preprogrammed actions based on this input.

4. The lighting and/or HVAC controller switches OFF or modulates loads according to the selected program.

5. When the security system is disarmed, corresponding signals to the integrated controllers again change their operating program back to an occupied-building mode.

Integration with Access Control Systems

A primary function of access control systems is the authentication of system users. Since access to the configuration, programming, and setting of alarm systems also requires authentication, it is common to integrate the system with access control devices, such as keypads and card readers. This reduces the number of installed devices between the two systems and may simplify the management of system rights for the two systems. For example, instead of having separate access codes or keycards for two different systems, users can use the same authentication to interact with both systems.

Upon successful authentication, the access control device signals the security system control panel that access to a certain area has been granted and the area will soon be occupied. This signal is accepted as a programmable input by the security system. **See Figure 8-29.** The security system then deactivates the alarm functions for that system partition. This type of integration is relatively simple and can be accomplished with many existing security systems.

Operating Conditions. The security and access control systems are integrated in a way similar to lighting and HVAC system integration, except that the security system control panel receives inputs rather than sending output signals. The input is typically a connection point programmed as a momentary digital input, which can be used with a variety of other ON/OFF devices. The control panel may require programming to recognize this input and designate which partition to arm/disarm according to the received signals.

Controlling Lighting or HVAC Systems

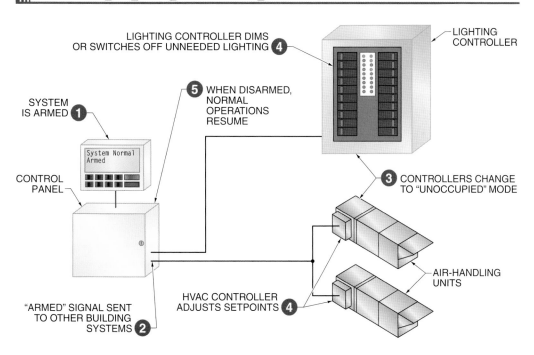

Figure 8-28. The armed/disarmed status of a security system can be used to control other building systems that are based on occupancy, particularly lighting and HVAC systems.

Integration with Access Control Systems

ACCESS CONTROL
KEYPAD

Access
Granted

1 2 3
4 5 6
7 8 9
0 CANCEL

DESIGNATED
PARTITION DISARMED ❸

System Normal
Partition 3
Disarmed

SECURITY SYSTEM
RECEIVES INPUT ❷

❶ KEYPAD SENDS SIGNAL
TO SECURITY SYSTEM

SECURITY SYSTEM
CONTROL PANEL

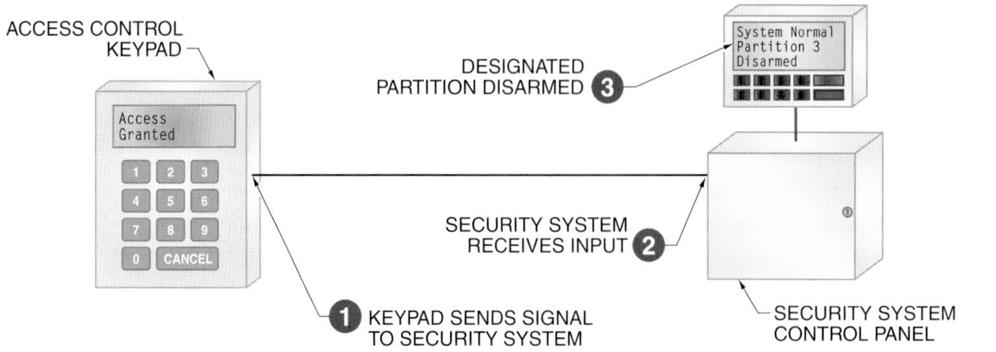

Figure 8-29. *Security and access control systems can share many of the same devices, which can be used to seamlessly integrate the operation of the two systems.*

Control Sequence. The control sequence for the arming and disarming of a security system via access control keypad is as follows:

1. Upon successful authentication at an access control device, the device sends a signal to the security system.
2. The security system receives the control input from the access control system at a programmable input connection.
3. The security system deactivates the designated partition or the entire security system.

Integration with VDV Systems

Voice-data-video (VDV) systems include several subsystems to handle audio, video, and information separately. VDV systems are commonly integrated with security systems because video and audio information can be very useful in security applications. Each subsystem can be activated or commanded by the security system for security-related functions. Video cameras deter many potential intruders who fear the risk of detection because the cameras record suspicious activity for evidence. Audio systems such as facility intercoms can be used for two-way conversations with security personnel and for broadcasting messages.

Operating Conditions. Some video systems monitor and record continuously: the cameras sweep large areas in their field-of-view to catch any suspicious activity. This requires

special video recording systems to continuously overwrite old, unneeded footage with new video, while saving important portions for later review. Other systems, however, only record on demand. The video cameras and recording systems can accept signals to move cameras to the necessary position and begin recording. The security system can command these actions based on inputs from sensing devices indicating an intruder or other security problem. **See Figure 8-30.** VDV systems are connected to programmable outputs of the security system control panel, which then send the necessary signals to control the VDV systems.

Control Sequence. The control sequence for integration with VDV systems is as follows:

1. A person under duress activates a panic button.
2. The security system activates sirens in the area and an alarm signal is received at the security station.
3. A programmable output from the security system sends control signals to other building systems so that the security station can remotely investigate and monitor the location of the panic alarm.
4. The intercom system receives a signal from the security system control panel to activate the two-way intercom unit closest to the panic alarm location. This allows security personnel to communicate verbally with people in the area.

Refer to
Quick Quiz®
on CD-ROM

5. The closed-circuit television (CCTV) system receives a signal from the security system control panel to reposition nearby cameras in order to focus on the location of the alarm. This allows the security personnel to notify responders of any activity in the area.

6. The digital video recorder receives a signal from the security system control panel to start recording real-time events from all cameras in the alarm area.

Many security system control panels allow customization of inputs and outputs for system size, design, or integration with other systems. For example, many include a number of freely programmable outputs that can be used to integrate with other automated building systems for special applications. Typically, these connections act like a relay, using contact closures to send digital (ON/OFF) signals. If not already built-in, some control panels work with expansion modules to provide additional programmable outputs.

Integration with VDV Systems

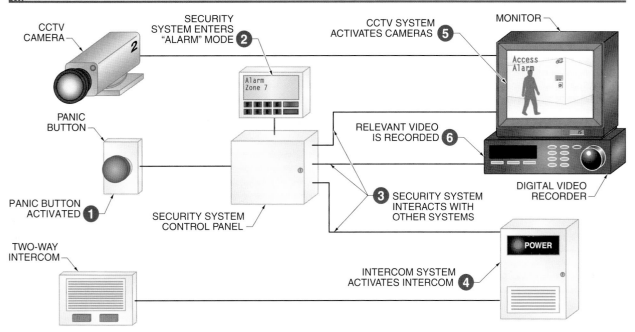

Figure 8-30. *Security systems can take control of audio and video devices to selectively monitor building areas for suspicious activity.*

Summary

- A security system is managed by a central control panel that interfaces with all other security devices and, as necessary, other building systems.

- Security system protection includes three different types: point-of-entry protection, specific-area protection, and spot protection.

- The method of notification is used by security system professionals to classify different system types.

- Many security systems installed in residential and small commercial buildings are local security systems.

- Supervisory security systems require on-site 24 hr monitoring, while central station security systems provide 24 hr protection without the facility employing its own 24 hr security personnel.

- A building typically includes at least several zones, defined by location, size, and security requirements.

- Zones can be programmed to behave in certain ways under certain conditions.

- Security systems can be placed into a variety of states of readiness, or modes.

- Changing a security system between different modes requires authentication.

- Partitioning allows the system to be subdivided into two or more separate systems using the same control panel and control devices.

- An alarm signal is initiated by various sensing devices that detect intrusion into the secured space.

- Trouble conditions do not require an alarm response by municipal authorities but do require a prompt response by building personnel to correct the problem quickly.

- The control panel is programmed with instructions for various scenarios and it makes decisions based on the status of the system and field devices.

- Security system circuits are primarily closed-loop circuits.

- A single end-of-line resistor circuit configuration monitors the circuit for wiring faults but cannot distinguish between an open due to wiring trouble and an open due to sensing-device activation.

- A double end-of-line resistor circuit configuration is able to distinguish an open-circuit condition from an alarm condition.

- Features incorporated into the security system control panel provide procedural solutions to prevent accidental false alarms.

- Security systems use many of the same types of sensing devices.

- Most security system sensing devices open a normally closed-loop circuit when triggered. This changes the electrical resistance of the circuit, causing the control panel to register an alarm.

- Most alarm notification devices are either audible, such as sirens or speakers, or visual, such as bright flashing lights or strobes.

- The security system must indicate the alarm status, operating mode, and programmed behavior of each zone to security personnel and provide a means with which to change the operating modes and programs.

- Since security systems may be integrated with access control systems, many access control devices can also be used as security system interface devices.

Definitions

- A *security system* is a building system that protects against intruders, theft, and vandalism.

- *Point-of-entry protection* is a type of security system protection that monitors the specific points that might allow entry or exit from a secure area.

- A *point-of-entry* is a potential opening in the envelope or perimeter of a building or area through which a person may enter.

- *Specific-area protection* is a type of security system protection that monitors only one defined location.

- *Spot protection* is a type of security system protection that monitors one particular object.

- A *local security system* is a self-contained security system that does not require full-time surveillance or off-premises wiring.

- A *supervisory security system* is a security system that is monitored continuously by on-site security personnel.

- A *central station security system* is a security system that is monitored by an off-site station.

- A *zone* is a defined area protected by a group of security system sensing devices.

- A *zone type* is a definition of the type of alarm a zone initiates and the corresponding reaction of the control panel.

- A *delay zone* is the zone type that allows a certain period of time to elapse before a triggered sensor initiates an alarm.

- An *instant zone* is a zone type that indicates an immediate alarm condition when a sensing device in the zone is activated.

- An *interior zone* is a zone type used to differentiate interior-sensing devices.

- A *stay/away zone* is a zone type that is programmed to ignore signals from its sensing devices when the system is armed, but when there is no exit from the premises.

- A *24-hour zone* is a zone type that is always active (never disarmed).

- A *24-hour supervisory zone* is a zone type that is always active and initiates a trouble signal.

- A *24-hour fire zone* is a zone type that is always active and initiates a fire alarm.

- A *null zone* is an unused zone.

- *Authentication* is the process of identifying a person and verifying their credentials.

- A *disarmed mode* is a security system mode in which the control panel ignores all inputs from sensing devices.

- An *armed (away) mode* is a security system mode that makes all security-sensing devices active.

- An *armed (stay) mode* is a security system mode that makes all perimeter-sensing devices active while signals from the sensing devices (that were defined as being stay/away zone types) are ignored.

- A *partition* is a portion of a security system that operates under a different program than the rest of the system.

- An *alarm* is a security system signal that indicates an intrusion or other alarm condition.

- A *trouble signal* is a signal that indicates a wiring or power problem that could disable part or all of the system.

- An *open-loop circuit* is a security system circuit with devices that are all normally open.

- A *closed-loop circuit* is a security system circuit with devices that are all normally closed.

- A *nonsupervised circuit* is a security system circuit of field devices that cannot monitor the integrity of the wiring.

- A *supervised circuit* is a security system circuit of field devices that can monitor the integrity of the wiring. This type of circuit requires end-of-line resistors.

- An *end-of-line (EOL) resistor* is a resistor installed at the far end of an initiating-device circuit for the purpose of monitoring the circuit.

- An *addressable circuit* is a circuit of addressable devices that communicate using digital signals.

- An *addressable device* is a device with a unique identifying number that can exchange messages with other addressable devices.

- A *false alarm* is an activation of an alarm without any evidence of an actual or attempted intrusion.

- *Voice verification* is a security system control panel feature that allows two-way communication between the central station and the facility where the security system is installed.

- *Fire alarm verification* is a security system control panel feature that requires two trips of a fire alarm-initiating device within a specified amount of time before it transmits a fire alarm signal.

- *Programmable abort* is a security system control panel feature that delays the transmission of an alarm signal for a short period to allow a manual cancellation.

- An *audible/visual exit delay countdown warning* is a security system control panel feature that audibly and visually indicates the amount of exit delay time remaining before the system is armed.

- An *exit delay* is the programmed amount of time between when a security system is commanded to be armed and when the control panel begins registering alarm signals from sensing devices.

- *Cross zoning* is a security system control panel feature that requires two predetermined zones to trip before an alarm signal is transmitted.

- *Swinger shutdown* is a security system control panel feature that disables a zone after repeated triggering during an armed period.

- A *sensing device* is a security system device that sends an input signal to the control panel when triggered by some security condition.

- A *door/window contact* is a sensing device that indicates whether a door or window is fully closed.

- An *acoustic glass break detector* is a sensing device that detects the noise (sound frequencies) of breaking glass.

- A *shock sensor* is a sensing device that detects the physical vibration of breaking glass.

- A *motion sensor* is a sensing device that detects movement within the coverage area of the detector.

- A *passive infrared (PIR) sensor* is a sensor that activates when it senses the heat energy of people within its field-of-view.

- A *microwave sensor* is a sensing device that activates when it senses changes in the reflected microwave energy caused by people moving within its field-of-view.

- A *panic/duress button* is a sensing device that causes an alarm signal when manually activated.

- A *floor mat* is a sensing device that detects the steps of a person walking on it.

- A *stress sensor* is a sensing device that measures a weight load over a specific area.

- An *environmental sensor* is a sensing device that detects abnormal conditions in the environment.

- A *notification device* is a device that alerts building occupants of an alarm condition.

- A *speaker* is a device that produces sound from electrical audio signals.

- A *siren driver* is a circuit that produces the frequency, volume, and pattern of a desired sound.

- A *strobe* is a visual notification device that flashes a bright light at regular intervals.

- An *automatic phone dialer* is a notification device that dials telephone numbers and communicates a trouble or alarm condition to the recipients.

- An *interface device* is a device that facilitates communication between a person and the security system for sharing information or controlling the system.

- A *keypad* is an interface device used to display the status and control the operation of the security system.

- A *biometric device* is an interface device that evaluates specific human physical characteristics for authentication.

- A *card reader* is an interface device that evaluates the magnetic or electrical characteristics of a special handheld keycard for authentication.

- A *key switch* is an interface device that arms or disarms a security system with a special key.

- A *wireless key fob* is an interface device that arms or disarms a security system from a handheld wireless transmitter.

Review Questions

1. What types of protections do security systems offer for different types of areas and objects?

2. How do local security systems, supervisory security systems, and central station security systems differ with relation to the roles of security personnel?

3. Why can it be important to define different zone types in a security system?

4. In what circumstances might partitioning be appropriate?

5. How do double end-of-line supervised circuits differentiate both open-circuit and short-circuit faults from sensing-device activation?

6. How do the cross zoning and swinger shutdown features reduce the incidence of false alarms?

7. Explain the differences between the detection methods of acoustic glass break detectors and shock sensors.

8. How can environmental sensors protect against property damage?

9. Describe how sirens and speakers are activated differently.

10. Why are security system arm/disarm signals sometimes the most accurate means for determining the unoccupied/occupied status of a building?

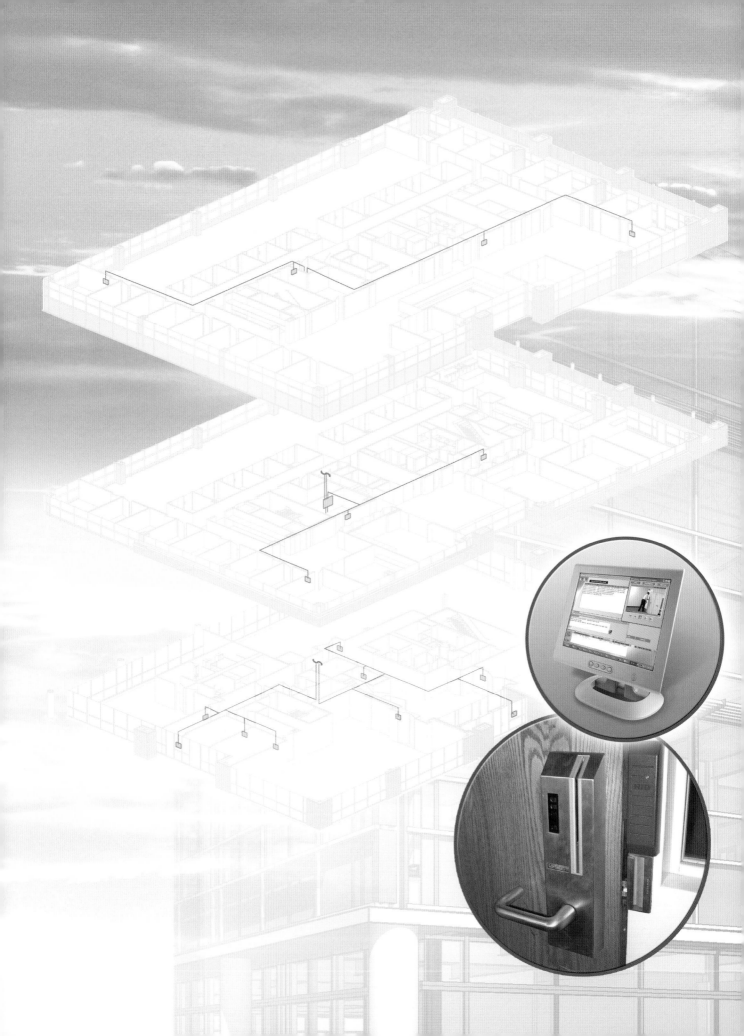

Chapter Nine

Access Control System
Control Devices and Applications

Access control systems can be very simple, providing basic features to restrict access to certain areas in a building. Alternatively, these systems can be very complex, with the ability to control and integrate with other building systems, such as security systems, CCTV systems, and lighting systems. Because the access control system is operated by software on a computer, it is particularly flexible for programming specific functions and control actions for various scenarios.

Chapter Objectives

- *Describe the primary functions of an access control system.*
- *Compare the common access control features available through software programming.*
- *Describe the different types of credential and barrier devices.*
- *Identify common strategies for integrating access control systems with other automated building systems.*

ACCESS CONTROL SYSTEMS

All buildings use some form of access control to prevent unauthorized access into the building. An *access control system* is a system used to deny those without proper credentials access to a specific building, area, or room. Access control provides security for persons and/or property by restricting access to a building or specific area to unauthorized persons through locked doors, fences, or some other form of barrier intended to block entry.

Access control also restricts unauthorized people from dangerous environments. These environments could include high-voltage areas such as open electrical vaults and chemical storage areas. Access control can also be used to prevent people from leaving a specific area such as patients in a nursing home or treatment facility or a baby nursery in a hospital.

Access control is commonly used for convenience applications. For example, a parking facility may charge a monthly fee for parking. Upon arrival at the parking facility, the person uses an access card to gain entry into the ramp. When leaving the parking ramp, the access card is used again to exit the ramp. Access control devices are also used in many mass transit applications for gaining access to the station through a turnstile. The same process is used to exit the transit station.

Castles had tall walls surrounded by deep ditches, often filled with water, as barriers. A drawbridge was lowered to allow access to authorized individuals known to the gatekeeper or those who spoke the correct password.

Access Control Methods

An access control system must have a method in place to grant or deny access to a specific area. The common methods utilized are human control, mechanical control, and intelligent control. A typical building may use one or more of these methods, depending on the level of security needed.

Human Control. *Human control* is a form of access control that requires human involvement. This can be a human physically opening a door or gate or placing a barrier to deny access to a certain area.

Guards provide flexibility due to their ability to react to special circumstances and document events. **See Figure 9-1.** Guards also receive special training that makes them more alert and aware of their surroundings. Though the costs associated with employing guards typically exceed the cost of other types of access control, the mere presence of a guard can deter most people from trying to gain entry into restricted areas. Common applications where guards are used include government buildings, hospitals, airports, mass transit stations, buildings with sensitive materials, international borders, and military checkpoints.

Guards

Gamewell-FCI

Figure 9-1. Guards provide a human control in applications that require on-the-spot decision making.

Mechanical Control. *Mechanical control* is a form of access control that uses a mechanical device to maintain a closed area. The most common form of mechanical control uses a lock and key. The lock-and-key method has been in use for many years and is still one of the most effective methods to protect the security of a building. In most buildings where electronic access control is used, locks are still installed on doors as an alternate method to gain access if there is a total system failure. A standard lock-and-key system is common in many commercial buildings and may have varying levels of security depending on the needs of the customer.

Key Systems

The most common type of access control is with mechanical lock and key systems. Though these devices are not controllable in an automation system, they have established some of the concepts about the control of access and hierarchies that are carried through to electronic systems. The various levels of key security include a single key, master key system, grandmaster key system, and great-grandmaster key system.

A single key unlocks one type of lock by one type of key. A facility may have multiple doors in a building keyed the same (using locks with identical keys) for convenience. A master key system, however, keys each door individually, but has a master key that can open all doors within the master key system. This provides two levels of access. This key system is commonly used in commercial buildings such as office buildings with several tenants. Each tenant possesses a key that will only open the door to individual office space, while the building management personnel can open any door within the master key system with the master key.

A grandmaster key is one level higher than a master key. In this system, a grandmaster key can open all doors within multiple master key systems. This type of system can be used in a building that has multiple floors or multiple divisions within an organization. For example, a two-story building has a master key system for each floor. The grandmaster key would open any door on either master key system, unless a specific door is not a part of either master key system.

The hierarchy concept can be extended as far as needed, resulting in great-grandmaster key systems and beyond. For example, a great-grandmaster key system will open any door within each grandmaster key system and master key system, unless the specific door is not part of the master key or grandmaster key systems.

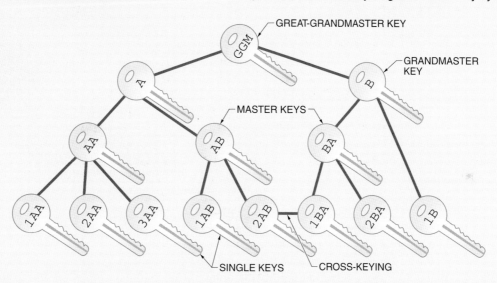

Sometimes a common entrance or door must be accessible by people with keys under different master key systems. This is accomplished by cross keying. Cross-keying is a keying method for master key systems that allows a lock cylinder to be opened with two or more different keys that would not normally operate together.

A *mechanical keypad* is a mechanical locking device that can be released by pressing a sequence of numbers on the keypad. **See Figure 9-2.** These mechanical keypads are relatively inexpensive and do not require power to operate. The mechanical keypad operates on principles similar to a combination lock. When the correct sequence of numbers, typically between two and six digits, is entered the lock is released.

Intelligent Control. Intelligent control involves electronics and computer systems to make control decisions. This is the type of system that can be integrated with building automation systems because the information and communication is based on digital signals. These are also the most flexible systems. Since all access information is stored electronically, it is easily changed, updated, and upgraded.

Mechanical Keypad

Figure 9-2. Mechanical keypads are an example of mechanical control of access to certain areas.

Access Control System Software

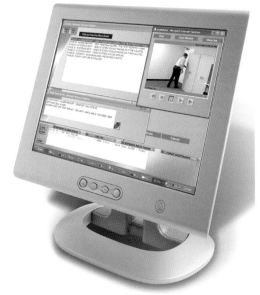

GE Security

Figure 9-3. Access control software monitors access control status, logs events, and integrates with other building systems.

Access Control System Software

Access control systems are conceptually very similar to key systems and use the same hierarchy of access rights. They also give the administrator many options regarding the system operation and level of security. These options are programmed into the access control system software.

Access control system software handles functions such as creating and maintaining authentication records, holding specialized information related to the operation of access control devices, logging events, and generating reports. **See Figure 9-3.** This information is typically stored by the software in a secure database. The administrator programs and manages the system with the users' access rights. *Access rights* are the collection of access control information about a user related to the user's access to certain areas, what times the user is allowed into those areas, and which doors are accessible. Additionally, one or more operators can be given the task of maintaining cards and entering minor programming details, such as time zone and holiday schedule.

The features programmed into most systems include card holders, time zones, access levels, access groups, momentary unlock, anti-passback, supervisor first, and conditional unlock. Though there are many other functions available in access control systems, these are the most common.

Card Holders. A *card holder* is the information identifying an access control system user and the user's access rights. A card holder is made up of four types of information:
* The identity of the user, such as their name and personal identifier.
* Information on their type of authentication, such as the type of card, access code, biometrics, or a combination of the technologies.
* The time periods when access is permitted, such as the time of day, days of the week, and weekends and holidays.
* The areas to which access is permitted, and the functional group the person is associated with.

Time Zones. A *time zone* is a specific range of time for an access control function. A time zone is typically assigned to a specific function, such as when a card holder is granted access. However, a time zone may have many other functions in a complex system. A single time zone may also be used for various card holders, as well as for control functions. Timed event schedules may be programmed to unlock doors between certain hours of the day or these events could be programmed to trigger relays to turn on certain lights at programmed times.

Access Levels. An *access level* is the level of security for a group or single card holder. An access level may define a high level of access to multiple doors or areas or it may define only one door or specific area.

Access Groups. An *access group* is a group of users with the same access rights. For example, a group called "maintenance" is defined, which grants access to a group of doors in a warehouse and the mechanical rooms of a building. **See Figure 9-4.** Multiple access groups may be assigned to a single cardholder to create different access strategies. For example, a maintenance supervisor may be part of both the "maintenance" access group and the "management" access group, allowing access to additional areas.

Momentary Unlock. *Momentary unlock* is an access control function that keeps a door unlocked for a certain amount of time after a valid authentication. A guard or administrator also has the ability to unlock a door momentarily for someone without proper credentials but who requires access to an area.

Anti-Passback. *Anti-passback* is an access control function that does not allow a credential to be used twice within a certain period or until the credential has been used to exit the area. This prevents users from passing an access card or other credential back to another user to be used again right away.

Supervisor First. *Supervisor first* is an access control function that allows no one into a secure area until a supervisor has made a valid read into that area.

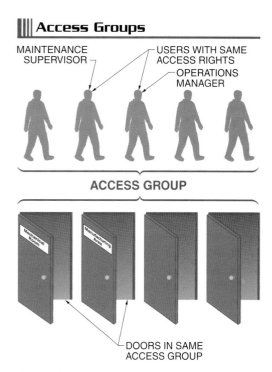

Figure 9-4. *An access group is a group of people with the same access rights to the same secure areas.*

Authentication

Authentication is the process of identifying a person and verifying their credentials. Authentication involves the collection of some type of information from the user, such as an access card identifier, memorized code, or fingerprint pattern. The information is then compared to information stored in the access control database for that person, which was generated and stored when the user was added to the system. **See Figure 9-5.** A match results in authentication. Once authenticated, the user is allowed to proceed with the requested action, such as entering through a doorway or accessing secure materials. A failed authentication results in a denial of the request. It may also trigger a security system response, such as sounding an alarm or alerting guards.

The primary advantage of an access control system is that it can be easily changed at any time. For example, if an access card is lost or stolen, the card can be deleted from the system and a new card issued, preserving the security of the restricted areas.

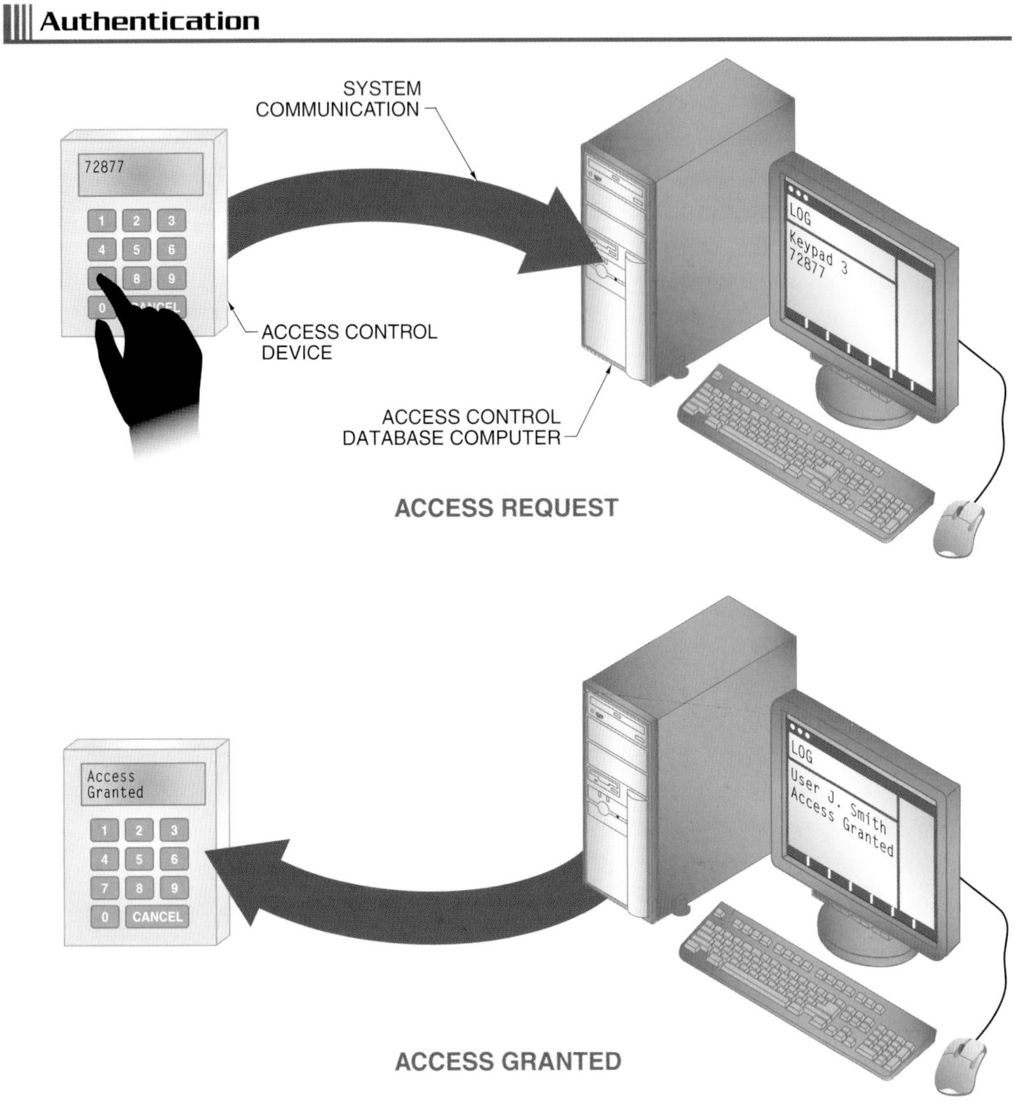

▌▌▌ Authentication

ACCESS REQUEST

ACCESS GRANTED

Figure 9-5. During the authentication of credentials, the access control device communicates with the computer with the access control database to confirm a match with the stored information.

Communications

There are many methods used in access control systems for communications, including digital, serial, and network communications. Digital communications are used to communicate between control panels and the server. Serial communications are commonly used by older systems, as well as some newer systems to communicate between a control panel and the server. Network communications are used to communicate with both a control panel and a server from anywhere inside or outside of a building.

ACCESS

Access is the characteristic that is monitored and controlled by the access control system. *Access control* is the ability to permit or deny the use of something by someone. Most systems control the access of people to certain building areas, such as file storage rooms, warehouses, offices, and vaults. Access control systems can also be used to protect small enclosures, such as jewelry cases or building control panels, from unauthorized users. These prevent theft or vandalism of valuable or critical items. Computer systems commonly use passwords to control access to sensitive electronic files. Some laptop computers and computer mice even include additional access control features, such as fingerprint scanners or card readers, or additional levels of access control. High-security applications may include multiple levels and types of access control.

ACCESS CONTROL SYSTEM CONTROL DEVICES

Effective access control systems can be composed of any number or combination of control devices. Both large and small access control systems use many of the same components. Common components include computers, control panels, credentials, barriers, door position switches, and request to exit devices. Though other components can be used within an access control system, these are the most common components.

Supervisory Computers and Control Panels

Access control systems must react to events and make intelligent decisions based on those events. A *supervisory computer* is a computer that stores access control database information and makes intelligent decisions for the system.

A *control panel* is a panel that contains electrical connections and control interfaces for a control system. Control panels connect to smaller control devices, such as keypads, and share their information with the supervisory computer. Control panels may be limited to this interface function or may be able to make access control decisions like the supervisory computer. In a large access control system, detection and action are managed electronically through control panels. These panels accept inputs from detection devices and issue instructions (outputs) to action devices. Control panels can also communicate with other control panels and computers throughout the system.

The two common configurations of supervisory computers and control panels in an access control system are the central database system and distributed processing. **See Figure 9-6.**

Honeywell International, Inc.

Some security systems can be armed or disarmed from a handheld keyfob device.

Access Control System Configurations

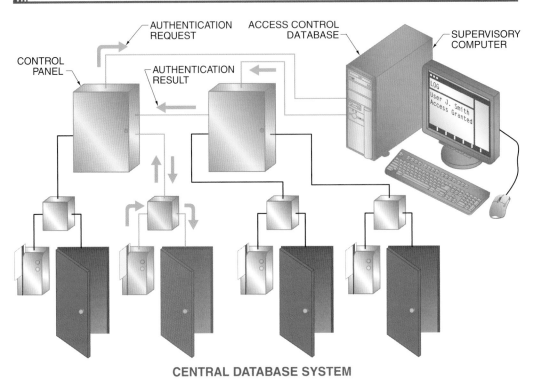

CENTRAL DATABASE SYSTEM

DISTRIBUTED PROCESSING SYSTEM

Figure 9-6. Depending on the access control system configuration, control panels may just pass along information or be involved in decision making.

Central Database Systems. A central database system requires a server or supervisory computer to make all of the intelligent decisions for the entire system. Control panels may be used, but only as connection points for the devices to communicate with the server. This design is an inexpensive system, but the entire system is dependent on the server. The system is completely disabled if the server goes down.

Distributed Processing Systems. In distributed processing, multiple devices can make intelligent decisions. A server or supervisory computer maintains the database information for the system. However, the control panels share many of the decision-making functions.

This method is more expensive to implement, but produces a robust system that tolerates some failures. The system remains at least partially operable in the event of a loss of system communications. The control panels buffer (store) the transactions and unbuffer (transmit) all transactions back to the server when system communications are reestablished.

Credentials

A *credential* is a method of authenticating an access request. A credential can be an access card that is inserted or held close to a reader, a biometric pattern, or a personal identification number (PIN) that is manually entered.

Access cards (also commonly referred to as badges) are typically the same size as credit cards. An *access card* is a plastic card encoded with some type of identification number that, when read by an access control device, allows access. **See Figure 9-7.** There are many different technologies for encoding access control cards, but the most common are barium ferrite, magnetic stripe, bar code, Wiegand, proximity, and smart cards. **See Figure 9-8.** Access cards can also be printed with user names and photos, allowing visual identification by security guards or other users. Some card technologies can be manufactured together into the same card, such as magnetic stripes and smart chips. These may be necessary for access to areas secured by different types of readers.

Access Card Credentials

Honeywell International, Inc.

Figure 9-7. Access cards are carried with the user and presented to a card reader for authentication.

A *reader* is a device that reads access cards or biometric patterns. Readers come in a variety of shapes and sizes and with features such as keypads and weatherproof housings. While readers are categorized primarily by the credential technology that they read, they are also referred to by the way they read it, such as touch, insertion, and swipe. Touch readers require that the access card touch a read surface of the reader. Insertion readers require that the access card be inserted into the reader, and then pulled out. Typically, only half of the access card enters the reader. Swipe readers require that an edge of the access card pass completely through the reader.

Barium Ferrite Credentials. Barium ferrite is a compound that can be used to magnetically store information in the form of bits. Barium ferrite cards contain up to 40 dots of this material, embedded within the card in a certain pattern. Each dot is magnetically polarized in one of two possible orientations, making a unique number for the card. The card is held up against a reader, which reads the orientations and locations of dots, identifying the card number.

||| Access Cards

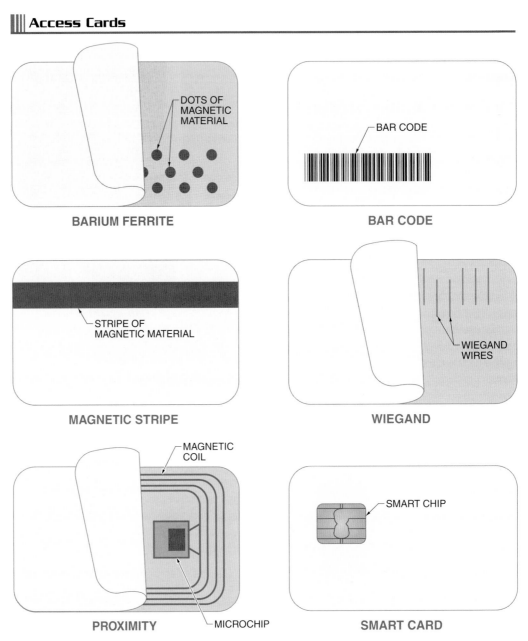

Figure 9-8. A variety of different technologies are used to embed access control information into access card credentials.

Because the magnetic dots are inside the card where they are not visible, this type of card is quite difficult to reproduce, making the cards relatively secure. However, barium ferrite cards are not commonly used with newer access control systems because the readers require a great deal of maintenance and also because strong magnetic fields around a card reduce its readability.

Bar Code Credentials. Bar code cards use a series of black lines of varying width to encode the identifying number. This is the same technology used in UPC labels on retail merchandise. However, a UPC label typically uses a 6-digit or 12-digit code, whereas an access card typically uses a much larger code. Code 39, which is also known as "code 3 of 9," is a code with up to 44 characters that includes any of the 225 ASCII characters. It can also include leading and trailing spaces, which allow two or more bar codes to be scanned as one long bar code.

Bar code access cards are typically read by swiping. As the bar code passes through the reader, a laser inside reads the light reflected by the white spaces between the lines and interprets this pattern of pulses as the identifying number.

This technology is among the least expensive to produce, but is also one of the easiest to duplicate, making this technology less secure than other technologies. To minimize the risk of duplication, a translucent patch is often placed over the bar code. If the card is photocopied, the duplicate shows only a black patch. Additionally, the bar code can be covered with an opaque film so that the bar code can only be read by a UV bar code reader.

Magnetic Stripe Credentials. Magnetic stripe cards are among the most common, most recognizable, and least expensive types of access cards to manufacture. These cards are commonly used for credit cards, hotel room keys, and even disposable transit tickets.

Magnetic stripe cards (commonly referred to as magstripe cards) contain a coated magnetic recording tape near the edge of the card. Similar to barium ferrite cards, magnetic stripe cards store information by the magnetic orientation of tiny magnetic particles in the stripe. The magnetic stripe contains three tracks of information. The way information is stored on the tracks varies, but there are some commonalities. Track 1 is typically used for information such as a name and or a title of a person. Track 2 is typically used for the access control identifying number. Track 3 is typically used for a facility code that can be used to increase the number of possible unique card numbers. For example, if this credential system has a range from 0001 to 9999, there are 10,000 possible card numbers. By adding facility codes, the total number of possible codes increases significantly.

Magnetic stripe readers are either swipe or insertion readers. They contain magnetic read heads that detect patterns on a magnetic stripe card when the stripe is passed across the read head. Different read heads are positioned to read different tracks. These patterns are then converted into the credential information.

The main drawback of magnetic stripe cards is that, as with barium ferrite, magnetic fields around the card can damage or destroy the information on the magnetic stripe. Also, constant wear on the stripe from the read heads may eventually damage the magnetic stripe, making the card difficult to read.

Wiegand Credentials. Wiegand access cards are embedded with wires made from a special alloy. The wires are laid out on the card in two parallel rows. Each of the wires represents a bit of information. The top row is a 0 bit and the bottom row represents a 1. The pattern of wires in the card represents the identifying number. The most common standard format contains 26 bits of information.

As the card passes through the magnetic fields in a Wiegand reader, each wire generates a pulse to the reader representing either a 1 or 0 bit. The reader senses the entire pattern of pulses and generates the identifying number.

Since the wire used to manufacture these cards is protected by a patent, they are very difficult to forge or reproduce, making them one of the more secure technologies.

Proximity Credentials. Proximity credentials require no actual contact between the access card and the reader, making them especially convenient. They only require the card to be present within a certain distance, or proximity, to the reader. **See Figure 9-9.**

||| Proximity Credentials

Honeywell International, Inc.

Figure 9-9. Proximity credentials require only that the access card be brought near the reader for authentication.

A proximity card has a magnetic coil and a microchip inside the card. The microchip can be either a read-only chip or a read/write chip, if the information must be reprogrammable. The proximity reader produces a magnetic field. When a proximity card gets close to the reader, the reader's magnetic field excites the coil in the card, providing power to the microchip. The microchip then transmits the card information back to the reader via the coil.

> The Wiegand protocol is a de facto communications standard that arose from the popularity of Wiegand card readers. The original Wiegand format included one 1 parity bit, 8 bits of facility code, 16 bits of ID code, and a stop bit, for a total of 26 bits.

The reading distance is dependent on the size of the reader and whether the card is an active card or a passive card. An active card contains a small lithium battery that increases the range of the proximity card, up to 100′. A passive card does not contain a battery and is limited in range to about 30″ or less, depending on the reader.

Smart Card Credentials. Smart card technology is the latest technology that is used in access control. A smart card has a microchip embedded either on the surface or inside the card. This chip can contain either non-volatile memory only or a microprocessor with volatile memory. Both types can store very large amounts of information, but a microprocessor type can carry out its own security functions such as encryption and mutual authentication.

A contact smart card is inserted into a reader to make physical contact with the sensors on the reader to transmit information. Alternatively, a contactless card works similarly to a proximity card. Coils in both the reader and the card provide power to the microcontroller and information is transmitted via radio frequency signals. This technology is a very secure technology and is likely to surpass the proximity card as the most commonly used card technology.

Biometric Credentials. Access control authentication has typically relied on credentials that a person carried, such as an access card, or memorized, such as an access code number. Biometric readers are newer technologies that utilize a person's unique characteristics to increase security and convenience. A *biometric reader* is an access control reader that analyzes a person's physical or behavioral characteristics and compares them to a database for authentication. Physiological biometric technologies measure a physiological characteristic such as a fingerprint or facial pattern. Behavioral biometric technologies measure a behavioral pattern such as speech or handwriting.

Physiological biometric technologies include fingerprint, hand geometry, facial recognition, iris scan, and retinal scan. Fingerprint technology is among the most accurate biometric technologies and analyzes the patterns of the skin on the fingertip. Hand geometry technology analyzes the shape of the hand. Facial recognition measures characteristics of a person's face when scanned by a camera. Iris and retinal scanning technology looks at specific characteristics of the human eye: the colored ring or the internal blood vessels, for instance.

Common behavioral technologies include voice recognition and signature recognition. Voice recognition analyzes unique vocal characteristics of a spoken phrase, such as air pressure and vibrations over the larynx. Random voice interrogation prevents a recording from being used to bypass the system by recording several possible spoken phrase templates for each person. Signature recognition measures the signature pattern, writing speed, and pen pressure used to create the signature.

Each type requires some type of body feature image or behavior data, which is then analyzed for a unique set of patterns. The patterns are compared to those stored in the access control database for that person. A sufficient match results in authentication.

Biometric authentication is processor intensive and more prone to error than access card authentication. However, biometrics are generally more secure, unique, difficult to falsify, and convenient for the user (since nothing has to be carried or memorized). **See Figure 9-10.** Also, newer technology continues to increase the accuracy and speed of biometric authentication.

Comparison of Biometric Credentials

Biometric	Universality	Uniqueness	Permanence	Collectability	Difficulty to Falsify
Fingerprint	Medium	High	High	Medium	High
Hand Geometry	Medium	Medium	Medium	High	Medium
Face	High	Low	Medium	High	Low
Iris	High	High	High	Medium	High
Retina	High	High	Medium	Low	High
Voice	Medium	Low	Low	Medium	Low
Signature	Low	Low	Low	High	Low

Figure 9-10. Biometric credentials are generally more secure than other technologies, though there are some differences between the common types of biometrics.

Electronic Keypads. An electronic keypad is the simplest type of electronic control. An *electronic keypad* is an authentication device that uses a manually entered numeric or alphanumeric code as a credential. **See Figure 9-11.** This electronic keypad typically has a pushbutton arrangement with digits 0 through 9, similar to a telephone, though some include additional special function keys. Most electronic keypads accept combinations of two to six digits and can store up to 99 access codes. When a correct access code is entered, a relay is triggered that releases a door or some other barrier. Many electronic keypads have the ability to report activity to a computer through a serial connection or downloadable event logs.

Common uses for stand-alone electronic keypads include locations such as daycare facility entrances and mechanical rooms. These keypads are an inexpensive solution for limited access control needs. However, this type of keypad is vulnerable to manipulation because all of the control components are located together at the entryway.

Mixed Credential Technologies. Mixed technology devices use a combination of two or more credential technologies in a single unit to increase the security of the authentication process. Mixed technologies can be a combination of an access card and keypad, access card and biometric, two or more biometric technologies, or any other combination of technologies. **See Figure 9-12.**

Electronic Keypad

GE Security

Figure 9-11. Electronic keypads are among the most common access control devices and do not require any physical credential such as an access card.

The most common biometric credential used for security systems is fingerprint identification. This type of credential is often combined with other biometrics for high-security applications. Authentication with multiple biometric technologies requires the different biometric readings to be taken at the same time to reduce the ability to falsify identification.

Mixed Credential Technologies

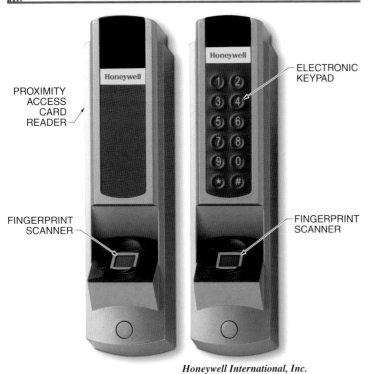

PROXIMITY ACCESS CARD READER

ELECTRONIC KEYPAD

FINGERPRINT SCANNER

FINGERPRINT SCANNER

Honeywell International, Inc.

Figure 9-12. Many access control devices require more than one type of credential to be presented and verified for authentication.

Barriers

Barriers clearly define the boundaries of an access-protected area, delaying entry into or through the protected area, and directing people or traffic to proper entry points. Barriers in access control systems include door closers, locks, gates, turnstiles, and mantraps. A *barrier* is a manmade obstacle strategically placed to prevent or limit access. Barriers are designed to discourage penetration. For example, a locked door stops a person from entering a secured area without the use of force.

Barriers provide security, but also slow the flow of traffic. This may be desirable or undesirable, depending on the situation. For example, access control on tollways moderates the speed of vehicular traffic, which increases safety. However, the railroad and trucking industries pioneered continuous-motion access control, allowing the vehicles to roll through checkpoints without stopping, which saves time and energy.

Access control barriers include a closing mechanism, an electrically activated lock, sensors (switches) that determine whether or not the door or barrier device is properly closed, and computerized control either in the locking device itself or in a nearby control panel.

Door Closers. A *door closer* is a device that pulls a door shut after an access has been gained. Door closers are springs with strong enough tension to pull doors completely shut after use, yet not so strong that it makes opening the door difficult or warps the door during normal use. Door closers are also used on other barriers. For example, a timer on an automatic gate motor reverses its direction after a specific amount of time.

Electric Strikes. An *electric strike* is an electrically operated lock that unlocks by moving the door strike. A door strike is the metal plate on a doorframe that prevents a door from opening without the latch-bolt being retracted. The electric strike, which requires power, is installed in the doorframe, while the non-powered latch is installed in the door. When activated, the strike moves out of the way, releasing the latch-bolt and allowing the door to open. **See Figure 9-13.**

Electric Strikes

MOVABLE DOOR STRIKE

Figure 9-13. An electric strike releases a locked door by allowing the strike plate to swing away from the bolt.

There are two configurations for electric strikes: fail-safe and fail-secure. A *fail-safe lock* is an electric door lock that is normally unlocked when unpowered. A *fail-secure lock* is an electric door lock that is normally locked when unpowered. This distinction is important when considering the state of the locking device if the power is lost to the facility. Local, state, and federal regulations also require certain doors or access points, such as emergency exits, to have fail-safe locks so that people can exit a building in the event of an emergency.

Magnetic Locks. A *magnetic lock* is a lock that uses magnetic force to hold a door closed. Magnetic locks are also called mag locks. A magnetic lock is installed on or in the doorframe and an electric lock armature is installed on the door. When electric current is applied to the magnetic lock, a strong magnetic force between the lock and the lock armature keeps the door closed. The holding force available for these locks varies from 500 lb to 3000 lb.

Power is required to keep the door locked, but not to keep it unlocked. Therefore, magnetic locks are fail-safe; if power is lost, the locks remain unlocked.

Electric Locks. An *electric lock* is a lock that uses a solenoid to actuate the bolt. The solenoid is energized or de-energized in order to lock or unlock the lock. The two basic types of electric locks are cylindrical and mortise locks. A cylindrical lock is a common doorknob arrangement. A mortise lock is housed in a rectangular metal unit that is embedded at the edge of the door. Since these devices are mounted in the door itself, and not the doorjamb, they require electricity to be run to the door, typically via a flexible cable of the secure side of the door.

An electric deadbolt lock can come in either a cylindrical or a mortise style and does not have a spring-loaded retractable latch like other locks. The electrically powered deadbolt is fitted into either the jamb or the door itself. When activated, it protrudes or swings into a mortised strike plate on the adjoining surface.

Gates. Gates control access of motor vehicles into a secure area. Some gates open to the side and some have an arm that moves up and down. **See Figure 9-14.** Both types of gates are operated by electric motor controls. If a gate requires a credential reader such as an access card reader, the gate does not operate until proper authentication is achieved.

Gates

Figure 9-14. Gates provide access control for large points of entry, such as driveways.

An access control system must release all door locks for emergency egress from a building upon receiving a fire alarm signal. The locking devices are controlled by the access control system, but the unlocking of the doors is commanded by the fire alarm system.

Turnstiles. Turnstiles are used to limit traffic flow by creating a narrow alley that allows only one person to enter at a time. There are two common types of turnstiles, manual and electronic. Manual turnstiles control traffic flow, but do not require credentials. These are common in applications such as transit stations. Electronic turnstiles are similar, but do require authentication before allowing a person to pass. If a person passes through the turnstile without proper credentials, an alarm is triggered.

Turnstiles help eliminate piggybacking by forcing people to enter an area individually. *Piggybacking* is the action of a person following directly behind another person to avoid presenting a credential for authentication.

Mantraps. A *mantrap* is a barrier configuration that holds a person in a small area until the door or gate just passed through is secured. **See Figure 9-15.** The mantrap is a common configuration in high-security areas such as prisons.

Mantraps

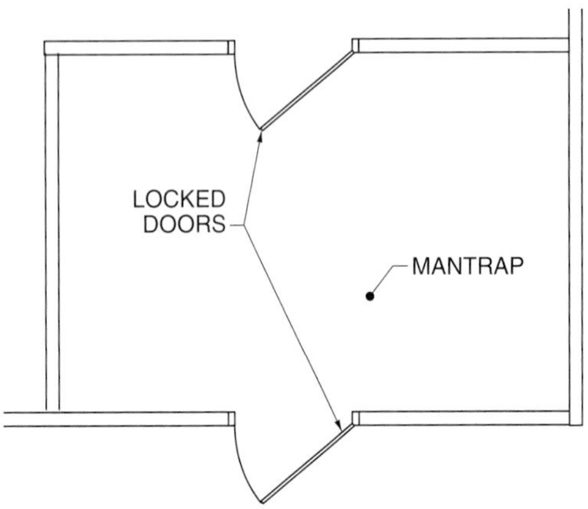

Figure 9-15. Mantraps allow access through a second door only when the first door is securely closed and locked.

Door Position Switches

Access control systems are also used to monitor the status of doors and other barriers with door position switches. A *door position switch* is a magnetic contact that indicates an abnormal condition of a door or barrier. These magnetic contacts are the same type of contacts used in security systems and function similarly. The door position switch indicates an abnormal condition to the access control database, which displays the information on a computer screen monitored by a guard. For example, if a door is propped open while an employee steps outside, after a specified amount of time (such as 10 seconds), the access control system displays an alarm condition or door open signal.

Request-to-Exit Devices

While entering a secure area requires authentication, sometimes exiting one does not. A *request-to-exit device* is a device that automatically authorizes an exit from a secure area without needing credentials or without causing an alarm condition. These devices are installed on the interior of the secure area and bypass the door position switch or report to the controller that an authorized exit is occurring.

Three common devices that are used for request to exit are manual pushbuttons, motion sensors, and touch-sensitive or mechanical switches. The manual pushbutton is mounted on the wall near the exit door. When pushed, it sends an input to the controller, which either bypasses the door position switch or triggers an output to release the door. A motion detector operates using PIR technology similar to that in lighting or security systems. The detector is mounted on the top of a doorjamb or near a doorway. **See Figure 9-16.** The detector typically points downward, focusing on a very small area around the door opening. It also has multiple relay contacts, which are used to trigger inputs and/or outputs on the controller. A touch-sensitive switch on a panic-bar-style lock allows free egress from a secure area with the push of the bar on the door.

ACCESS CONTROL SYSTEM CONTROL APPLICATIONS

An access control system can be used to control aspects of other building systems either by choice or requirement (due to local, state, or federal codes). Regardless of the reason, the access control database helps the system to make intelligent decisions based on events and/or inputs.

⫼ Request-to-Exit Devices

REQUEST-TO-EXIT PUSH BUTTON ⅃

Figure 9-16. *Pushbuttons are a common type of request-to-exit device.*

⫼ Lockdown

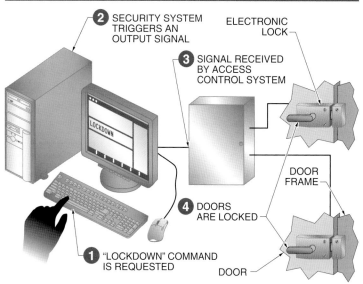

② SECURITY SYSTEM TRIGGERS AN OUTPUT SIGNAL

ELECTRONIC LOCK

③ SIGNAL RECEIVED BY ACCESS CONTROL SYSTEM

DOOR FRAME

④ DOORS ARE LOCKED

① "LOCKDOWN" COMMAND IS REQUESTED

DOOR

Figure 9-17. *Lockdown is an integration of the security and access control systems that automatically locks all the electrically actuated door locks in a building.*

Most access control systems have multiple control relays built into the control panels to allow for driving external systems. These relays typically have normally open and normally closed contacts, which allow for any wiring configuration that may be needed. In addition, the database software controlling the relays can be programmed to pulse, latch, or time the activation of a relay.

Lockdown

Access control and security systems have been integrated for many years, mostly because these two systems utilize many of the same devices. However, recent innovations in intelligent control have made even more advanced integration between these two systems possible. Schools and hospitals are two applications that often integrate security systems and access control systems.

Operating Conditions. Typically a signal from the security system is sent to the access control system to command an action. The access control system responds to the input by sending an output to a controlled device based on information within the database. **See Figure 9-17.** Common examples of this type of integration include lockdown situations at schools and hospitals. A command by the security system instructs the access control system to lock all of the doors.

Control Sequence. The control sequence for an access control system to control the lockdown of a building is as follows:

1. A "lockdown" command is entered into the security system by manually pressing a special button on the security keypad.
2. The security system control panel receives the signal and triggers a programmable output signal to an input on the access control panel.
3. The access control panel receives the input signal and decides on the required action based on its program.
4. The access control panel sends an output signal to all locking devices, locking down the building until a reset signal has been sent.

Closed-Circuit Television (CCTV) Control

A closed-circuit television (CCTV) system can provide video documentation of the circumstances of an alarm condition or unusual event. However, until CCTV systems were integrated with security systems or access control systems, the chances of catching problems required either luck or reviewing of long durations of previous recordings. Integrating a CCTV system with an access control system provides an easier method of recording and documenting problems. For example, a person attempting to enter a restricted area without the proper access rights or a credential can be recorded for future review.

Operating Conditions. Similar to the lockdown sequence, a command signal is shared between the access control system and the other system. In this scenario, a signal from the access control system is sent to the CCTV system to command it to begin recording. **See Figure 9-18.** This simple scenario, however, assumes that a camera is pointed at the necessary area.

Control Sequence. The control sequence for an access control system to control the operation of a CCTV system is as follows:

1. A person presents a false credential to a card reader.
2. The reader checks the database to verify the credential and reports an invalid read to the control panel.
3. The control panel sends an alarm condition to the supervisory computer.

Refer to Quick Quiz® on CD-ROM

4. The supervisory computer displays the alarm condition on a guard monitor and sends a signal to the CCTV system's video recording device.
5. The video recording device displays the video signal from the indicated camera on an alarm monitor and begins recording.

Lighting and HVAC System Control

An access control system can be integrated with a building automation system to control many different types of systems. Since the access control system contains a database to make intelligent decisions based on certain inputs, as well as detailed action items, it is well suited for integration with automated building systems such as lighting and HVAC control systems.

Operating Conditions. Integration features are accomplished primarily through programming in the access control database. The program ties an authentication request to a certain output from the access control panel. The output signal is sent to the control panel of another system, which causes an action on the system based on this input. **See Figure 9-19.** This feature can also use any access control software feature, such as time zones, to further enhance the interaction.

▌▌▌ Closed-Circuit Television (CCTV)

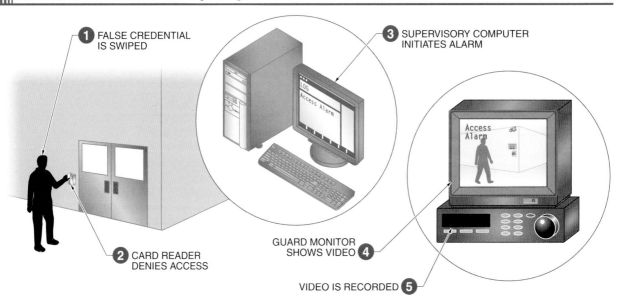

Figure 9-18. Access control systems can command a CCTV system to record video of invalid access attempts.

Lighting and HVAC Control

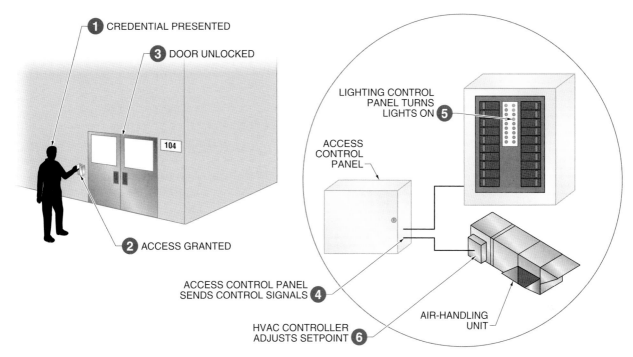

Figure 9-19. Access control systems can be used to activate lighting and HVAC systems when a person enters a secure area.

Control Sequence. The control sequence for an access control system to control lighting and HVAC systems is as follows:

1. A person presents a credential to a reader.

2. The reader authenticates the credential with the access control database.

3. The door is unlocked or released, permitting entry.

4. The access control panel sends signals to the lighting control system and HVAC control system.

5. The lighting control panel receives the signal and executes a preprogrammed action, such as turning ON a certain bank of lights near the access control device.

6. The HVAC controller receives the signal and executes a preprogrammed action, such as raising a temperature setpoint.

The unlocking of a door for access can also be used to control other building systems related to the accessed building area.

Summary

- Access control provides security for persons and/or property by restricting unauthorized persons' access to a building or specific area.

- Access control also restricts unauthorized people from entering dangerous environments.

- Human control involves physically opening a door or gate or placing a barrier to deny access to a certain area.

- The most common form of mechanical control uses a lock and key.

- When a correct access code is entered on an electronic keypad, a relay is triggered that releases a door or some other barrier.

- Access control system software handles functions such as creating and maintaining authentication records, holding specialized information related to the operation of access control devices, logging events, and generating reports.

- A time zone is typically assigned to a specific function, such as when a card holder is granted access.

- Multiple access groups may be assigned to a single card holder to create different access strategies.

- Authentication for a particular person is accomplished within an access control system by the information contained in the database.

- The primary advantage of an access control system is that it can be easily changed at any time.

- High-security applications may include multiple levels and types of access control.

- Access control systems must react to events and make intelligent decisions based on those events.

- Control panels may be limited to an interface role or may be able to make access control decisions like the supervisory computer.

- A distributed processing system is more expensive to implement than a central database system, but is a robust system that tolerates some failures.

- A credential can be an access card that is inserted or held close to a reader, a biometric pattern, or a personal identification number (PIN) that is manually entered.

- Readers are categorized by access card technology and the way they read a card, such as touch, insertion, and swipe.

- Magnetic stripe cards are among the most common, most recognizable, and least expensive types of access cards to manufacture.

- Proximity credentials require no actual contact between the access card and the reader, making them especially convenient.

- Biometric readers are newer technologies that utilize a person's unique characteristics to increase security and convenience.

- Biometric methods are generally more secure, unique, difficult to falsify, and convenient for the user.

- Barriers clearly define the boundaries of an access-protected area, delaying entry into or through the protected area, and directing people or traffic to proper entry points.

- Access control barriers include a closing mechanism, an electrically activated lock, sensors (switches) that determine whether or not the door or barrier device is properly closed, and computerized control either in the locking device itself or in a nearby control panel.

- Most access control systems have multiple control relays built into the control panels to allow for driving external systems.

Definitions

- An *access control system* is a system used to deny those without proper credentials access to a specific building, area, or room.

- *Human control* is a form of access control that requires human involvement.

- *Mechanical control* is a form of access control that uses a mechanical device to maintain a closed area.

- A *mechanical keypad* is a mechanical locking device that can be released by pressing a sequence of numbers on the keypad.

- *Access rights* are the collection of access control information about a user related to the user's access to certain areas, what times the user is allowed into those areas, and which doors are accessible.

- A *card holder* is the information identifying an access control system user and the user's access rights.

- A *time zone* is a specific range of time for an access control function.

- An *access level* is the level of security for a group or single card holder.

- An *access group* is a group of users with the same access rights.

- *Momentary unlock* is an access control function that keeps a door unlocked for a certain amount of time after a valid authentication.

- *Anti-passback* is an access control function that does not allow a credential to be used twice within a certain period or until the credential has been used to exit the area.

- *Supervisor first* is an access control function that allows no one into a secure area until a supervisor has made a valid read into that area.

- *Authentication* is the process of identifying a person and verifying their credentials.

- *Access control* is the ability to permit or deny the use of something by someone.

- A *supervisory computer* is a computer that stores access control database information and makes intelligent decisions for the system.

- A *control panel* is a panel that contains electrical connections and control interfaces for a control system.

- A *credential* is a method of authenticating an access request.

- An *access card* is a plastic card encoded with some type of identification number that, when read by an access control device, allows access.

- A *reader* is a device that reads access cards or biometric patterns.

- A *biometric reader* is an access control reader that analyzes a person's physical or behavioral characteristics and compares them to a database for authentication.

- An *electronic keypad* is an authentication device that uses a manually entered numeric or alphanumeric code as a credential.

- A *barrier* is a manmade obstacle strategically placed to prevent or limit access.

- A *door closer* is a device that pulls a door shut after an access has been gained.

- An *electric strike* is an electrically operated lock that unlocks by moving the door strike.

- A *fail-safe lock* is an electric door lock that is normally unlocked when unpowered.

- A *fail-secure lock* is an electric door lock that is normally locked when unpowered.

- A *magnetic lock* is a lock that uses magnetic force to hold a door closed.

- An *electric lock* is a lock that uses a solenoid to actuate the bolt.

- *Piggybacking* is the action of a person following directly behind another person to avoid presenting a credential for authentication.

- A *mantrap* is a barrier configuration that holds a person in a small area until the door or gate just passed through is secured.

- A *door position switch* is a magnetic contact that indicates an abnormal condition of a door or barrier.

- A *request-to-exit device* is a device that automatically authorizes an exit from a secure area without needing credentials or without causing an alarm condition.

Review Questions

1. What are the typical ways in which access control systems are used?

2. What are the typical functions of access control software, and who is responsible for maintaining the database?

3. Describe the process of authentication.

4. What are the two common roles of a control panel in an access control system?

5. What are the different ways in which a reader can read an access card?

6. What types of information are typically stored on a magnetic stripe access card?

7. How does biometric authentication compare with access card authentication?

8. How are barriers also used to limit the flow of traffic?

9. What is the function of a request-to-exit device?

10. How are access control systems typically connected to other building systems for integration?

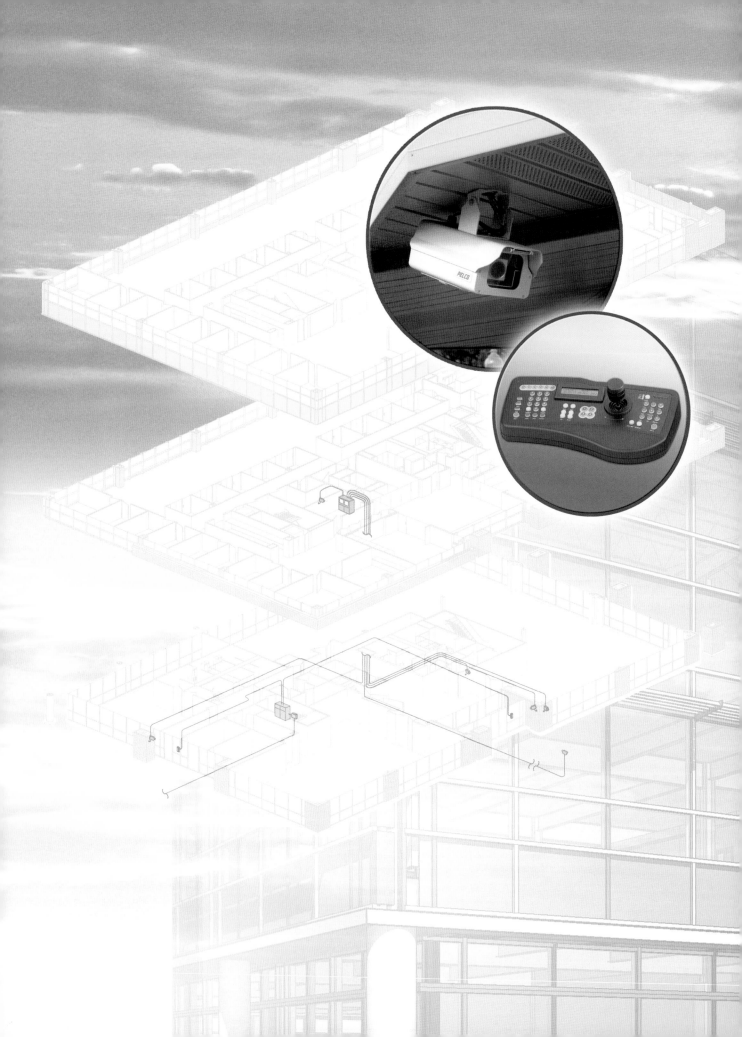

Chapter Ten

Voice-Data-Video (VDV) System
Control Devices and Applications

Voice-data-video (VDV) systems are the primary communication networks in a building. They include telephony, computer, and closed-circuit television (CCTV) systems. These systems are sometimes used as the media for the transmission of control signals, but in some applications, can be controlled themselves. Common control scenarios include seizing a telephone line for emergency use, transmitting control signals with telephony tones, and repositioning CCTV cameras based on information about events provided by other building systems.

Chapter Objectives

- *Describe the different types of VDV systems.*
- *Evaluate how VDV control information can be used to integrate VDV systems with other building systems.*
- *Compare the features and advantages of analog and digital video equipment.*
- *Describe common control applications involving VDV systems.*

VOICE-DATA-VIDEO (VDV) SYSTEMS

A *voice-data-video (VDV) system* is a building system used for the transmission of information. A transmitting device converts the information into a time-varying signal made up of electrical or optical pulses. VDV systems are particularly useful for sharing information that could not have otherwise been shared because of distance. This signal is carried over a transmission medium, such as electrical conductors or radio frequency waves, to a receiver. The receiver monitors the transmission medium for signals and converts them back into useable information. For two-way communication, each end device must act as both a transmitter and a receiver. These devices are known as transceivers.

Transmission signals are limited by distance. The signal gradually degrades until it is no longer useful. The limits depend on the signal power, encoding, and the transmission medium. For greater distances, intermediate devices are used to repeat (amplify) and route a signal between its origin and destination.

VDV systems encompass several systems, each designed specifically for transmitting a certain type of information. **See Figure 10-1.** Telephony systems provide the infrastructure and control of point-to-point audio signals, which are typically used for voice transmission but can also be used for data communication. Data systems are any system connecting electronic equipment that exchanges data with other equipment. The most common example of this is a local area computer network, such as those within office buildings or schools. Closed-circuit television (CCTV) systems manage the transmission of video signals in a relatively small system, such as within a building or campus.

Telephony Systems

Early telephony systems used manually operated switchboards to connect a caller with the intended receiver. The switchboards were staffed with telephone operators who used patch cables and plug connectors to make connections between telephones. A *telephone* is a device used to transmit and receive sound (most commonly speech) between two or more people. These switchboard stations were located in a telephone exchange building. A telephone exchange could include hundreds of switchboards and operators to handle the telephone traffic. Long-distance telephone calls required multiple connections at different telephone exchanges to make a continuous connection between the caller and receiver. This was the beginning of what is now known as the public switched telephone network (PSTN).

The modern telephone system works in much the same way, but the connections are now made electronically. This makes the system faster, more accurate, and more reliable. Telephone exchange buildings are still large facilities, though they now handle millions of simultaneous calls. **See Figure 10-2.** Also, the electronics technology in telephone systems has made small-scale exchanges feasible for local systems, such as in offices. These provide many of the same features and capabilities for private telephone systems as the major exchanges do for the PSTN.

▎▎Voice-Data-Video (VDV) Systems

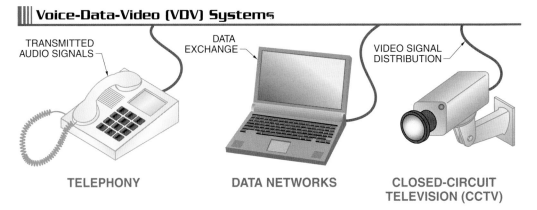

TRANSMITTED AUDIO SIGNALS

DATA EXCHANGE

VIDEO SIGNAL DISTRIBUTION

TELEPHONY

DATA NETWORKS

CLOSED-CIRCUIT TELEVISION (CCTV)

Figure 10-1. Voice-data-video (VDV) systems include telephony, data network, and closed-circuit television (CCTV) systems.

History of Telephony

The first commercially available electrical communications system was the telegraph, which used Morse code to send and receive information over great distances. With the invention of the telephone, two-way voice communication was possible. These devices changed varying sound pressure (voice) into mechanical energy with the movement of a diaphragm. The mechanical energy was then converted into analog electrical signals. The receiving telephone reversed the process to reproduce the sounds at the other end.

The telephone was followed closely by the invention of the radio, which also used Morse code to send and receive signals. The vacuum tube later made voice transmission over radio possible. Modulations schemes such as amplitude modulation (AM) and frequency modulation (FM) improved the quality and clarity of radio voice transmission. Throughout radio's heyday, amateur radio operators (hams) are credited with most of the advancements in radio technology. Although surpassed by the use of the cell phone and digital communications such as instant messaging and email, amateur radio still has a strong following and still offers very reliable communication services during major telephone system outages, such as those caused by natural disasters.

||| Public Switched Telephone Network (PSTN)

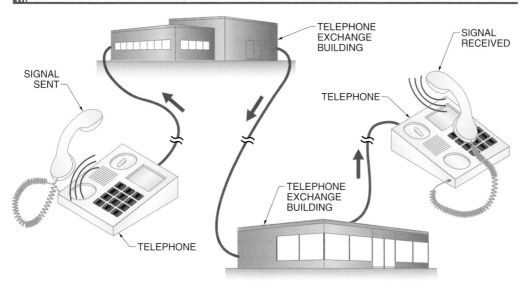

Figure 10-2. The public switched telephone network (PSTN) is a network of telephone exchanges that can be connected together to make a continuous circuit for each telephone call.

Key Systems. A *key system* is a multiline telephone system in which each telephone has direct access to each outside (PSTN) line. In a key system, the switching between connections occurs at the telephone. These telephones are distinguished by a pushbutton for each line. Key systems originally had electromechanical relays, which were eventually replaced by integrated circuits. A key system may also include single-line telephones, but each can only connect to one outside line.

Key system telephones are typically limited to a relatively small number of lines. Key systems do not include an intelligent switching system, which connects and disconnects calls based on programmed logic. Therefore, most key systems do not have the ability to make decisions about the use of pooled or shared outside telephone lines.

Private Branch Exchanges. The *private branch exchange (PBX)* is a telephone exchange that serves a small, private network. Improved electronics led to the development of separate switching devices that could handle more lines and calling features than a key system for a private telephone network. Given their similarity to PSTN exchanges, though on a smaller scale, these devices became known as private branch exchanges. PBXs are also known as private automatic branch exchanges (PABX) or electronic private automatic branch exchanges (EPABX).

PBX switches are at the heart of voice communications for small networks. **See Figure 10-3.** Like a computer, a PBX includes a central processing unit (CPU), which runs software to intelligently manage the network. With its programmed logic, the PBX controls local in-house dialing, local calls outside the premises, and long distance calls. All calls are supported by the switch for the duration of the call. PBXs include programming to minimize the probability of getting a busy signal when connecting to an outside line.

▌▌▌ Private Branch Exchanges (PBXs)

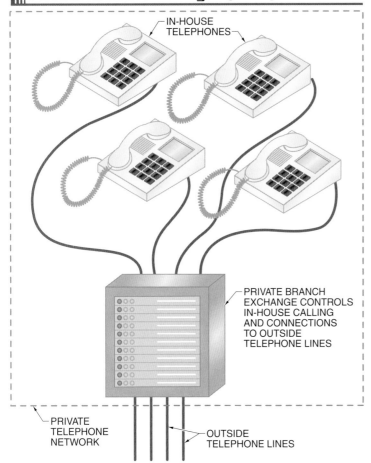

Figure 10-3. *Private branch exchanges (PBXs) connect in-house telephones together within a private network and manage the telephones' access to the PSTN.*

The PBX has two main functions. First, it establishes a connection between two telephones and maintains it for as long as required without interruption or interference. Also, it maps a dialed number to a physical telephone and ensures that the telephone is not already busy. The conference feature establishes connections between four to eight parties, including both outside lines and PBX extensions. Dual-tone multifrequency (DTMF) cards are needed for dialing out and receiving the touchtone signals dialed by each telephone.

A PBX may have additional features available, either built-in or with expansion cards. Station message detail recording (SMDR) provides information and reports on dialed numbers, times, and durations of calls made and received.

Voice-over-IP Systems. *Voice over Internet Protocol (VoIP)* is the transmission of audio telephone calls over data networks. This system makes use of IP networks, such as the Internet and/or local intranets, instead of the traditional telephone network. VoIP has the capability to deliver new services that combine voice and data. VoIP is becoming increasingly popular because it offers many traditional telephone services, such as long-distance calling, at a reduced cost.

Data networks transmit information differently than the traditional telephone network. A dedicated circuit is required to make a telephone call on the PSTN. IP networks, on the other hand, transmit the call in individual packets of information. **See Figure 10-4.** Packets from many different calls and data communications are transmitted over the same connections. IP devices along the way are able to distinguish and route each packet individually. The packets for a particular telephone call are then reassembled at the receiver and converted back into audio signals. With data networks becoming increasingly common, VoIP allows users to streamline their communications requirements by maintaining only the data network, instead of two separate systems.

Moving data on IP networks is efficient, but one of the challenges presented by IP telephony is connecting the devices that turn voice into IP packets and vice versa. Specialized telephones, Internet voice gateways, telephone systems, and other devices bridge the gap between the IP network and the sound of a voice. For a unified platform for VoIP equipment from different manufacturers, the telecommunications industry has established standards such as the widely adopted Session Initiation Protocol (SIP). IP networks use the Internet Protocol, and VoIP devices use SIP.

Closed-Circuit Television (CCTV) Systems

A *closed-circuit television (CCTV) system* is a system of video capture and display devices where the video signal distribution is limited to within a facility. CCTV systems are used primarily for security and surveillance applications, though they are also used for in-house television programming, robotics, live video of special events within a facility, remote process control, safety management, or other applications. Security applications require either full-time monitoring by security personnel or recording of video for later playback.

Security applications of CCTV systems include both overt and covert installations. Overt installations rely in part on the deterrence effect of highly visible CCTV cameras. That is, sometimes the possibility that someone is watching or recording their actions deters would-be criminals from committing criminal acts. Some facilities take advantage of this effect by installing dummy cameras to create or support the appearance of a secure area. Many security surveillance equipment manufacturers even sell nonfunctioning but identical-looking versions of their cameras for this reason. Covert installations, however, hide the existence of the surveillance system. Since the cameras cannot be seen by the would-be criminal, this reduces the chance that the camera's view will be avoided, and the subsequent crime will likely be recorded.

All CCTV systems consist of lenses, cameras, and monitors. The image is focused by the lens and converted by the camera into an electrical signal. This signal is then distributed to one or more locations where it is converted back into a visible image by a monitor. Most CCTV systems also include a video recording device, especially if the system is not continuously monitored. **See Figure 10-5.**

The simplest CCTV system is one camera connected to one monitor, though most

systems include several devices. Sophisticated multicamera CCTV systems include an array of monitors, recorders, and matrix switching devices. Optional features for high-security applications include video motion detection and facial recognition software, which can trigger an alarm for all or a specific group of people within the field-of-view of a camera.

||| Voice-Over-IP Systems

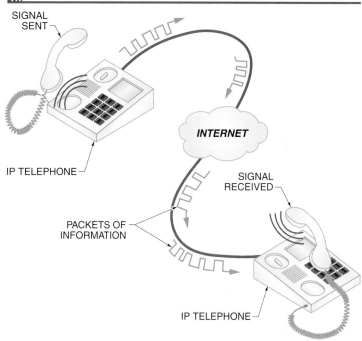

Figure 10-4. Voice-over-IP (VoIP) systems use special telephones that convert voice audio into packets of digital information to be transmitted over the Internet.

||| Closed-Circuit Television (CCTV) Systems

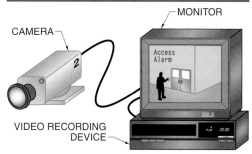

Figure 10-5. The simplest CCTV system consists of just a camera and a monitor, although most systems also include a recording device.

Paging Systems

A paging system is a VDV system used for broadcasting audio announcements to all personnel within a certain area. The area may be an individual building floor or an entire campus of many buildings. A paging system can be used for a variety of reasons. Informational messages reach everyone quickly and simultaneously and replace paper memos. Emergency messages can broadcast an alarm and initiate an evacuation.

Page messages are used to quickly locate personnel and direct them to where they are needed, increasing business efficiency. Background music or sound masking "white noise" may be broadcast on the system to increase productivity and conversation privacy. Almost any application requiring the broadcasting of audio signals can make use of a paging system.

There are two types of paging systems: central-amplified systems and self-amplified systems. Both types include an audio source, amplifiers to enhance the input audio signal, and speakers to reproduce the paging signal. An audio amplifier increases low-power audio signals to a level suitable for driving speakers. Both may also use interface devices to connect the paging system to the telephone system (if required). The primary difference between the two systems is the location of the amplifiers.

Central-Amplified Paging Systems

A central-amplified system is also known as a high-power or 70 V system. The central amplifier limits the voltage supplied to the speakers, regardless of the amplifier's power capacity. Each speaker in this system includes a step-down transformer that is used to convert the high-voltage/low-current amplifier signal of the central paging amplifier to the low-voltage/high-current signal the speakers use. The speaker transformers can be tapped at different power levels, providing more or less power to the speaker, depending on its necessary coverage. Central-amplified systems can power a large number of speakers with extremely long connections.

Central-amplified systems are cost-effective for large systems and offer excellent audio performance. Central amplifiers also provide features such as extra microphone inputs, tone control, automatic level control, automatic music muting during a page, and "night ringer" signal distribution.

Self-Amplified Paging Systems

A self-amplified (24 V) paging system is known as a low-power, a 24 V, or a distributed system. Self-amplified speakers each contain a built-in, miniature amplifier that directly drives the speaker. Each speaker is supplied with both 24 VDC to power the internal amplifier and a low-level audio signal. All self-amplified speakers have volume controls to adjust output level.

Self-amplified systems are cost effective for small systems (generally up to six speakers or four horns). However, they do not typically offer all of the sophisticated features of most central-amplified systems and do not perform as well in combined music and paging systems. Some self-amplified speakers have inadequate heat sinking and are not recommended for constant use.

Amplifiers

Amplifiers for paging systems may include connections for a variety of audio sources, either built-in or by way of changeable input modules. Depending on the source, input connections may require different types of connectors, such as RCA connectors, XLR connectors, or simple screw terminals.

Auxiliary Inputs. The auxiliary input (AUX) is the most common type of input for paging and connects to most music sources, such as CD players and tuners. AUX inputs have a high input impedance so that they do not put too much load on the source equipment's output. Shielded cable must be used with this type of input in order to avoid inducing noise into the system.

Telephone Input. The telephone input (TEL) is compatible with page outputs of telephone systems. The telephone input is a 600 Ω, transformer-coupled input that matches the impedance of the telephone port, electrically isolates the amplifier from the PBX or key system, and provides a balanced input with a great deal of noise immunity. Telephone inputs do not have to be shielded, but shielding provides more noise immunity. Higher noise immunity allows the amplifier to be located much further away from the source equipment than an unbalanced input would allow.

Microphone Input. The traditional paging amplifier input is the microphone input (MIC). When connected properly, a microphone can be hundreds of feet away from the amplifier and still provide a clear audio signal. Microphone cable is always shielded. The input requires three connections: two for the balanced signal and one for the shield ground.

Speakers

There are three major types of speakers: ceiling speakers, wall baffles, and horns. Ceiling speakers are typically installed in drop ceilings, which are common in offices and stores. Speakers are typically spaced a distance approximately equal to twice the ceiling height, which provides smooth coverage that works well with both voice and music. Wall baffles are used when ceiling speakers are not practical. Wall baffles have similar output characteristics as ceiling speakers except that they are designed to project forward rather than downward. Wall baffles are common in spaces with regularly spaced pillars or posts, which simplifies the mounting of these types of speakers. Horns are common for outdoors, extreme environments (such as freezers), high-noise areas, and indoors where large area coverage is necessary.

Sound Masking

Sound masking is an application of paging systems that broadcasts a subtle background sound into an environment. By reducing the intelligibility of speech, this increases conversation privacy and reduces noise-induced stress.

Sound masking works by changing the dynamic range of sound in an environment. Typically, the dynamic range is large, meaning that there is a large difference between the lowest (ambient) sound level and the highest (conversation) sound levels. Offices are often perceived as noisy because of a large dynamic range, even if the overall sound level is moderate. The higher sound level of normal speech is easily heard against a backdrop of the lower ambient background sound.

Sound masking subtly raises the ambient background sound level, reducing the environment's dynamic range. This effectively masks unwanted noise, makes speech unintelligible (creating privacy), and makes the environment more acoustically comfortable.

VDV SYSTEM CONTROL INFORMATION

The control aspects of VDV systems depend on the system. Telephony systems cannot be changed in their operation, though they can be seized for use by other systems. They also provide a means to transmit control signals. Operators can change setpoints by dialing into the system. This provides limited control functionality, but can be particularly useful as a back-up method if other means of remote access are unavailable. Control of video systems, such as CCTV systems, typically involves changing the field-of-view and recording a video feed for investigation and documentation purposes.

Dual-Tone Multifrequency (DTMF) Tones

Dual-tone multifrequency (DTMF) signaling is a method used to transmit numbers with a pair of tones. Almost all modern telephone systems are capable of DTMF signaling. DTMF signaling uses two tones to represent each key on a numeric keypad. **See Figure 10-6.** A total of eight tone frequencies are used, in combinations of two per number. The frequencies were chosen and standardized to avoid problems from harmonics which can cause unreliable signals. For example, no frequency is a multiple of another, the difference between any two frequencies does not equal any of the frequencies, and the sum of any two frequencies does not equal any of the frequencies.

DTMF Tones				
Frequency†	1209	1336	1477	1633
697	1	2	3	A
770	4	5	6	B
852	7	8	9	C
941	*	0	#	D

† in Hz

Figure 10-6. Each telephone keypad digit corresponds to a pair of DTMF tones.

With the typical layout of telephone keys, tones are assigned to rows and columns. The two tones for any key are the tones assigned to its row and column. When a key is pressed, the two tones are generated and signaled simultaneously. For example, pressing the "8" button generates the tones 852 Hz and 1336 Hz.

DTMF signaling is primarily used for signaling the dialed telephone number to the telephone exchange, but can also be used for other purposes. A common use is to input caller responses to recorded or interactive telephone menu options. DTMF signaling can also send signals to remotely control automated building systems, such as security and CCTV systems, over dial-up lines.

Camera Field-of-View

The video feed from a camera is largely determined by the choice of the camera and the rest of the video signal distribution network along with the installation. The image from the camera and viewed on a monitor is usually unchangeable. However, there are some instances where the field-of-view can be controlled, either by the CCTV system itself or with commands from other building systems. Systems can be designed with video distribution devices that can be used to automatically or manually switch between the feeds from different cameras.

More importantly, however, CCTV systems can use special cameras that can be controlled to point in a different direction and zoom in and out of a field-of-view. **See Figure 10-7.** This powerful capability can be controlled manually by security personnel at a security station, using a special keypad joystick, or automatically by other building systems. The commands to control these functions are standardized electrical signals.

VDV SYSTEM CONTROL DEVICES

Control devices in VDV systems are divided between telephony and CCTV systems. Each type involves the conversion of electrical signals into some other useable form. Telephony control devices convert electrical signals into audio signals. CCTV control devices convert light into electrical signals or video signals.

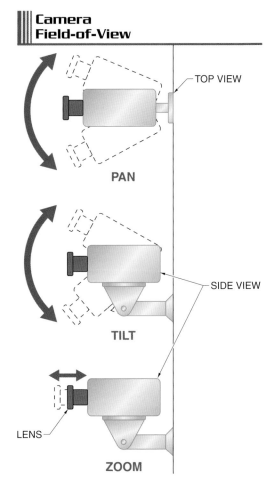

Figure 10-7. *Control of the field-of-view of a camera involves changing its position and zoom level.*

Telephones

A telephone is an end point in a telephony connection. It is one of the most common devices in the world, in both residential and commercial settings. Telephones initiate and terminate telephony connections and manage the conversion between audio signals and electrical signals.

The base of the telephone manages the making and receiving of calls. **See Figure 10-8.** The signaling equipment consists of a bell to alert the user of incoming calls and a dial to enter the phone number for outgoing calls. When a handset is picked up, releasing the switch hook, the telephone goes into active or "off hook" state. This puts a resistance across the telephone line, causing current to flow. The telephone exchange detects the DC current, sends a dial-tone sound signal, and gets ready to receive a dialed number. Dialed numbers are signaled with DTMF codes generated in the telephone. The exchange reads the outgoing phone number, makes the desired connection, and alerts that line.

Telephone Base

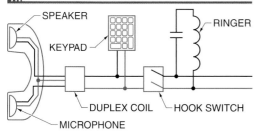

Figure 10-8. *A simplified circuit of a telephone base shows how the position of the hook switch changes the operation of the telephone.*

When a phone is inactive or "on hook," its bell or other alerting device is connected across the line through a capacitor, indicating to the exchange that it is on hook. Only the bell is electrically connected. When a call is received, the exchange applies a high-voltage pulsating signal, causing the alerting device to ring. When the called party picks up the handset, the hook switch disconnects the bell, connects the audio parts of the telephone, and puts a resistance on the line, confirming that the phone has been answered and is active. The parties are now connected and can converse using the audio parts of their telephones. When either party puts their handset on the cradle, current ceases to flow in the line, signaling to the exchange to disconnect the call.

The audio conversion parts of the telephone are in the handset and consist of a transmitter (microphone) and a receiver. **See Figure 10-9.** The transmitter, powered from the telephone line, outputs an electric current that varies in response to the sound pressure waves of the voice. The resulting analog electric current is transmitted along the telephone line to the telephone at the other end. Then it is fed into the coil of the receiver, which is a miniature speaker. The analog electric current causes the coil to move back and forth, reproducing the sound pressure waves of the transmitter.

Telephone Handset

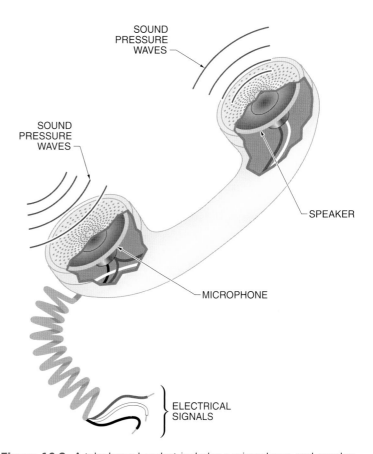

Figure 10-9. *A telephone handset includes a microphone and speaker that convert sound pressure waves to and from electrical signals.*

Modems

A *modem* is a device that modulates (changes) an analog carrier signal into digital information for transmission over a medium. On the other end of the transmission, another modem demodulates this same carrier signal to remove the coded information for its intended use. The term "modem" comes from the combination of the terms "modulate" and "demodulate." There are a wide variety of modulation schemes used to encode binary data, depending on the carrier and the transmission media.

The most familiar use of a modem is to change digital information into audio signals so that it can be transmitted over telephone lines. This type of modem is known as a voice modem or dial-up modem. There are digital modems used in

ISDN, DSL, ADSL, T1, and T3 digital telephone carrier circuits. There are also broadband, cable, wireless, and fiber-optic modems.

Modems are characterized by speed as well as modulation technique. Two measures of modem speed are bit rate and baud rate. *Bit rate* is a measure of the number of binary bits transmitted per unit of time, typically a second. *Baud rate* is a measure of the number of symbols transmitted per unit of time, typically a second. In some systems, a bit and a symbol are the same, but in most, they are not. **See Figure 10-10.** Since a symbol can include more than two states, it can carry more information than a bit. For example, if a system transmits information as one of four possible states, and its baud rate is 600, its equivalent bit rate would be 1200 bits/s (since two bits would be required to represent four states).

Modem Speed

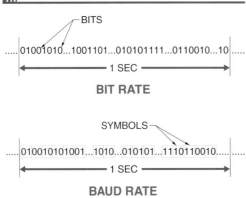

Figure 10-10. If the transmitted symbols are greater than one bit in length, then a bit rate is greater than a corresponding baud rate.

Facsimile Machines

A *facsimile machine* is a device used for transmitting image information over telephone lines. A facsimile, or "fax," machine typically consists of a scanner, a modem, and a printer combined into a single package. Fax machines are usually connected to a dedicated telephone line. This makes it easier to dial a fax machine directly if the other lines are connected to a PBX.

A scanner converts a printed document into a digital image. This scanned image is buffered (stored) for transmission as soon as scanning

is complete. An internal modem dials the telephone number of the receiving fax machine. When the receiving fax machine goes off hook, the two modems negotiate a two-way conversation. The transmitting fax machine sends the image as data over a telephone line to another device, and the printer at the far end produces a copy of the transmitted document.

Similarly, a computer with a fax-capable modem and software can also act as a fax machine. The process is nearly identical, however, computers provide the added feature of being able to transmit electronic documents directly from the computer without scanning.

Camera Lenses

A *lens* is a clear optical element for focusing a pattern of light, or image, onto a smaller area. A camera lens functions in a similar way as the human eye, collecting the reflected light from the target subject and focusing it onto the camera's image sensor. **See Figure 10-11.** An iris or diaphragm controls the amount of light reaching the image sensor. The lens is focused by moving it slightly forward or backward in relation to the image sensor. Exposure and focus can typically be adjusted manually, but automated systems may require these adjustments to be performed by servomotors, which can accept electronic control signals.

Camera Lens Function

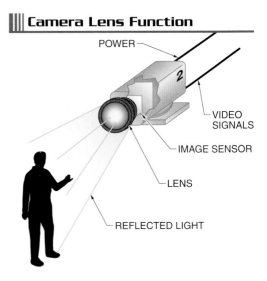

Figure 10-11. A camera lens focuses light reflected from an object within its field-of-view on an image sensor, which converts the light information into electrical signals.

Lenses used for most video applications have fixed focal lengths and fixed fields-of-view. Fixed focal length lenses are commonly known as wide-angle, medium-angle (also known as a normal lens; comparable to the normal field of view of the human eye), and narrow-angle or telephoto lenses. **See Figure 10-12.**

Camera Lens Focal Length

WIDE-ANGLE

MEDIUM-ANGLE

NARROW-ANGLE

Figure 10-12. The focal length of a camera lens determines how much area is covered by the camera.

Cameras

A *camera* is a device that converts an image into a time-varying electrical signal. The camera's image sensor scans the image focused by the lens by measuring the varying light intensity on the image sensor either line by line for analog cameras, or pixel by pixel for digital cameras. **See Figure 10-13.** In the case of a color camera, this is done three times: once each for red, green, and blue.

Cameras are broadly categorized as analog or digital cameras. Either type can be color or black and white. There are also special types of cameras for certain applications. For example, thermal infrared (IR) cameras sense infrared wavelengths, which are used to detect people in very low light or no-light conditions.

Modern cameras use solid-state image sensors known as charge-transfer devices (CTDs). These are available in three types depending on the manufacturing technology: charge-coupled device (CCD), charge-priming device (CPD), and charge-injection device (CID). Another sensor type introduced recently to the security market is the complementary metal-oxide semiconductor (CMOS) sensor. By far the most popular sensors used in security cameras are CCD and CMOS sensors.

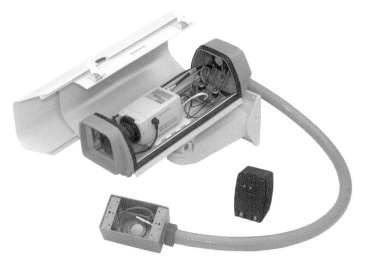

Honeywell International, Inc.

For outdoor applications, CCTV cameras are housed in rugged and waterproof enclosures.

||| Video Image Signals

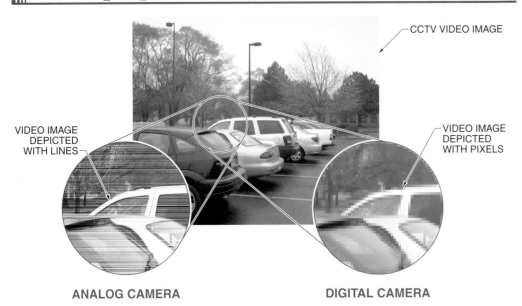

Figure 10-13. Analog video image signals are composed of lines of information, while digital video image signals are composed of pixels.

Both analog and digital CCTV cameras use one of these image sensors. The method of processing the image by the camera and what type of signal the camera outputs determine whether the camera is digital or analog.

Analog Cameras. Most CCTV applications for security and surveillance use a CCD sensor and analog transmission of composite video over coaxial or unshielded twisted pair (UTP) cables. A *composite video signal* is an analog video signal encoded with all of the component parts of a video signal. These component signals are encoded (combined) in the camera, sent over the media, and separated back out into their component level signals for viewing on the monitor.

The component video signal contains (for monochrome signals) horizontal line synchronization pulse, black level, luminance level (grayscale), and field synchronization pulses. Color signals contain light intensity and color information, so hue (color or tint), color intensity (saturation), and color sync pulses (color burst signals) are added to the horizontal and field synchronizing signals. Signals from an analog camera are typically recorded as an analog signal.

When recording an analog camera signal to a digital recorder (DVR or PC), the composite signal goes through an analog to digital (A/D) conversion. To view this digital signal on an analog monitor, the digital video signal goes through a digital to analog (D/A) conversion back to a composite video signal.

Digital Cameras. Digital CCTV cameras have built-in digital signal processing (DSP) and other digital enhancements. Typically, this enhanced video output signal is transmitted as an analog signal. (The exception is a network camera, which connects directly to an IP network and transmits video information in digital packets. These are also known as IP cameras or Internet cameras.) Analog signals can be transmitted farther than digital signals, so most CCTV signals are analog.

Digital cameras offer better image quality than analog cameras because of their higher signal-to-noise (SNR) ratio and offer other features not possible on an analog camera. For example, digital cameras offer video motion detection (VMD), electronic zoom, automatic backlight compensation, automatic iris and shutter control, in-camera scene processing (automatic processing and compensation of the bright and dark parts in a scene), and time and date stamp.

Camera Mounts

Cameras are installed on sturdy mounts that hold the camera securely and facilitate the routing of power and video cables. While surveillance systems can easily switch between different views of an area provided by different cameras, many cannot actually move the camera. Some cameras, however, are installed on mounts (or include mounts) that allow the camera to move to and zoom in on certain areas. A *pan-tilt-zoom (PTZ) camera* is a camera that can be remotely controlled to change its field-of-view. Panning is side-to-side movement. Tilting is up-and-down movement. Zooming is in-and-out movement.

PTZ commands can be issued from remote devices, even those in other building systems, to control the field-of-view of the camera. The command signals are standardized into a protocol and transmitted via separate conductors. PTZ controllers are commonly included on keypads in security stations. **See Figure 10-14.** When not being actively controlled by a guard, PTZ controllers can be programmed to sweep the camera over a wide area.

Monitors

Video information is displayed on monitors. A *monitor* is a device that converts video signals into monochrome or color motion images. Monitors are necessary for viewing live video and video played from recording devices. The simplest video system shows one camera feed per monitor. However, as this setup can become cumbersome when using multiple cameras, video signals are frequently manipulated to display multiple video feeds on a single monitor.

A multi-image monitor allows a split-screen display of 4 (quad), 9, or 16 video feeds simultaneously. **See Figure 10-15.** A greater number of feeds makes each one smaller, possibly requiring a larger monitor. A multi-image monitor, or its matrix switchers, can

also be connected to a sensor to switch the screen to a single feed. This is used to view video from a camera that has detected motion in its field-of-view.

PTZ Camera Controller

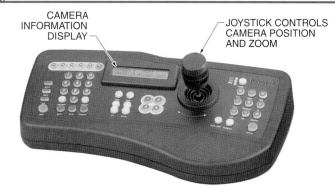

CAMERA INFORMATION DISPLAY

JOYSTICK CONTROLS CAMERA POSITION AND ZOOM

Honeywell International, Inc.

Figure 10-14. The position and zoom level of a pan-tilt-zoom (PTZ) camera is controlled by a PTZ controller keypad.

Multi-Image Monitor

GE Security

Figure 10-15. CCTV systems with many cameras typically display multiple video feeds on a monitor simultaneously.

Analog Monitors. An analog monitor consists of a cathode ray tube (CRT), video amplifier, CRT deflection circuit, and video-processing circuitry used to separate video and synchronizing signals. A monochrome CRT is composed of a heated cathode, electron gun, and a glass enclosure with a phosphor coating on the inside of the viewing side. A color CRT has three electron guns used to display each of the primary colors (red, green, and blue), with corresponding phosphor coatings.

Video recorders may include connections for digital alarms signals, which are used to switch to particular video feeds and/or change recording speed.

Digital Monitors. Digital monitors, commonly known as digital flat-screen or flat-panel monitors, consist of three major types: liquid crystal display (LCD), plasma, and organic light-emitting diode (OLED). The image on a digital display is composed of a rectangular grid of individual pixels. *Resolution* is the number of pixels displayed on a monitor screen. Depending on the monitor and its screen settings, there are 72 or 96 pixels per inch on the display. *Color depth* is the number of possible colors for each display pixel. Color depth is represented in bits per pixel (bpp). For example, 8 bpp yields 256 colors ($2^8 = 256$), 16 bpp yields 65,536 colors ($2^{16} = 65,536$), and 24 bpp yields 16,777,216 colors ($2^{24} = 16,777,216$), which is known as "true color."

Display resolution is the maximum number of pixels that can be displayed on a monitor screen. Display resolution is the product of the number of horizontal columns and vertical lines. The horizontal number is stated first. For example, display resolution of 640×480 is 640 horizontal pixels by 480 vertical pixel lines.

Recorders

Even when a guard is employed to watch security monitors for suspicious activity, there is no guarantee that a problem will be noticed. It is impossible for a person to have their full attention on multiple video images simultaneously, especially for hours at a time. Also, if there is an intruder or other problems, it may be necessary to have documented evidence of the event. Therefore, CCTV systems for security applications usually include some type of video recording device.

Videocassette Recorders. A *videocassette recorder (VCR)* is a video storage device that records information on magnetic tapes. VCRs allow real-time 30 fps (frames per second) video recording and a convenient method of handling and storing tapes. However, for security applications requiring continuous recording, this would fill a tape in a few hours. To solve this problem, the time-lapse videocassette recorder (TLVCR) records 24 hours of real-time video, or up to 40 days (960 hours) of extended time-lapse video, on the same T160 cassette tape. **See Figure 10-16.**

TLVCRs include alarm activation that switches from time-lapse mode to real-time mode when an alarm signal is received. This balances efficient use of tape with detailed recording of important events. Optional features of a TLVCR include a built-in camera switcher, sequence or interval recording of multiple cameras on one tape, remote control via RS-232, time and date generator, and interface connections for other devices, such as an access control system.

VCRs have their drawbacks. They require frequent tape changes, such as once in every 24 hours and tape storage can be problematic and costly. The magnetic-read heads require constant head cleaning and can wear out or become misaligned. Drive belts stretch and need replacing. Repeated use of the same tape degrades the recorded-image quality. All of these problems have been eliminated in the digital video recorder.

Time-Lapse Recording

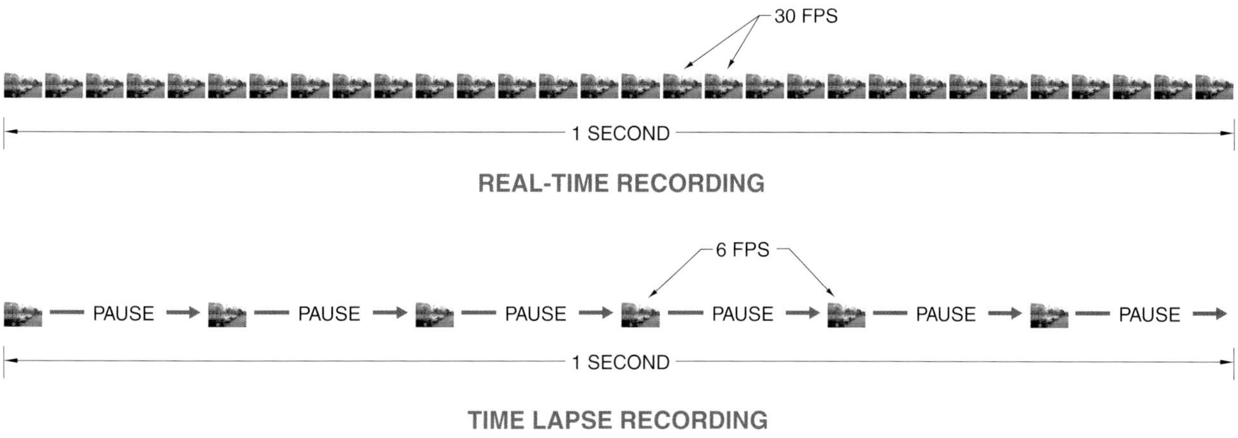

30 FPS

1 SECOND

REAL-TIME RECORDING

6 FPS

PAUSE → PAUSE → PAUSE → PAUSE → PAUSE → PAUSE →

1 SECOND

TIME LAPSE RECORDING

Figure 10-16. Time-lapse recording records video images at a fraction of the frame rate of real-time recording.

Digital Video Recorders. A *digital video recorder (DVR)* is a video storage device that records information on digital memory disks. Video images are initially stored on hard disk drives, making the video information very easy to work with. Video from a DVR can also be easily copied to DVDs, CDs, digital audiotapes (DAT), or flash memory devices for long-term storage or backup.

DVRs have several advantages over VCRs. DVRs record higher resolution and full-frame video information with little noise. (VCRs record fields, which are half of a frame each, and can introduce noise into the recorded image.) DVRs can record at lower resolutions and in time-lapse modes to conserve memory and automatically switch to full-resolution real-time recording when an alarm is received. However, during any of these recording options, the video image quality remains the same. DVRs provide higher image quality with no distortions during pause, single frame advance, fast-forward, and reverse. An event can be found and played back in an instant, as opposed to advancing through a VCR tape. Unlike VCR tapes, DVR recordings can be copied and played repeatedly with no reduction in image quality.

DVRs come in many forms, from cards that plug into personal computers, to multiple-channel recorders, to network video recorders. **See Figure 10-17.** The basic DVR is a single-channel device, which is the logical replacement for a TLVCR. It has controls that are the same as those used on a VCR, making it easy to install and use as a replacement.

A multiplex DVR provides recording capabilities for 4 to 16 cameras, but not in real time. It must time-share recording time with each camera. The number of cameras affects the time allotted for each camera. Since DVRs record at 60 fps, the number of frames per second that can be recorded for each camera is 60 fps divided by the number of cameras. For example, a system with 4 cameras would record 15 fps for each camera. However, alarm inputs to the multiplexing DVR can trigger real-time recording of the alarming camera.

Network DVRs

Honeywell International, Inc.

Figure 10-17. Network DVRs record digital video information on storage media that is accessible via the local network.

Hard disk drives will eventually fail, but the mean time between failures is very high compared to VCRs and magnetic videotapes.

Alternatively, multichannel DVRs are capable of recording in real time at 60 fps for up to hundreds of camera channels simultaneously. Multichannel DVRs are limited only by the amount of storage space the system can handle. Systems like these are critical for certain large applications with special security needs, such as casinos, which can have up to 1500 cameras or more.

Most current multiplexing and multichannel DVRs are triplex types, meaning that they can simultaneously display, play, pause, fast forward, or rewind any video information without interrupting the current recording. Some DVRs also allow remote viewing via a network connection using a web browser.

VDV SYSTEM CONTROL APPLICATIONS

Telephone systems and CCTV systems function independently, but can be manipulated to serve the purposes of other building systems. This is especially true for high-priority applications such as security and fire protection.

Fire Alarm/Security System Line Seizure

Connections to a fire/security central monitoring system may be directly connected with a dedicated telephone line from the panel to the monitoring station or service. This connection is always available and the monitoring station can monitor the fire alarm or security system in real time for alarm or trouble conditions.

Some CCTV cameras, or their video feeds, can be controlled in building automation applications.

A less expensive option is to have the fire alarm and/or security system dial out to the central monitoring service over one of the regular voice telephone lines. In order to accomplish this, the fire or security system must seize the telephone line, dropping any call that may be in progress and dialing out to the central monitoring service. **See Figure 10-18.**

Operating Conditions. Line seizure is a built-in feature on fire alarm and security system control panels. This gives the panels priority over a telephone line when an alarm has been triggered. The telephone connection must be made upstream of any other telephone or device connected to the same line.

When the fire and/or security control panels are not directly connected to the central monitoring service, a special telephone jack is required between the incoming voice line and all of the telephones that will be connected downstream to that voice line. This standard jack is known as an RJ-31X. **See Figure 10-19.** Upon insertion of the 8-position modular plug into the RJ-31X jack, mechanically held shorting bars connecting the telephone wiring to the facility equipment are lifted.

Control Sequence. The control sequence for seizing a telephone line is as follows:

1. An alarm panel has an alarm or trouble condition that requires dialing the central monitoring station.

2. A relay on the panel's circuit board opens the connection between the telephone wiring and the facility's equipment. This disconnects all of the telephone equipment downstream of the alarm panel. The switching of the relay also puts the alarm panel in an "off hook" condition.

3. When the alarm panel senses dial tone, it dials a preprogrammed telephone number to the central monitoring station.

4. When the central monitoring station picks up the call, the alarm panel transmits the information about the alarm or trouble condition.

Fire Alarm /Security System Line Seizure

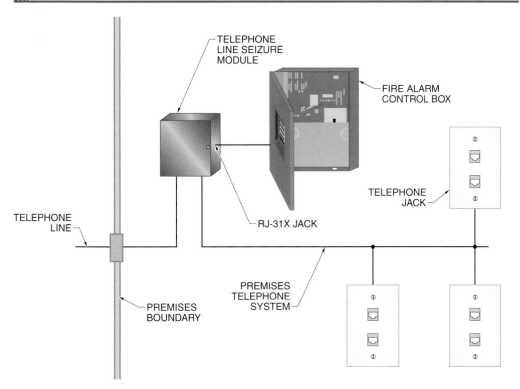

Figure 10-18. Telephone line seizure devices are installed upstream of any other telephone devices within the premises so that if necessary, an emergency control panel can take control of the line.

Telephone Line Seizure Jacks

Figure 10-19. Telephone line seizure modules use special jacks that open the connection between the premises wiring and the telephone line, inserting the dialer into the circuit.

DTMF Control Signaling

Many building automation systems are connected to the Internet for remote control and monitoring. As a backup, many systems still allow the user to dial in over telephone lines and affect control of a device or system. This application uses DTMF tones to transmit control information from any touchtone telephone to a building system.

One popular use of DTMF signaling is with door and gate access control. **See Figure 10-20.** This is an application where someone wishing to gain access through a certain door or gate must first contact someone on the premise via a dedicated call box. That person can then command the door or gate to open with DTMF codes generated by their telephone.

Operating Conditions. DTMF control signaling requires the device or system to have an onboard dial-up modem and built-in DTMF decoder. Remote control can also be achieved with stand-alone DTMF controllers that include relays. For door or gate control, some type of access control device listens for a certain DTMF code on the telephone line and authorizes passage if it is received.

▌▌▌ DTMF Control Signaling

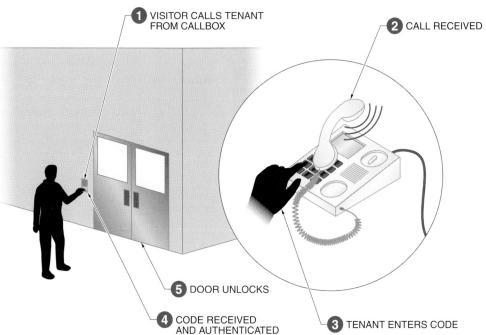

1 VISITOR CALLS TENANT FROM CALLBOX

2 CALL RECEIVED

5 DOOR UNLOCKS

4 CODE RECEIVED AND AUTHENTICATED

3 TENANT ENTERS CODE

Figure 10-20. DTMF signaling can be used to send control information across telephone lines.

Control Sequence. The control sequence for controlling a system with DTMF codes is as follows:

1. The person trying to gain access calls a tenant from a dedicated call box near the locked door.
2. The tenant receives the call on their normal telephone.
3. The tenant allows access by entering an access code on the telephone's keypad.
4. The access control device receives the access code as DTMF tones and receives authentication.
5. The access control device releases the door lock.

Before DTMF signaling, telephone systems employed pulse signaling to dial numbers, which rapidly disconnects and connects the calling party's telephone line. As the dial spins, the pulses sound like a series of clicks. The exchange would count the pulses to determine the telephone number of the desired destination. This system was limited in range by distortion and other technical problems.

PTZ Camera Control

Pan-tilt-zoom (PTZ) cameras allow a wide variety of possible integration applications with other building systems. Nearly any type of event can be recorded accurately by directing a PTZ camera to focus on a certain area. Very common examples include surveillance of areas near access control devices to ensure that no tampering of the devices occurs and to investigate invalid authentications. **See Figure 10-21.**

Operating Conditions. A surveillance system may include just one or two PTZ cameras that augment a system of several other fixed cameras. These PTZ cameras are placed so that they can view the entire secure area. The DVRs, monitors, and cameras are all controlled from a guard position with the use of a controller keypad. The DVR records video from multiple cameras continuously. When not being controlled by the guard, a PTZ camera is programmed to slowly "tour" the secure area. An alarming output on an access control device is used to control the field-of-view of the PTZ camera.

PTZ Camera Control

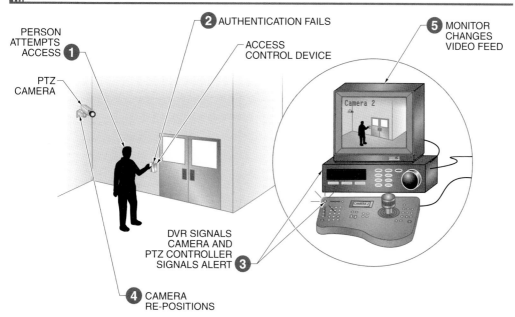

Figure 10-21. PTZ cameras can be controlled by other building systems, such as the access control system, to monitor and document events.

Control Sequence. The control sequence for controlling a PTZ camera to cover an event is as follows:

1. A person attempts to gain access through a door by swiping an access card.
2. Authentication fails and the access control device sends an alarm to the CCTV system DVR via an access control panel.
3. The DVR responds by sending a signal to the PTZ camera with new position commands.
4. The PTZ camera immediately moves to a position preprogrammed to center its field-of-view on the door.
5. The DVR records the video feed from the PTZ camera and switches the monitor view to the PTZ camera.
6. The PTZ controller keypad alerts the guard to the event with its built-in annunciator.

Refer to Quick Quiz® on CD-ROM

A VDV system is also known as a low-voltage system to distinguish its cabling infrastructure from a higher voltage power infrastructure. Low-voltage cabling may also be known as structured cabling and is subject to different, often less stringent, code requirements for installation.

Summary

- A transmitting device converts information into a time-varying signal made up of electrical or optical pulses.

- Telephony systems provide the infrastructure and control of point-to-point audio signals, which are typically used for voice transmission, but can also be used for data communication.

- Closed-circuit television (CCTV) systems manage the transmission of video signals in a relatively small system, such as within a building or campus.

- Data systems are any system connecting electronic equipment that exchanges data with other equipment.

- With its programmed logic, the PBX controls local in-house dialing, local calls outside the premises, and long distance calls.

- Voice-over-IP (VoIP) systems make use of the Internet and/or local intranets instead of the traditional telephone network to transmit audio information for telephone calls.

- VoIP allows users to streamline their communications requirements by maintaining only the data network, instead of two separate systems.

- CCTV systems are used primarily for security and surveillance applications.

- Security applications of CCTV systems include both overt and covert installations.

- Dual-tone multifrequency (DTMF) signaling is primarily used for signaling the dialed telephone number to the telephone exchange, but can also be used for control purposes.

- CCTV systems can use special cameras that can be controlled to point in a different direction and zoom in and out of a field-of-view.

- Control devices in VDV systems are divided between telephony and CCTV systems.

- The most familiar use of a modem is to change digital information into audio signals so that they can be transmitted over telephone lines.

- A camera lens collects the reflected light from the target subject and focuses it onto the camera's image sensor.

- The camera's image sensor scans the image focused by the lens by measuring the varying light intensity on the image sensor either line by line (for analog cameras) or pixel by pixel (for digital cameras).

- Cameras are broadly categorized as analog or digital cameras.

- Pan-tilt-zoom (PTZ) commands can be issued from remote devices, even those in other building systems, to control the field-of-view of the camera.

- Video signals are frequently manipulated to display multiple video feeds on a single monitor.

- CCTV systems for security applications usually include some sort of video recording device to fully monitor all video feeds and document any unusual events.

- Time-lapse (TL) VCR records 24 hours of real-time video, or up to 40 days (960 hours) of extended time-lapse video recording, on the same T160 cassette tape.

- In comparison with VCRs, DVRs record higher resolution and full-frame video information with less noise.

- Multiplex DVRs provide recording capabilities for 4 to 16 cameras, but not in real time.

- Multichannel DVRs are capable of recording in real time at 60 fps for up to hundreds of camera channels simultaneously.

- A fire or security system can seize a telephone line in an emergency, dropping any call that may be in progress and dialing out to the central monitoring service.

- As a back-up method, many systems still allow a user to dial in over telephone lines and affect control of a device or system with DTMF code signaling.

- An alarming output on an access control device may be used to control the field-of-view of the PTZ camera.

Definitions

- A *voice-data-video (VDV) system* is a building system used for the transmission of information.

- A *key system* is a multiline telephone system in which each telephone has direct access to each outside (PSTN) line.

- The *private branch exchange (PBX)* is a telephone exchange that serves a small, private network.

- *Voice over Internet Protocol (VoIP)* is the transmission of audio telephone calls over data networks.

- A *closed-circuit television (CCTV) system* is a system of video capture and display devices where the video signal distribution is limited to within a facility.

- *Dual-tone multifrequency (DTMF) signaling* is a method used to transmit numbers with a pair of tones.

- A *telephone* is a device used to transmit and receive sound (most commonly speech) between two or more people.

- A *modem* is a device that modulates (changes) an analog carrier signal into digital information for transmission over a medium.

- *Bit rate* is a measure of the number of binary bits transmitted per unit of time, typically a second.

- *Baud rate* is a measure of the number of symbols transmitted per unit of time, typically a second.

- A *facsimile machine* is a device used for transmitting image information over telephone lines.

- A *lens* is a clear optical element for focusing a pattern of light, or image, onto a smaller area.

- A *camera* is a device that converts an image into a time-varying electrical signal.

- A *composite video signal* is an analog video signal encoded with all of the component parts of a video signal.

- A *pan-tilt-zoom (PTZ) camera* is a camera that can be remotely controlled to change its field-of-view.

- A *monitor* is a device that converts video signals into black and white or color motion images.

- *Resolution* is the number of pixels displayed on a monitor screen.

- *Color depth* is number of possible colors for each display pixel.

- *Display resolution* is the maximum number of pixels that can be displayed on a monitor screen.

- A *videocassette recorder (VCR)* is a video storage device that records information on magnetic tapes.

- A *digital video recorder (DVR)* is a video storage device that records information on digital memory disks.

Review Questions

1. Briefly explain the primary roles of a private branch exchange (PBX).

2. What are some of the advantages of a Voice-over-IP (VoIP) system?

3. How can overt closed-circuit television (CCTV) systems help avoid crime?

4. How does a telephone communicate with the public switched telephone network (PTSN) exchanges to place a call?

5. How does a pair of modems exchange information over a telephone line?

6. How are many digital cameras analog in signal output?

7. How is a pan-tilt-zoom (PTZ) camera used to dynamically change the field-of-view of a video feed?

8. What are some of the advantages of digital video recorders (DVRs) over videocassette recorders (VCRs)?

9. How does a fire or security system control panel seize a regular telephone line for emergency use?

10. How is DTMF signaling implemented to control devices in other systems?

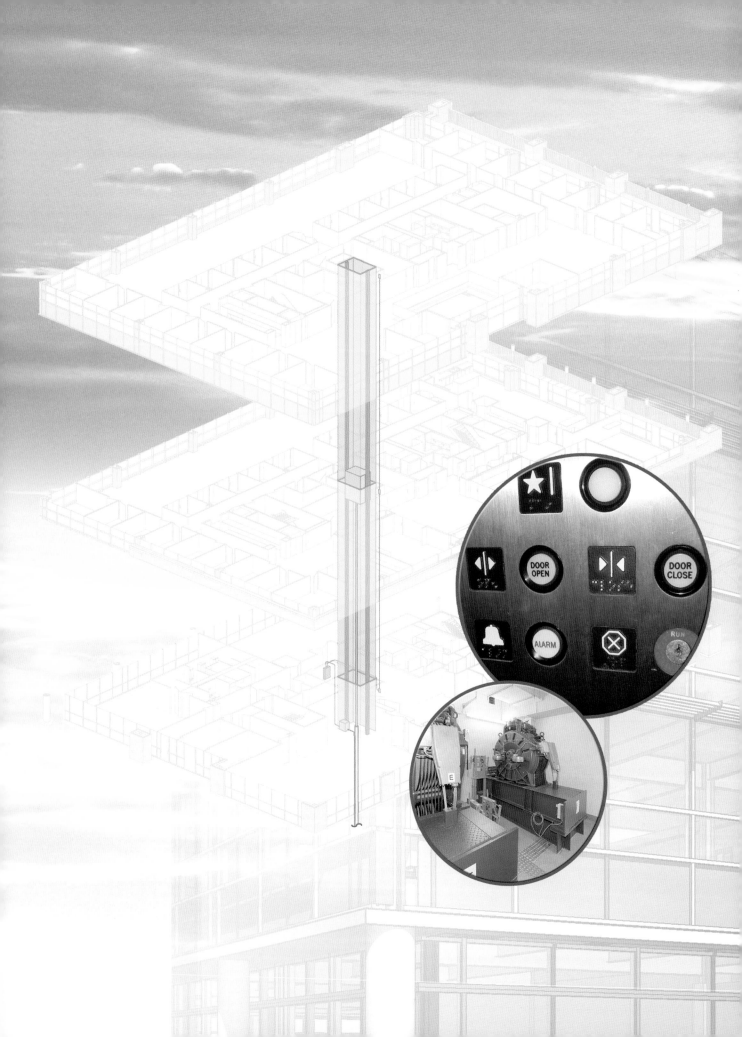

Chapter Eleven

Elevator System
Control Devices and Applications

Without elevators, building height would be limited to just a few floors. Elevators provide safe and efficient transportation for people, materials, and equipment vertically between building floors. The normal operation of an elevator system is managed internally by the elevator controller, which connects to elevator system-specific control devices. However, many elevator systems can also receive inputs from other building systems, such as the fire alarm and access control systems, to override the normal operation for special applications.

Chapter Objectives

- Compare the different types of elevators by their function and mechanical operation.
- Describe the basic operation of common elevator system safety features.
- Differentiate between the normal control of elevators by the elevator controller and the control inputs to the controller from other building systems.
- Describe common elevator system control applications that involve other building systems.

ELEVATOR SYSTEMS

A *conveying system* is a system for the transporting of people and/or materials between points in a building or structure. Conveying systems may operate horizontally, vertically, or even diagonally. **See Figure 11-1.** The most common type of conveying system in commercial buildings is the elevator.

Conveying Systems

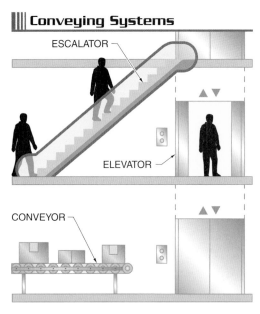

Figure 11-1. Conveying systems include systems to automatically transport people and/or materials between areas of a building.

Escalators are a common conveying system in shopping malls, airports, and transit stations, where they provide efficient transportation for passengers.

Many commercial buildings are several stories high, which allows for a large amount of space in a relatively small footprint. However, this also requires a means to move people and materials quickly and efficiently between floors. This is accomplished with one or more elevators. An *elevator* is a conveying system for transporting people and/or materials vertically between floors in a building.

Functional Elevator Types

Passengers and/or materials ride within the elevator car, which can vary in size and design, depending on the type of elevator. The car moves vertically in either direction within a narrow shaft and can stop at certain points along the shaft to allow for loading and unloading. The stops are determined by requests made by both the riding and the waiting passengers.

Beyond the basic function of an elevator, there are different types of elevator systems that specialize in certain types of operations. The functional types of elevators include passenger, service, freight, dumbwaiter, and construction elevators. **See Figure 11-2.**

Passenger Elevators. Passenger elevators are the most common type of elevator in typical multistory buildings. A *passenger elevator* is an elevator primarily intended to transport people. The capacity of passenger elevators typically ranges from 2000 lb to 5000 lb, in 500 lb increments. The vertical speed of a passenger elevator varies depending on the number of floors the elevator services. In a building with ten floors or fewer, the elevator likely travels 500′ per min (fpm) or less, while in a taller building, the elevator may travel up to 2000 fpm.

Elevators may be assigned specific functions in order to improve efficiency. For example, it is very inconvenient to a person wanting to reach a high floor for the elevator to stop at many lower floors on the ride up. An *express elevator* is an elevator specifically intended to service a single floor or a specific group of floors. Express elevators can be used for dedicated access to a penthouse or other special areas within the building, such as a restaurant at the top of an office building.

Elevator Types

PASSENGER ELEVATORS

Kone, Inc.

SERVICE ELEVATORS

Kone, Inc.

FREIGHT ELEVATORS

Waupaca Elevator Company, Inc.

DUMBWAITERS

CONSTRUCTION ELEVATOR

Figure 11-2. While passenger elevators are the most common type of elevator, other types of elevators are specifically designed for other functions.

Very tall buildings often use express elevators to serve specific groups of floors. This system may also incorporate local elevators for access between the floors within a group. For example, a certain express elevator may travel only between the ground floor and the 50th floor. To reach the 54th floor, a passenger travels first to the 50th floor, and then takes a local elevator that serves only the 50th to 59th floors. Despite the multiple trips and dedicated elevators in this express elevator system, it is very efficient and typically saves much of the passenger's time.

Service Elevators. A *service elevator* is an elevator reserved exclusively for the use of building service personnel. These elevators are used by maintenance technicians, cleaning crews, and building managers. In many instances, gaining access to these elevators requires a key or access card. In some buildings, these elevators can access floors that the passenger elevators cannot, such as basements. Otherwise, service elevators operate the same as passenger elevators.

A residential elevator is a passenger elevator for a one-family home. They are the same as commercial elevators, except that they typically serve far fewer floors. Residential elevators are particularly useful for the disabled and are becoming increasingly popular and affordable.

Other Conveying Systems

Conveying systems include other means of moving people and/or materials through a building besides elevators, such as escalators, conveyors, and moving walkways.

Escalators

An escalator is a conveying system similar to a moving set of stairs that transports people diagonally between floors in a building. Escalators are common in commercial buildings such as department stores, malls, transit stations, and airports. The escalator has a few significant advantages over the elevator. First, there is no wait to board an escalator. Also, the escalator can typically handle as many people as can comfortably fit onto it, whereas the elevator has weight limitations. However, unlike elevators, escalators are practical for spanning only two to four floors. They also consume a large amount of space.

An escalator is a continuous set of steps connected to a conveyor belt that is driven by an electric motor. The steps have rollers on each side that ride along a track to maintain their horizontal position at all times. At the top and bottom, the steps flatten to allow a smooth transition for entry and exit from the escalator. Escalators are fully reversible, allowing for ascending or descending travel. A handrail travels at the same speed as the steps, providing passenger stability. A typical escalator travels at between 90 fpm and 180 fpm, with most operating at approximately 100 fpm. This speed has proven to be an efficient speed with easy transitions for stepping onto and off the escalator.

Safety features built into escalators provide safe passage for the riders. Lighting illuminates the entrance and exit points of the escalator. Step lines are lines painted on the steps to make the edge more visible, thus directing passengers to the proper standing locations. Impact switches in the combplate detect objects caught in the escalator and automatically shut down the escalator. Skirt brushes prevent foot entrapment in the step/skirt. Emergency stop buttons are installed at one or both ends of the escalator to immediately stop the escalator in the event of an accident. Safety instructions are posted near an escalator to clearly indicate precautions for escalators. Handrail speed devices are used to make sure the handrail operates at the same speed as the escalator or moving walkway. The dedicated safety zone clearly identifies that the areas around the top and bottom exit/entry points are free of clutter or anything else that may create a hazard to the safe operation of escalators and moving walkways.

Conveyors

A conveyor is a system for transporting materials horizontally. Conveyors are used in many industrial facilities, such as for raw material handling, manufacturing processes, and assembly lines. However, there are also many conveyor applications in commercial buildings, including shipping and receiving, security checkpoints, airport baggage handling systems, and retail checkout stations. Most conveyors are controlled for ON/OFF, speed, and/or direction (forward and reverse). Specialized conveyor systems may also incorporate scanners to inspect items on the conveyor and devices to remove selected items from the conveyor.

Most conveyors consist of a continuous belt driven by a motor. The belt is usually supported by free-spinning rollers held in a frame that carry the weight of the items on the belt without affecting horizontal travel. Some conveyors, though, do not include belts. A gravity conveyor is an unpowered conveyor that provides a low-friction path for items moving due to gravity or external forces. These conveyors consist of only a series of free-spinning wheels or rollers so that even heavy items move easily along the conveyor with a gentle push. By installing the conveyor at a shallow angle, the force of gravity moves items down the slope automatically. A powered roller conveyor is a variation of this beltless conveyor that drives some or all of the rollers along the conveyor to push items mechanically.

Most conveyors are typically designed for moving materials and are not recommended for moving people because of safety issues. A moving walkway, however, is a special type of conveyor designed specifically for moving people. These passenger conveyors are primarily used in airports and transit stations so that people can travel down long terminals much faster than they can normally walk. The most common moving walkway is flat, though they can be inclined up to 12°. The typical speed for a moving walkway is approximately 130 fpm, though faster speeds, up to a maximum 180 fpm, are becoming more popular.

With many moving parts, all conveyors include safety features to protect those working near or traveling on them. Basic safety features include guardrails to prevent accidental contact with pinching and electrical shock hazards, and emergency shutoffs to de-energize the conveyor instantly. Many moving parts of conveyors can also be noisy, but significant improvement has been made in noise reduction to protect the hearing of those nearby.

Freight Elevators. A *freight elevator* is an elevator primarily intended to transport equipment or materials. Passengers are typically prohibited from riding freight elevators. A typical freight elevator will be located on or near the loading dock of the building. The interior of a freight elevator car is generally plain and clad with rugged materials to tolerate the abuse of frequent loading and unloading of freight. The doors may operate either automatically or manually. A freight elevator is typically much larger than a passenger elevator and can handle loads from 10,000 lb to 15,000 lb.

Dumbwaiters. A *dumbwaiter* is an elevator for transporting very small loads, primarily consisting of domestic items. Dumbwaiters are used to transport food, laundry, dishes, and other items in restaurants, hotels, and Victorian-style residences. Dumbwaiters are too small for passengers and may not include the safety features required for passenger service, so they are restricted to only carry materials. Older dumbwaiters were operated by pulling ropes by hand, but newer dumbwaiters are similar to other modern elevator designs. However, they are smaller in scale. Dumbwaiters usually have a weight capacity of about 750 lb to 1000 lb.

Construction Elevators. Elevators in new buildings are typically not ready for operation until well into the construction process. However, the elevator function is extremely useful during the early phases of construction, so temporary construction elevators are sometimes built.

A *construction elevator* is a temporary elevator for the transportation of materials and workers during the building construction. This type of elevator is usually anchored to the exterior shell of the building itself. The construction elevator can transport tools, equipment, and supplies that would be difficult or impossible to carry up multiple flights of stairs. They can also be used to safely bring construction debris down to ground level for disposal.

For safety reasons, construction elevators are strictly regulated and cannot be erected until the building is complete enough to serve as a solid anchoring structure. When construction elevators are not available, tower cranes with personnel baskets can be used, but these are much slower.

Elevator Mechanical Operation

Elevators can be categorized by the design used to move the elevator car vertically. Most elevators have either traction or hydraulic designs. Specific elevator design and operation may vary between manufacturers, but most elevator systems share common parts and features with others of its type.

Traction Elevators. The most common type of elevator is the traction elevator. A *traction elevator* is an elevator system that raises and lowers the elevator car with cables operated by an electric motor. **See Figure 11-3.** Most designs use a set of several thick steel cables. One end of each cable attaches to the top of the elevator car and the other end wraps over a drive sheave (pulley) and attaches to the top of a counterweight. The sheave is approximately 2′ to 4′ in diameter and has grooves that grip the cables to move them up or down. The sheave is driven by an electric motor, either directly or through a geared transmission. Geared traction elevators use a worm and worm-gear-type reduction unit, which allows for precise control of the elevator position, though the vertical speed is limited. A typical geared traction elevator can travel at speeds of 125 fpm to 500 fpm. A typical gearless traction elevator can travel at speeds greater than 500 fpm.

Most traction elevators require a machine room. A *machine room* is a space directly above an elevator shaft to house the motor, sheave, and elevator controls. **See Figure 11-4.**

A double-deck elevator car is one elevator car stacked on top of another elevator car, which allows two consecutive floors to be serviced simultaneously. A scenic elevator car has a glass enclosure that allows passengers to see out as they are traveling to their destination. In some installations, the scenic elevator car rides on the exterior of the building.

Traction Elevators

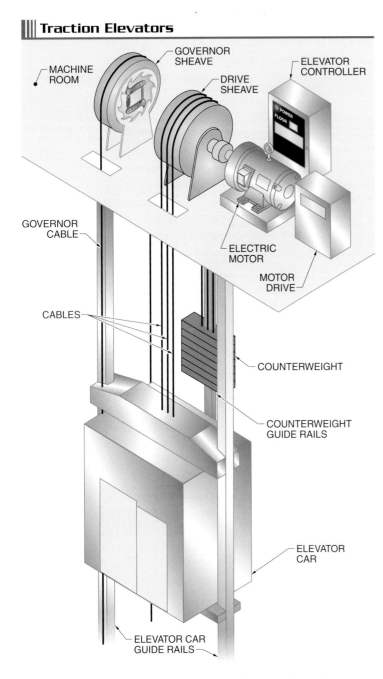

Figure 11-3. Traction elevators are raised and lowered through the elevator shaft by cables operated by electric motors.

The size of the elevator machine room is determined by elevator codes, electrical codes, and manufacturer specifications. The elevator code mandates the minimum dimensions, the electrical code mandates the required space around the electrical equipment, and the manufacturer specifications mandate minimum dimensions based on the equipment to be installed.

Machine Rooms

Kone, Inc.

Figure 11-4. Most of the equipment needed to operate a traction elevator is located in a machine room above the elevator shaft.

The counterweight reduces the amount of energy needed to raise and lower the elevator car by approximately balancing the loads on either side of the sheave. As the elevator car is raised, the counterweight is lowered, and when the elevator car is lowered, the counterweight is raised. A typical counterweight will weigh approximately the weight of the elevator car plus 40% to 50% of the capacity of the elevator car. For example, if the weight of the elevator car is 1000 lb and the rated capacity of the elevator is 3000 lb, the counterweight will weigh from approximately 2200 lb to 2500 lb. Both the elevator car and counterweight ride up and down the elevator shaft on their own set of guide rails, preventing them from swaying due to shifting loads and starting and stopping.

Innovations in elevator design have also produced machine room-less elevators. A *machine room-less elevator* is a traction elevator system using special materials and improved electric motors that require little space, eliminating the need for the machine room. **See Figure 11-5.** This elevator design uses flat polyurethane-coated steel belts that are considerably smaller than the traditional cables, allowing the sheave to be much smaller. Reducing the size of the sheave allows the elevator drive components to be installed directly in the elevator shaft. Additionally, the controls can be installed in a more convenient area of the building. The machine room-less elevator system is suitable for applications between 2 to 30 stories and is approximately 40% more energy efficient than comparable traditional traction elevator systems.

Hydraulic Elevators. A *hydraulic elevator* is an elevator that lifts the elevator car from below using a hydraulic ram. **See Figure 11-6.** A *hydraulic ram* is a piston that is driven into or out of a hollow cylinder by fluid pressure. An electric motor pumps fluid into the cylinder, causing the elevator car to rise. When the car reaches the destination, the pump stops and the pressure in the cylinder holds the elevator in position. Excess hydraulic fluid is kept in a reservoir located in the elevator machine room on the lowest floor and a valve controls the amount of fluid into or out of the reservoir. To descend, the valve is opened to allow hydraulic fluid to return to the reservoir. Once the car falls to the requested level, the valve is closed and the car remains in position. A typical hydraulic elevator can travel at speeds up to 150 fpm.

Traditional hydraulic elevators require the hydraulic cylinder to be located underground to a depth equal to the highest level the elevator will reach, which usually limits the applications to two to seven floors. Using one or more telescoping cylinders, which nest multiple cylinder sections together, allows for shallower holes. A holeless elevator utilizes two relatively small telescoping cylinders attached directly to the sides of the steel elevator car supports, significantly reducing (though often not completely eliminating) the need for space below the lowest elevator level. **See Figure 11-7.** Holeless elevators are particularly suitable for areas with high water tables or for existing building construction.

The roped hydraulic elevator is a special design that uses both traction and hydraulic elevator technologies. Steel cables are attached indirectly to the elevator car and two telescoping plungers. This type of elevator extends the available rise of a holeless elevator to approximately 60′.

Machine Room–Less Elevators

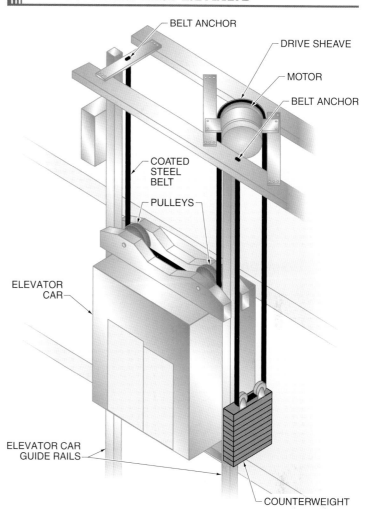

Figure 11-5. Machine room-less elevators use a design that reduces the size of the traction equipment enough so that it does not require a machine room.

Kone, Inc.

The drive equipment for a machine room-less elevator takes up very little space.

Hydraulic Elevators

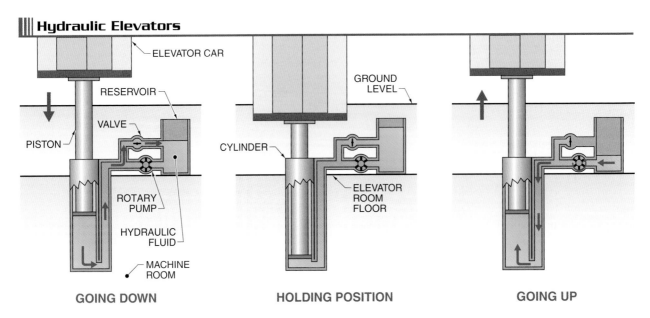

GOING DOWN **HOLDING POSITION** **GOING UP**

Figure 11-6. Hydraulic elevators are raised and lowered by pistons filled with fluid.

Holeless Hydraulic Elevators

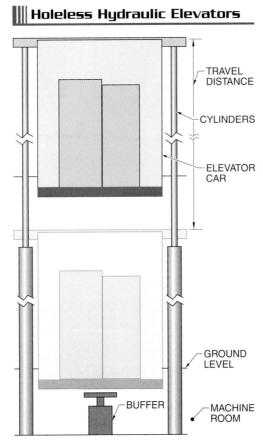

Figure 11-7. Holeless elevators reconfigure the hydraulic pistons to the side of the elevator, reducing the amount of equipment space needed directly beneath the elevator shaft.

Elevator Safety Features

Since most elevators transport passengers and the vertical height presents a great potential for injury, elevator system safety features are of primary importance. In fact, while the basic operating principle of an elevator is thousands of years old, elevators did not become common and practical until adequate safety features were implemented. In the decades since then, many of the basic safety features, such as leveling, emergency brakes, door openers, and car buffers, have not changed significantly. There are many additional safety features, but the basic features have the biggest impact on safety in the elevator industry.

Elevator Leveling. As passengers or materials get on and off an elevator, the load changes, which could cause the elevator to rise and fall slightly. This is a tripping hazard to the passengers crossing the threshold of the elevator car. Automatic leveling is a safety feature built into modern elevator systems to prevent this from happening. The elevator controller constantly monitors the load in the elevator car. If there is a substantial change in the load, the controller automatically adjusts the elevator up or down to ensure the elevator car floor stays level with the building floor, keeping the transition smooth.

Emergency Brake System. During normal operation, the electric motor or hydraulic system (depending on the elevator type) is responsible for stopping and holding elevator cars at their destinations. The emergency brake system stops the elevator car if its speed becomes excessive, most likely in the event of a catastrophic equipment failure that causes the car to fall. The brake system includes a governor and a brake system on the elevator car.

The governor system is built on a separate sheave located in the machine room. The elevator car is connected to a loop of cable that goes around the governor sheave and another sheave at the bottom of the elevator shaft. As the elevator car travels up and down, the cable turns the governor sheave accordingly. **See Figure 11-8.** A pair of hooked flyweights (weighted arms) is attached to the face of the governor sheave. In normal operation, the spinning flyweights remain near the center of the sheave because the rotation is relatively slow. However, if the elevator speed reaches the maximum allowable speed, the fast-spinning flyweights extend outward due to centrifugal force and the arms hook onto stationary ratchets. This locks the governor sheave, jerking on the governor cable connected to the elevator car brake system, which engages the emergency brakes.

The brakes on the elevator car use brake pads that squeeze up against the guide rails to slow and stop the car. The pads are mounted on sets of wedge-shaped blocks on each side of the car. **See Figure 11-9.** Normally, there is a gap between the brake pads and the guide rail. However, when the governor cable jerks suddenly to a stop, a linkage in the elevator shaft is actuated that pulls up on the car's brake assemblies, wedging the brakes tightly up against the guide rails. The linkage is designed to brake both sides of the car at the same time and with the same force, bringing the car to a controlled stop.

Hydraulic elevators use fluid-filled cylinders to push elevator cars to higher levels.

||| Governor Sheave

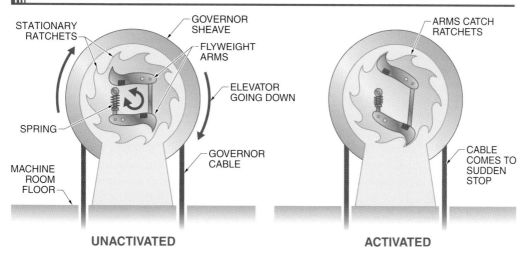

Figure 11-8. The governor sheave engages the emergency brake system if the elevator falls too rapidly.

▌▌▌ Elevator Brakes

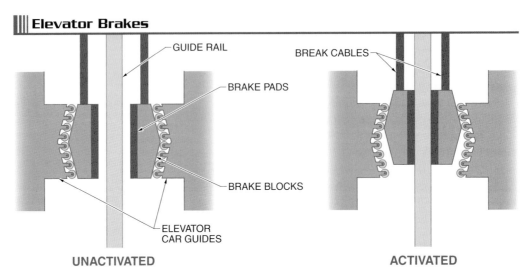

GUIDE RAIL

BREAK CABLES

BRAKE PADS

BRAKE BLOCKS

ELEVATOR CAR GUIDES

UNACTIVATED

ACTIVATED

Figure 11-9. When engaged, emergency elevator brakes wedge up against the rails guiding the elevator car, slowing it to a stop.

History of Conveying Systems

Humans were using elevators as early as the 3rd century BC. These elevators were basic systems used to hoist materials for construction and were powered by water wheel, animal power, or human power. By the mid 19th century, elevators were commonly used in factories, mines, and warehouses to lift equipment and materials. Though these elevators were very useful, they lacked safety features. If the cable or rope broke, there was nothing to prevent the elevator from falling to the ground floor. Therefore, they could not be used for transporting people.

In 1852, Elisha Otis developed a brake system that revolutionized the elevator industry. This safety feature was introduced and demonstrated by Elisha Otis himself in 1853 at the World's Fair in New York. This safety feature finally made passenger elevators practical and the first one was installed in New York in 1857. Within 20 years, there were over 2000 elevators being used by department stores, factories, hotels, and office buildings throughout the United States.

The moving staircase was patented in 1892 by Charles A. Wheeler and the first one was built by Jesse W. Reno in 1897 at New York's Coney Island as an amusement ride. He sold the machine to the Otis Elevator Company in 1899. Also in 1899, Charles D. Seeberger joined Otis Elevator Company, bringing with him the word "escalator," which is a combination of "scala" (Latin for "steps") and "elevator." The Otis Elevator Company installed the first escalator for public use at the 1900 Paris Exhibition.

Conveyors date back to the late 1700s. Early conveyors used belts made from leather, canvas, or rubber that traveled over a wood platform. These early conveyors moved products such as grain over very short distances. The first roller conveyor patent was received by Hymle Goddard of the Logan Company in 1908. The roller conveyor did not gain much popularity until 1919, when the automotive industry started using conveyors in production. In the 1920s, conveyor popularity increased as more factories used them for mass production and coal mines used them to transport coal in underground mines.

Door Operating System. The door operating system is used to open two different sets of doors: the elevator car doors and the elevator shaft doors. Both sets of doors must operate correctly for safe operation of the elevator system. An open door to the elevator shaft without the elevator present is a serious safety hazard because someone could easily fall into the open elevator shaft.

The elevator car doors are opened with an electrical motor and linkage assembly mounted to the top of the elevator car. **See Figure 11-10.** When the car reaches the correct position, the motor turns 180° in one direction, moving a pair of arms that pulls the doors open. Each set of doors rides along a set of rails and may consist of two separate panels.

Door Operating System

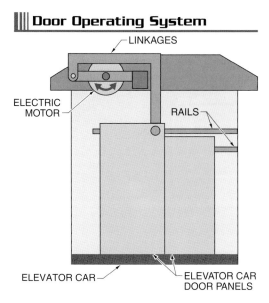

Figure 11-10. When the elevator car is in the correct position, the door operating system unlocks and pulls open both the car doors and the elevator shaft doors.

A mechanism on the car doors unlocks the elevator shaft doors and pulls them open at the same time. This safety feature prevents an elevator shaft door from opening when no elevator is present. Elevator cars also include sensors to detect people in the elevator car doorway, preventing the doors from closing until the area is clear.

Elevator Car Buffer. If all other safety features fail and the elevator car reaches the bottom of the elevator shaft, an elevator car buffer helps minimize damage and injuries by absorbing some of the impact force. An *elevator car buffer* is a piston and cylinder arrangement designed to cushion the fall of an elevator car or counterweight. One or two buffers are placed in the bottom of the elevator shaft to act as shock absorbers in the event the elevator falls to the lowest level of the building. For traction elevators, there are also buffers to cushion the descent of falling counterweights.

CONVEYANCE

A *conveyance* is a means by which people and materials are transported, such as an elevator. Much of an elevator system's operation is managed directly by an elevator controller. The control of elevator car speed, stops, direction, leveling, door opening, and safety features is accomplished by the controller, which uses its own control devices, such as call buttons and electric motors. This is the case regardless of whether the facility implements a building automation system. Therefore, these variables are not directly controllable by devices in other building systems.

However, the controller may be added to a building automation network to share information with and receive external inputs from other building systems. This integration can involve sending additional call signals to the elevator controller, controlling access with elevators, and changing the elevator's operating mode. **See Figure 11-11.** For example, an access control system may request an elevator car to a certain floor, or a fire alarm system disables an elevator system for occupant use for safety reasons.

Elevator Controllers

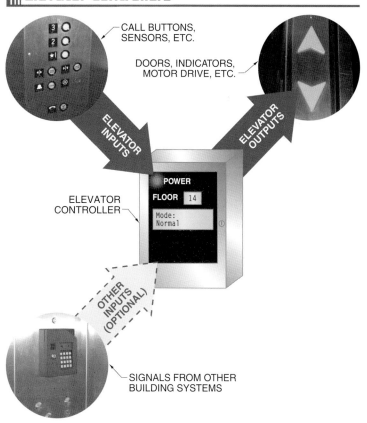

Figure 11-11. The elevator system can respond to signals from other building systems to add calls, control access, and change elevator operating modes.

Calls

A *call* is a request for an elevator to stop at a certain floor to either pick up or drop off passengers. Calls are typically activated from inside the elevator, where riding passengers press buttons corresponding to the desired drop-off floors, or from an elevator lobby on any floor the elevator serves, where waiting passengers typically press buttons for the desired direction (up or down). The elevator controller constantly monitors the car position and the calls from all sources and controls the elevator's direction and stops according to a preprogrammed algorithm. **See Figure 11-12.** An *algorithm* is a sequence of instructions for producing the optimal result to a problem. In this case, the optimal operation of the elevator minimizes the waiting times for all the passengers riding and waiting for the elevator.

For single-elevator systems, the control algorithm is very simple:

1. Continue in the same direction as long as there are remaining calls for/from floors in that direction.
2. Stop first at floors with waiting passenger calls in the same direction of travel.
3. If there are no more calls for one direction, change direction.
4. If there are no more calls at all, stop and become idle.

The number and distribution of calls determines the amount of time necessary to fulfill a specific call, so waiting times can vary significantly. However, this algorithm is very effective at minimizing the average waiting time for all elevator users.

For systems with multiple elevators, the idea is similar, but the algorithm becomes more complicated because there are more elevator cars that can be dispatched. Sophisticated elevator systems may also allow waiting passengers to specify the desired floor, instead of only the direction. The controller analyzes this more detailed information and plans dynamic travel routes for each elevator. By being able to group floor requests together, even if the directions of travel are not all the same, the system further reduces waiting time. Waiting passengers are then directed to the particular elevator that will fulfill their call request, which may be different from the elevator serving other waiting passengers.

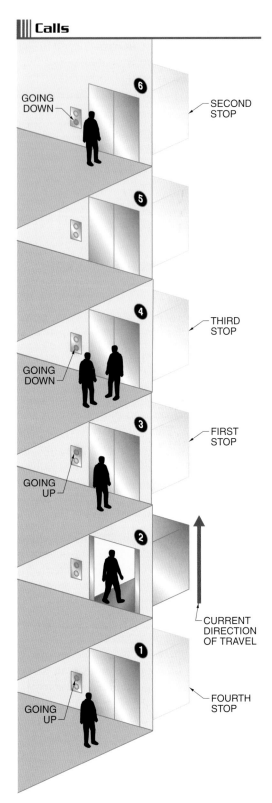

Calls

Figure 11-12. The normal operating algorithm of an elevator system determines the best sequence of stops in order to minimize passengers' waiting time.

Building automation systems can be used to add call signals to the elevator controller based on other system actions. For example, the elevator may be called to a certain floor to pick someone up when the security system is armed at the end of the day, or when the access control system admits a person to the building. In these cases, the call signal is sent to the controller from other control devices, but typically has the same priority as those from lobby call buttons. The controller adds this call to other received call signals and stops at the requested floor according to the normal operating algorithm. It is possible, however, for some controllers to be programmed to give higher priority to calls from certain sources. This would cause them to deviate from the normal algorithm to travel to the high-priority floors immediately.

Access

Many buildings include areas that are not intended to be accessible by the general public. Basements and mechanical areas, for example, are typically accessible by only maintenance personnel. High-security floors, such as penthouses, secure storage areas, and private offices, may also require restricted access that affects the operation of an elevator system. There are two common methods of managing access with elevator control systems.

Large buildings may segregate these special areas by having dedicated elevators that can only serve those areas from a central location, such as the lobby. For example, a dedicated service elevator may travel only to floors with heavy mechanical equipment, such as basements and rooftops. Access to those areas is controlled by controlling access to the elevators. In this situation, instead of a standard call button, the elevator is called to the waiting passenger when a correct authorization is given via some type of access control device.

For example, a key or access code may be required to call the elevator. Once the passenger is in the elevator, the call buttons inside the car can be used to call a certain floor. However, if the elevator serves more than one secure area, the system may restrict passengers to certain authorized floors. For example, one

access code may authorize the passenger to travel only to the basement, while another access code may authorize unlimited travel, including penthouses.

Smaller buildings with fewer elevators may not want to dedicate an entire elevator for special-access situations. If the secure areas require access only occasionally, then this elevator would be idle much of the time, wasting this building resource. Instead, one or more of the elevators providing regular service to unsecured floors may include a means to also transport authorized passengers to secure floors. In this situation, the access control device is located inside the elevator. **See Figure 11-13.** In addition to selecting the desired secure floor, a passenger must also use the necessary key, code, or other authorizing means, depending on the access control device. Only when the passenger has confirmed his or her authorization will the elevator travel to the secure floor. This system is commonly used for lower-security applications, however, since it is relatively easy for unauthorized passengers riding in the same elevator car to also gain access to secure areas.

Elevator Access Control Devices

Figure 11-13. Elevators can be used to control access to certain areas of a building by either controlling the access to the elevator or controlling where the elevator can stop.

Access control with elevator systems may be implemented with the elevator controller alone, or integrated with the building's access control system. Integration involves sharing authorization information between the two systems, but allows a single central database to be maintained for access privileges for building personnel.

Special Operating Modes

For the majority of time, an elevator system operates in a normal mode, and the controller moves the elevator car according to the regular algorithm. However, for reasons of efficiency, convenience, service, or safety, the elevator system can be placed into special operating modes. These modes give different priorities to the typical control inputs, or respond to completely different inputs altogether. The most common special operating modes include up peak mode, down peak mode, independent service mode, inspect mode, and fire service modes.

These special modes can be activated in a variety of ways. Some require manual activation, such as by a keyswitch. Some may be activated by an automated means within the elevator controller. For example, the controller may be programmed to switch to a different mode by analyzing the pattern of passenger calls and their locations, or it may be programmed to switch modes at certain times of the day. **See Figure 11-14.** Alternatively, some special elevator operating modes can be activated by systems outside of the elevator system, such as an access control system or fire alarm system. This can be the case if these systems are integrated together in a building automation network to share control signals.

Up Peak Mode. Elevator usage typically changes throughout the day. For example, in the morning, most calls originate from the building's entry floor by arriving employees who need to get to their place of work on upper floors. *Up peak mode* is an elevator system operation mode that stations all available elevator cars at the building entry floor in anticipation of many upward traveling passengers. Up peak mode operates the elevator more efficiently (minimizing passenger waiting times) during these busier times of the day by stationing elevator cars at certain floors in anticipation of calls.

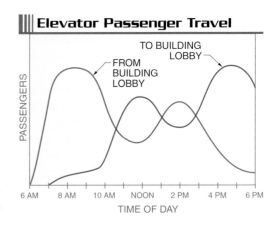

Figure 11-14. Some special elevator operating modes help the elevator system serve passengers more efficiently during certain periods of the day.

Down Peak Mode. Down peak mode is the opposite of up peak mode. In the late afternoon, most calls are for downward travel. While they probably come from a variety of floors, some calls may be clustered at certain times from certain floors. For example, the employees on the 10th floor may leave work at 4 PM, while the employees on the 15th floor leave at 5 PM. *Down peak mode* is an elevator system operation mode that stations all available cars on a certain building floor in anticipation of many downward traveling passengers. Alternatively, the cars may be sent to the highest building floor to wait so that they can travel down through all the floors to pick up the many waiting passengers wishing to descend.

Independent Service Mode. *Independent service mode* is an elevator system operation mode that releases the elevator from control by the central elevator controller. This mode is used when someone needs uninterrupted use of an elevator, such as a company moving from one floor of a building to another floor of the building. When placed in this mode, the elevator no longer responds to other calls until it is placed back to normal operating mode. It remains parked at a floor with its doors open until a user selects a different floor from inside the elevator car. If the building has a single elevator, this mode disables all call buttons. If the building has more than one elevator, the calls are serviced by the other elevators while one is in independent service mode.

Inspect Mode. *Inspect mode* is an elevator system operating mode in which the elevator ignores all control inputs. The inspect mode essentially disables the elevator and is used for elevator maintenance. All calls are redirected to other elevators. This mode protects the maintenance personnel from the hazards of unexpected movement of the car while working on the elevator, though lockout/tagout precautions on the elevator power supply are also taken as a safety measure. This mode requires a special key for activation.

Fire Service Modes. The operation of elevators during a building fire is a critical life safety decision. Elevators can become dangerous during building fires and occupants must be prevented from using an elevator for evacuation. Both a fire and a fire suppression system can damage the elevator motors, doors, cables, controls, and emergency brakes, potentially trapping elevator occupants in an inoperable and unsafe elevator. However, elevators can be useful for firefighters to move around the building quickly to look for fire hazards and trapped occupants. These two considerations correspond to the phase one and phase two special operating modes, which can be activated in the event of a fire in the building.

Phase one mode is an elevator system operating mode in which the elevator car automatically moves to a designated floor during a fire. This mode is required by code that mandates an emergency recall function. The elevator system receives notification of an alarm from the fire protection system, which initiates phase one operating mode. The elevator car moves to a certain recall floor (determined by code requirements), ignoring all normal call signals. If the fire alarm system indicates a fire hazard on that floor, such as alarms from smoke detectors, then the elevator moves to a predesignated alternate recall floor. The elevator car doors open so that any riding passengers can exit the elevator, and then the doors close, remaining closed while in phase one mode.

If the fire department personnel decide that the elevator is safe to use under the circumstances, they can manually override the otherwise disabled elevator. *Phase two mode* is an elevator system operating mode that enables rescue personnel to manually operate an elevator that is in phase one mode. This mode is activated by fire department personnel with a special key switch inside the elevator car. **See Figure 11-15.** While in this mode, automatic functions such as doors opening and closing do not operate normally. For example, to open the door, the door open button must be pressed and held until the door is fully open. Likewise, the door close button must be pressed and held until the door closes completely. Once the door is closed completely, the elevator will move to the requested floor. If the firefighter wants the elevator to remain on a specified floor, the hold position on the key switch is used. When the firefighter wishes to return to the recall floor, the phase two keyswitch is simply turned off.

The elevator shutdown (shunt trip) function is a separate function that automatically shuts down the elevator before water from sprinklers in the elevator machine room or elevator shaft affects the essential safety features of the elevator, such as brakes. Once an elevator has been disabled by the shunt trip, it cannot be used in either fire service mode.

Phase Two Mode Keyswitch

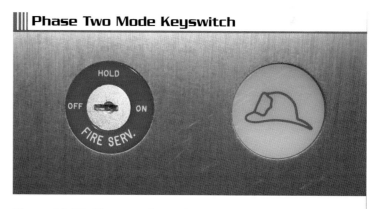

Figure 11-15. *Phase two fire service mode functions are activated by a keyswitch, typically inside the elevator car.*

Phase two mode requires manual use of elevator car buttons for every door opening and closing action. This is in case there is a fire near the elevator doors on the arrival floor. If the firefighter can feel the heat, then he or she can choose not to open the door.

Direction indicators are often found inside elevator cars, but must always be in the lobbies. Direction indicators are sometimes etched with or shaped like arrows. They may also use the convention that the "up" direction indicator light is green and the "down" indicator light is red. Also, most elevators chime to indicate the traveling direction of an arriving elevator, usually in conjunction with the indicator lights. Universally, one chime is for up, two is for down, and none indicates an elevator that is free.

ELEVATOR SYSTEM CONTROL DEVICES

Similar to the operation of fire alarm systems and security systems, the elevator system is managed by an elevator controller that is connected to its own control devices, with or without an automation system in the building. When integrated with other building systems for automated operation, however, the elevator controller and related devices must then interact with other control devices.

Elevator Controller

The dedicated elevator controller makes all decisions on elevator stops and directions during normal operation, so a building automation system does not need to manage normal elevator system operations. Elevator controllers include a means to accept inputs from sensors and call buttons, microprocessors to make control decisions based on operating algorithms and modes, and a means to directly control output devices, particularly the electric motor or drive that moves the elevator. **See Figure 11-16.** The controller adjusts the speed of the elevator as it approaches a destination floor, determines where to send an elevator car next, changes its direction, continuously monitors the position of each elevator car, and dispatches elevator cars to certain locations where peak usage is likely.

When integrated with other building systems, the elevator system requires a controller that can communicate with other control devices. This typically requires specialized or programmable inputs. For example, a special input connection is reserved for

alarm signals from the fire alarm system. When the controller receives a signal on that connection, it immediately initiates phase one mode. Programmable inputs, however, may be used in a variety of ways, depending on the desired application. For example, an input connected to an access control device at the building entrance may be programmed to add a call to the elevator controller, as if a waiting passenger pressed a call button in the building lobby.

Elevator Controller

ELEVATOR CONTROLLER   *Kone, Inc.*

Figure 11-16. Elevator controllers are connected to numerous elevator-specific control devices to operate the elevator system independently.

Information from the elevator system such as calls, previous stops, current location, service requirements, and load can also be shared with other building systems. This information may be used as inputs for the control of other systems, or may simply be monitored by security and/or maintenance personnel.

Elevator controllers are connected to various call buttons throughout the building. A *call button* is a user-operated pushbutton input device that sends a call signal to an elevator controller. Call buttons for riding passengers are located inside the elevator car and correspond to each floor accessible by that elevator. Call buttons for waiting passengers are located on each floor served by the elevator and typically correspond only to the desired direction of travel (up or down). Call buttons are fairly consistent among all elevators in both function and appearance, regardless of manufacturer. **See Figure 11-17.**

The controller maintains connections with the moving elevator car through the traveling cable. The *traveling cable* is a cable that provides power, control, communication, and other wiring to the elevator car. This cable travels up and down the shaft with the elevator car. At the car's farthest possible position from the controller, the traveling cable is almost completely extended. As the car moves closer, the traveling cable becomes slack and the extra length is taken up to avoid tangling.

Telephones

All elevators require a method to communicate with the outside world in the event of an elevator problem, particularly if the elevator becomes stuck between floors. Therefore, elevator cars include a dedicated telephone that automatically connects to either on-site or off-site security or maintenance personnel. A telephone may be mounted within a closed cabinet in the elevator car or the telephone line may be connected to a digital communicator. The telephone connection may be activated by picking up the receiver or pushing a "call for assistance" button. **See Figure 11-18.**

ELEVATOR SYSTEM CONTROL APPLICATIONS

With specialized or programmable inputs, elevator system controllers can be integrated with other building systems in practically any way desired, especially in a fully automated building with multiple systems that may provide input signals. It is required that elevator systems to be integrated with the fire alarm system to initiate phase one mode, also called emergency elevator recall. Another common application integrates access control devices with elevators to limit access to certain building areas to authorized persons.

Call Buttons

LOBBY CALL BUTTON　　ELEVATOR CAR CALL BUTTONS

Figure 11-17. Call buttons are the typical method for passengers to select their desired travel direction or floor.

Elevator Car Telephones

Figure 11-18. Telephones or similar two-way communication devices are installed inside elevator cars for emergencies.

Emergency Elevator Recall

Emergency elevator recall is a critical life safety application that integrates an elevator system with a fire alarm system to completely override normal operations. Signals from the fire alarm control panel (FACP) cause the elevator controller to initiate phase one mode, which recalls all elevator cars to a designated floor. **See Figure 11-19.** Once there, the elevators are shut down, unless they are used by firefighting personnel in phase two mode.

⫿⫿ Emergency Elevator Recall

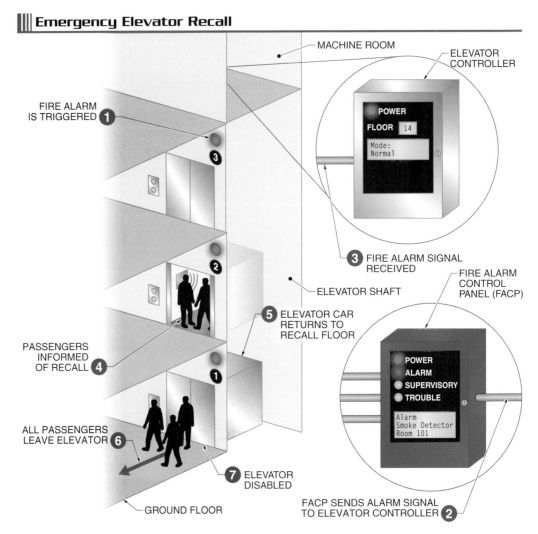

Figure 11-19. In emergency fire situations, the fire alarm control panel signals to the elevator controller to recall the elevator car(s) to a designated floor.

Operating Conditions. In order for the elevator controller to respond to fire alarm signals, an input connection on the controller is connected to an output connection on the FACP. Terminals may be provided on both the elevator controller and FACP that are dedicated for this application. Some elevator systems may also connect the dedicated telephone line to a security station or a prerecorded message to inform the riding passengers of the alarm, direct them to exit the elevator, and warn waiting passengers not to enter the elevator as the riding passengers exit. This option may require special features on the controller or integration with the security system.

Control Sequence. The control sequence for emergency elevator recall is as follows:

1. The building's fire alarm system registers a fire alarm from its initiating devices.
2. The FACP relays the signal to its outputs, one of which is connected to the elevator controller.
3. The elevator controller receives the fire alarm signal and interrupts its normal service to switch to phase one mode.
4. Since some passengers may already be riding the elevator, the controller may connect the elevator car's telephone line to building personnel or a message recording to give instructions.

5. The elevator controller operates the motor to return the elevator car to a recall floor. If the fire alarm signal indicates fire hazards on the primary recall floor (typically the ground floor), the elevator controller directs the car to a different floor.

6. The elevator controller signals the elevator car to open the elevator doors so that passengers can exit the elevator car and then close the doors.

7. While in phase one mode, the controller ignores all normal call signals, temporarily disabling the elevator system.

Automatic Elevator Call

For convenience, elevators can be automatically called to certain floors in anticipation of waiting passengers before any call buttons request a call. For example, when a person enters a multi-story building, it is likely that they will need to travel to a higher floor. Instead of waiting until a call button is pressed in the building lobby, the elevator can be already waiting to pick up the passenger by the time the passenger arrives at the elevator lobby.

Operating Conditions. This anticipation of elevator calls requires inputs from other building systems, such as access control or security systems, that can monitor entry into the areas near the elevators. **See Figure 11-20.**

Control Sequence. The control sequence for automatically calling an elevator from the access control system is as follows:

1. A person entering a secure building or elevator lobby area uses the necessary passcode or access card to authorize entry.

2. The access control device communicates with the access control database to check the person's credentials.

3. If the authorized user is verified, an access control device sends a waiting passenger call signal to the elevator controller. The call is specific to the floor where the access control device is located.

4. The elevator controller operates the motor to send the elevator car to the appropriate floor to pick up the passenger.

5. The elevator car opens its doors and waits for the passenger to enter.

Automatic Elevator Call

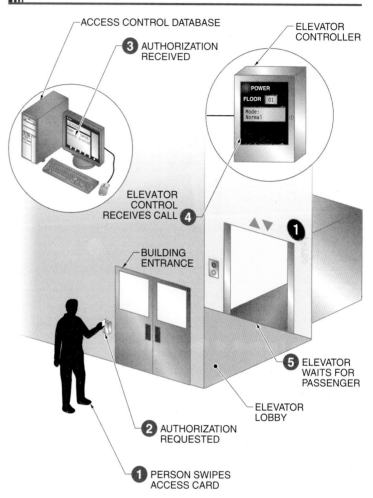

Figure 11-20. Inputs from access control systems can be used by elevator controllers to anticipate waiting passengers and automatically call elevators to certain floors.

Elevator Access Control

Integration with access control systems is commonly used for applications that require authorization to access certain floors. This can be done either in the elevator lobby or inside the elevator car. **See Figure 11-21.** In this application, a call for the desired floor or direction is sent to the elevator controller only when the access control device verifies an authorized user. **See Figure 11-22.**

Refer to
Quick Quiz®
on CD-ROM

Access Control of Elevator

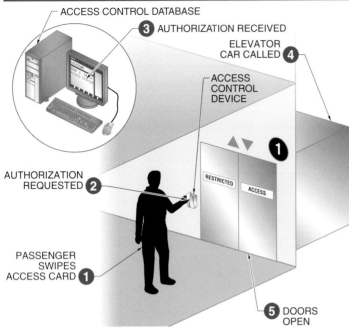

Figure 11-21. Access control systems can be used to admit only authorized personnel to elevators serving secure building areas.

Access Control of Secure Floors

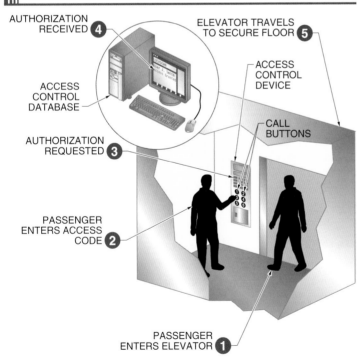

Figure 11-22. When integrated with access control systems, elevators can be used to secure individual building floors by requiring authorization for certain floor calls from inside the elevator car.

Operating Conditions. In order to control access to certain building floors with the elevator, the elevator controller must be integrated with an access control device. This could be accomplished with a stand-alone device connected only to the elevator controller, but typically this access control device is associated with the building's access control system to share a common database of authorization information, such as users, access codes, and authorized areas. In this application, the access control device notifies the elevator controller that the passenger's call request may be fulfilled based on the authorization results by forwarding on the pending call signal.

Control Sequence. The control sequence for controlling access to an elevator is as follows:

1. A waiting passenger uses the necessary access card to authorize use of the restricted elevator.
2. The access control device communicates with the access control database to check the passenger's credentials.
3. If the authorized user is verified, an access control device in the lobby sends a waiting passenger call signal to the elevator controller.
4. The elevator controller operates the motor drive system to send the elevator car to the lobby to pick up the passenger.
5. The elevator car admits the passenger and travels to any floor served by the elevator that is selected by the passenger from within the elevator car.

Control Sequence. The control sequence for controlling access to certain floors from inside an elevator is as follows:

1. A waiting passenger calls and enters the elevator car as normal.
2. If the desired floor is secure and requires authorization, the call for the desired floor is not sent to the elevator controller until the passenger uses the necessary access code or card.
3. The access control device communicates with the access control database to check the passenger's credentials.

4. If the authorized user is verified, the access control device in the elevator car sends the call signal for the desired floor to the elevator controller.

5. The elevator controller operates the motor drive system to send the elevator car to the desired floor.

> A special traction elevator design does not use the traditional counterweight, but instead has an elevator car at each end of the cables. The two elevators move synchronously in opposite directions, serving as each other's counterweight. This design is useful for special applications with high ride volume.

Summary

- The most common type of conveying system in commercial buildings is the elevator.

- The functional types of elevators include passenger, service, freight, dumbwaiter, and construction elevators.

- Most elevators have either traction or hydraulic designs.

- In a traction elevator, the counterweight reduces the amount of energy needed to raise and lower the elevator car by approximately balancing the loads on either side of the sheave.

- Most traction elevators require a machine room, though if smaller drive components can be installed directly in the elevator shaft, the result is a machine room-less design.

- Traditional hydraulic elevators require the hydraulic cylinder to be located underground to a depth equal to the highest level the elevator will reach, which usually limits the applications to two to seven floors.

- A holeless elevator utilizes two relatively small telescoping cylinders attached directly to the sides of the steel elevator car supports, virtually eliminating the need for space below the lowest elevator level.

- The emergency brake system stops the elevator car if its speed becomes excessive, most likely in the event of a catastrophic equipment failure that causes the car to fall.

- Both sets of elevator doors must operate simultaneously.

- The control of elevator car speed, stops, direction, leveling, door opening, and safety features is accomplished by the controller, which uses its own control devices, such as call buttons and electric motors.

- The elevator controller may be added to a building automation network to share information with and receive inputs from other building systems.

- The elevator controller constantly monitors the car position and the calls from all sources and controls the elevator's direction and stops according to a preprogrammed algorithm.

- A building automation system may send a call signal to the elevator controller with the highest priority, which overrides all other calls and immediately directs the elevator car to a certain floor, usually the lobby.

- Access control with elevator systems may be implemented with the elevator controller alone, or integrated with the building's access control system.

- For the majority of time, an elevator system operates in a normal mode and the controller moves the elevator car according to the regular algorithm.

- Some special elevator operating modes can be activated by systems outside of the elevator system, such as an access control system or fire alarm system, if these systems are integrated together in a building automation network.

- Elevator systems are managed by controllers, which include microprocessors to control every aspect of elevator operation.

Definitions

- A *conveying system* is a system for the transporting of people and/or materials between points in a building or structure.

- An *elevator* is a conveying system for transporting people and/or materials vertically between floors in a building.

- A *passenger elevator* is an elevator primarily intended to transport people.

- An *express elevator* is an elevator specifically intended to service a single floor or a specific group of floors.

- A *service elevator* is an elevator reserved exclusively for the use of building service personnel.

- A *freight elevator* is an elevator primarily intended to transport equipment or materials.

- A *dumbwaiter* is an elevator for transporting very small loads, primarily consisting of domestic items.

- A *construction elevator* is a temporary elevator for the transportation of materials and workers during the building construction.

- A *traction elevator* is an elevator system that raises and lowers the elevator car with cables operated by an electric motor.

- A *machine room* is a space directly above an elevator shaft to house the motor, sheave, and elevator controls.

- A *machine room-less elevator* is a traction elevator system using special materials and improved electric motors that require little space, eliminating the need for the machine room.

- A *hydraulic elevator* is an elevator that lifts the elevator car from below using a hydraulic ram.

- A *hydraulic ram* is a piston that is driven into or out of a hollow cylinder by fluid pressure.

- An *elevator car buffer* is a piston and cylinder arrangement designed to cushion the fall of an elevator car or counterweight.

- A *conveyance* is a means by which people and materials are transported, such as an elevator.

- A *call* is a request for an elevator to stop at a certain floor to either pick up or drop off passengers.

- An *algorithm* is a sequence of instructions for producing the optimal result to a problem.

- *Up peak mode* is an elevator system operation mode that stations all available elevator cars at the building entry floor in anticipation of many upward traveling passengers.

- *Down peak mode* is an elevator system operation mode that stations all available cars on a certain building floor in anticipation of many downward traveling passengers.

- *Independent service mode* is an elevator system operation mode that releases the elevator from control by the central elevator controller.

- *Inspect mode* is an elevator system operating mode in which the elevator ignores all control inputs.

- *Phase one mode* is an elevator system operating mode in which the elevator car automatically moves to a designated floor during a fire.

- *Phase two mode* is an elevator system operating mode that enables rescue personnel to manually operate an elevator that is in phase one mode.

- A *call button* is a user-operated pushbutton input device that sends a call signal to an elevator controller.

- The *traveling cable* is a cable that provides power, control, communication, and other wiring to the elevator car.

Review Questions

1. Which common types of elevators prohibit carrying passengers?

2. What is the function of the counterweight in a traction elevator?

3. Explain the operation of a hydraulic elevator.

4. How does the speed of the elevator car cause the emergency brakes to engage?

5. Why is it a necessary safety feature to have both sets of elevator doors open simultaneously?

6. Summarize the normal control algorithm for single elevator systems.

7. Explain the two common methods of controlling access to certain areas of a building with elevator systems.

8. In what ways can special elevator operating modes be activated?

9. What is the difference between phase one mode and phase two mode?

10. How are elevator controllers interfaced with other building system control devices?

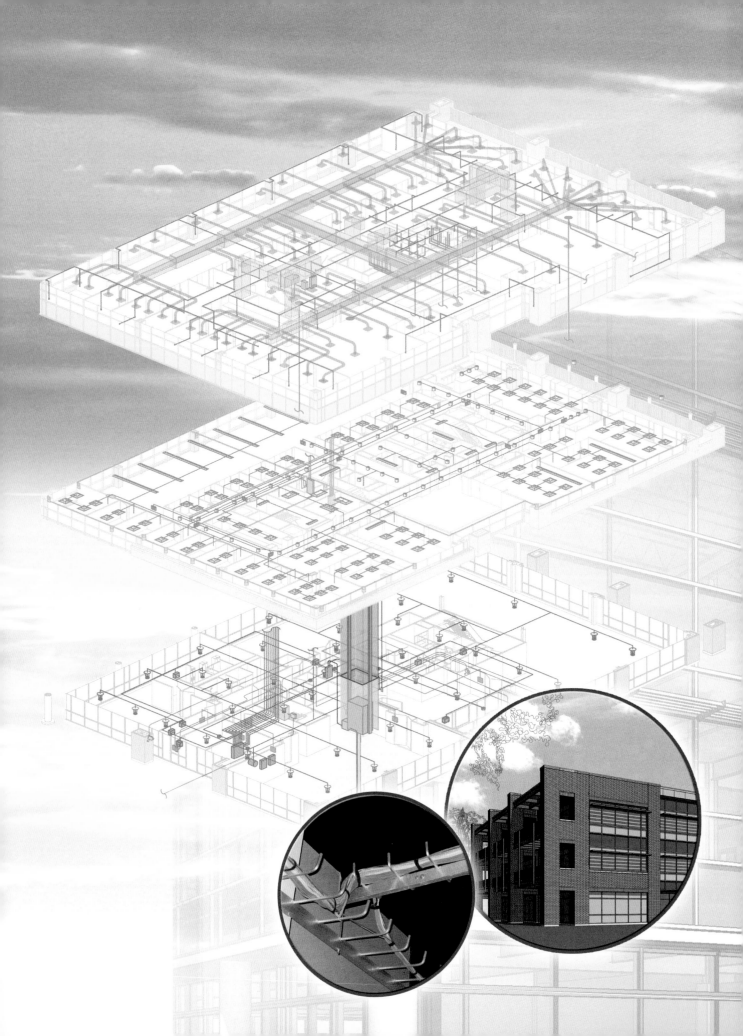

Chapter Twelve

Automated Building Operation

The number of potential ways in which various building systems can be integrated is practically limitless. Most automated buildings include only a few applications. Some include many, but there are always additional possibilities. A walk-through of an example of an automated building highlights only some of the common ways in which building systems are automated. There are also different protocol systems that can be used to achieve automation goals. Specific building and occupant requirements ultimately dictate the best choice of controls for a facility.

Chapter Objectives

- *Identify the ways in which controls for different building systems interact in an integrated building automation system.*
- *Compare the possible applications of building automation systems for different facility sizes.*
- *Describe the role of building automation in LEED® certification.*
- *Evaluate some of the possible control applications for building automation systems.*

BUILDING AUTOMATION SYSTEM OPERATION

Building automation systems use the abundant sharing of system information to provide a highly functional and flexible "living" environment for the building occupants. This exchange occurs on an infrastructure that connects the control devices for HVAC, lighting, fire, security, and other systems together on a common network. **See Figure 12-1.** Additionally, with energy management capabilities, these systems can reduce overall energy use and costs.

▌▌▌Modern Commercial Buildings

Figure 12-1. Building automation systems are becoming increasingly common in modern commercial buildings.

Although the initial cost for a building automation system is typically higher than for conventional construction, total life-cycle costs are generally lower due to energy savings, reduced maintenance costs, and increased value to potential tenants. The cost of implementing a building automation system is minimized when it is incorporated into new construction. Retrofitting a building for automation is usually also possible, but at a greater expense, inconvenience, and potential loss of system flexibility.

System Interdependencies

Building systems have been self-contained and independent in past generations. With multiple building systems being connected together into a "whole building" automation system, control devices become somewhat dependent on each other. The actions of one system may affect another. For instance, a controller may use sensor information from other networked devices in order to complete its sequences and share its output data for other devices to use. Critical interactions must be carefully designed for reliability, but the advantage is a significant reduction in redundancy. For example, only one outside air temperature sensor may be needed if its data is shared with all of the controllers that require this information.

Control Device Interoperability

A major concern with traditional automation systems is that they are typically limited to a single building system or equipment manufacturer. This severely limits the potential ability of automation to involve multiple systems seamlessly and reliably.

Most modern automation devices are built on a framework that allows complete compatibility and reliable and accurate communication between nearly any building automation devices, regardless of the manufacturer or the system in which they are found. For example, information from an occupancy sensor can be used to control both the lighting and the HVAC functions within a room. This interoperability of the system components allows for nearly limitless flexibility in system design and component sourcing. *Interoperability* is the ability of diverse systems and devices to work together effectively and reliably.

The communication requirements of such an interconnected system include a robust network infrastructure. **See Figure 12-2.** It is becoming increasingly common for building automation systems to merge with traditional information technology (IT) infrastructures. The sharing of these networks requires the IT professionals to take an active role in building automation system operations. Integration with Internet standards even allows the scope of a building automation system to extend beyond the walls of the building.

▌▌▌▌ Building Automation and Information Technology Systems

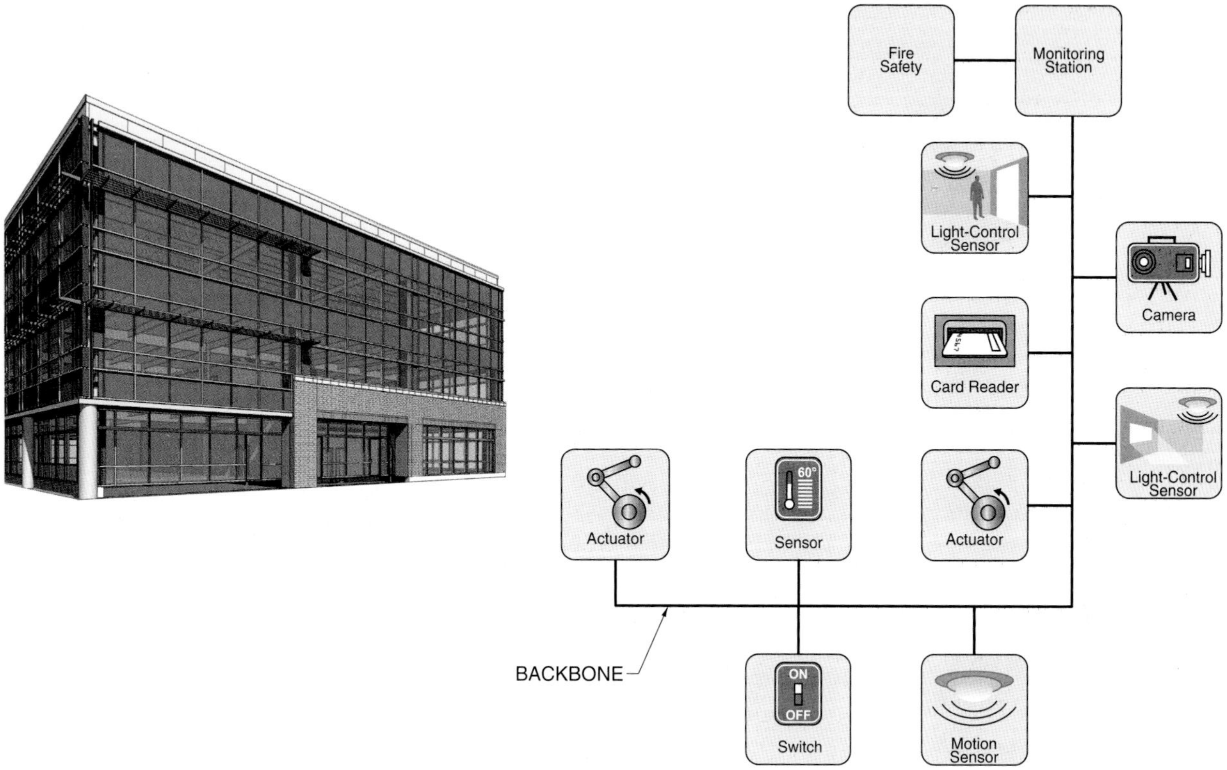

Figure 12-2. Building automation systems often use IT networks similar to those used for computer systems.

Location Applications

The flexibility of building automation systems is evident in the wide variety of possible applications. Building automation system location applications can be classified into building-wide applications, campus-wide applications, and worldwide applications. **See Figure 12-3.**

Building-Wide Applications. Building-wide applications include those operations within a discrete facility. The most common implementations of building automation systems are building-wide. The scope of the system extends no farther than that of the building systems themselves. The systems are monitored, controlled, and modified only from within the building.

Campus-Wide Applications. With a suitable communications network, automation systems can be extended beyond a single building. Campus-wide applications include those operations between and within a group of buildings in relatively close physical proximity.

For example, a campus consisting of several buildings such as a library, student union, classroom buildings, and athletic building may be managed by an automation system consisting of devices installed throughout the campus. The same systems managed as a building-wide application can be managed as a campus-wide application. This is especially important for campuses with centralized plant facilities, such as boilers and chillers. The campus-wide automation system can manage these resources in the most effective way across the entire area served by these plants.

Another example of a campus application is an airport. Facility systems in areas such as terminals, attached hotels, and the administration building can be managed from an automation system. Direct control and system monitoring is critical for addressing homeland security issues.

Building Automation System Location Applications

CONTROL OF EQUIPMENT
AND OPERATIONS WITHIN
DISCRETE BUILDING

GENERATOR

BUILDING-WIDE

CONTROL BETWEEN AND
WITHIN BUILDINGS IN
RELATIVE CLOSE PROXIMITY

CAMPUS-WIDE

CONTROL BETWEEN
AND WITHIN
DISTANT BUILDINGS

WORLDWIDE

Figure 12-3. With a suitable network infrastructure, building automation systems can extend beyond the walls of a single building.

Building-wide applications use control devices throughout the entire building.

Worldwide Applications. The worldwide Internet is a readily accessible network for applications beyond local areas. Many automation system device manufacturers include network tools to be able to access and control automation function from anywhere in the world.

Most worldwide automation applications involve the monitoring of system operations remotely. System technicians, maintenance personnel, or building owners can access building operation via an Internet connection and either special software or a standard web browser. Some systems even allow authorized users to change system information, such as operating modes or setpoints, remotely. Remote monitoring is particularly helpful when outside consultants are needed to troubleshoot system problems. Some issues may be resolved by monitoring system information via the Internet, thus reducing the need for expensive on-site visits.

If the application warranted it, systems could also be designed to automatically share control information between distant locations, though this would require careful design. For example, upon activation of a security alarm at a satellite office, personnel in the headquarters building could remotely control the surveillance CCTV system to investigate the trouble.

Automation systems can be used to monitor a steam plant remotely, allowing some routine monitoring and control functions to be performed by off-site personnel. **See Figure 12-4.** The boiler system's microcomputer burner control system (MBCS) is the local controller that analyzes data from boiler-specific sensors and accessories and makes the necessary data available for other controllers or remote monitoring via the building automation system. The remote monitoring of boilers or other potentially dangerous equipment does not replace the need for qualified personnel in close proximity to the facility. However, it provides additional options for monitoring and control.

▎Remote Monitoring of Boiler Plant

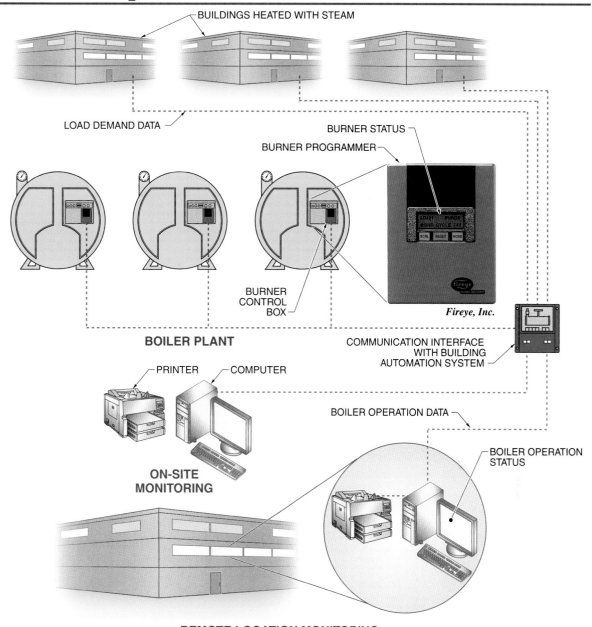

BUILDINGS HEATED WITH STEAM

LOAD DEMAND DATA

BURNER STATUS

BURNER PROGRAMMER

Fireye, Inc.

BURNER
CONTROL
BOX

BOILER PLANT

COMMUNICATION INTERFACE
WITH BUILDING
AUTOMATION SYSTEM

PRINTER   COMPUTER

BOILER OPERATION DATA

BOILER OPERATION
STATUS

**ON-SITE
MONITORING**

REMOTE LOCATION MONITORING

Figure 12-4. Remote access to a building automation system allows off-site personnel to monitor and control the building systems.

BUILDING AUTOMATION SYSTEM EXAMPLE

Building automation systems are increasingly popular in both new construction and existing buildings. The most common automation applications are related to HVAC and life safety operations. In some buildings, other applications such as security, access control, and other building functions are also included. Most buildings do not integrate all possible building systems. However, an overview of a building automation system example can be used to emphasize several applications that are possible with existing technology. **See Figure 12-5.**

Example Building Exteriors

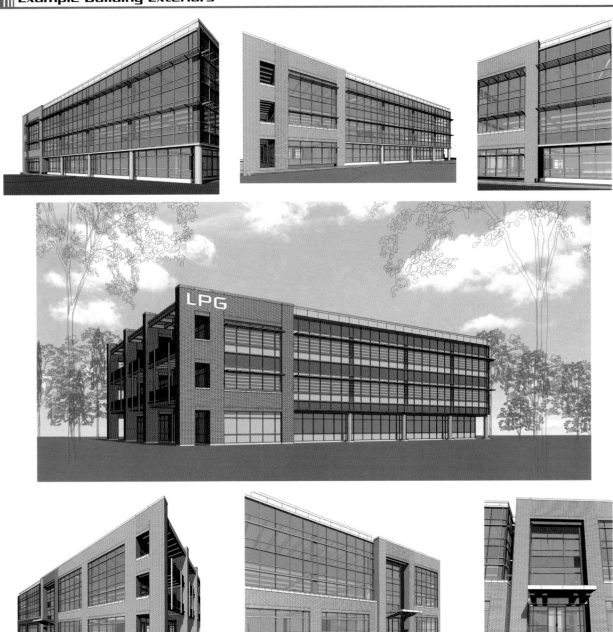

Figure 12-5. A walk-through of an example of an automated building highlights some of the possible automation applications.

Building Owner Profile

The Lincoln Publishing Group (LPG) is a 100-employee organization that publishes several trade magazines, journals, training products, and government documents. The company works with government officials, authors, advertisers, and other vendors lo-cated throughout the country. A variety of media is used in these products, including print, audio, and video. Some publications involve homeland security and classified government documents.

Because many of the publications cover technology and its applications in industry and government facilities, the building was designed with many automation features that showcase technology and energy savings. Additionally, the building was designed as a sustainable building by including provisions for energy efficiency, reducing maintenance, and enhancing occupant well-being and performance. The building was also designed and built to be Leadership in Energy and Environmental Design® (LEED®) certified by the U.S. Green Building Council.

Building Description

A conscious decision at the early planning stage was made by the management of the company to construct a building with sustainable building features. Input was solicited and received related to features and amenities that provide efficiency as well as comfort in the working environment. In addition, future expansion for increased production capacity was considered in the facility requirements.

The building site is 4.25 acres of previously farmed land in an area that has seen limited development. The area is zoned as office/research/light commercial. Because the property is not part of a subdivision, improvements for road, utilities, sewer, and water detention were required inside and adjacent to the property boundaries. Grading was completed to facilitate the flow of water into a dry bottom detention area and away from the property by earth swales. **See Figure 12-6.** A crushed rock walking path around the detention area provides an opportunity for moderate outdoor exercise. Landscape lighting is provided on the path in intervals.

> The Leadership in Energy and Environmental Design® (LEED)® Green Building Rating System™ is a method of quantifying the measures taken during the construction and operation of a building to save energy and use sustainable materials. Buildings are rated as Certified, Silver, Gold, or Platinum buildings.

||| Site Plan

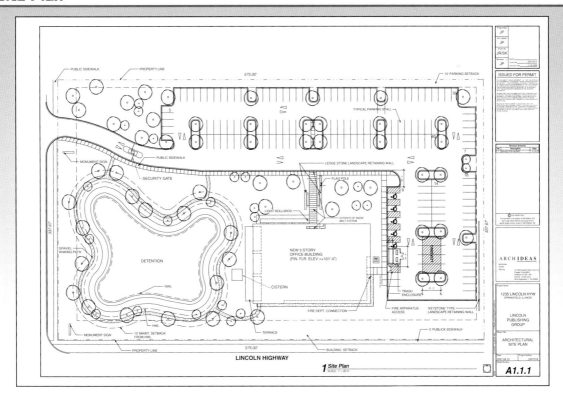

Figure 12-6. Some building automation features, such as security and outdoor lighting equipment, are located outside of the actual building.

An engine generator is a combination of an internal combustion piston engine and a generator mounted together to produce electricity. The engine burns a compressed fuel-air mixture to provide the mechanical power, and the generator converts mechanical energy into electricity by means of electromagnetic induction.

The building has 45,000 sq ft of total usable space on three floors. It has a structural steel core and glass curtain walls with aesthetic brick facades. Building orientation and architectural features provide shading for maximum lighting and heating benefits from the sun. A patio is located off the west wall of the first floor. Balconies are provided above the patio on the second and third floors. A sidewalk heating system circulates hot water from a boiler through heat exchanger loops in the sidewalk

to prevent slippery conditions from ice and snow in the winter. A soil moisture sensor and temperature sensor activate the system. A 400 kW diesel-powered generator serves as a back-up power source for the fire pump, elevator, computer servers, and other critical load circuits. **See Figure 12-7.** Critical loads are grouped into a common panelboard.

Each floor has 10′ ceilings to maximize natural light penetration and an open environment. Package HVAC units are located centrally on the roof. A trunk and branch HVAC system provides conditioned air into building spaces. The supply air is distributed by branches extending from the trunk (main supply duct) to each register. Airflow from the register is controlled by the vanes on register panels. Return air is drawn through grills back into the return-air ductwork.

Back-Up Engine Generator

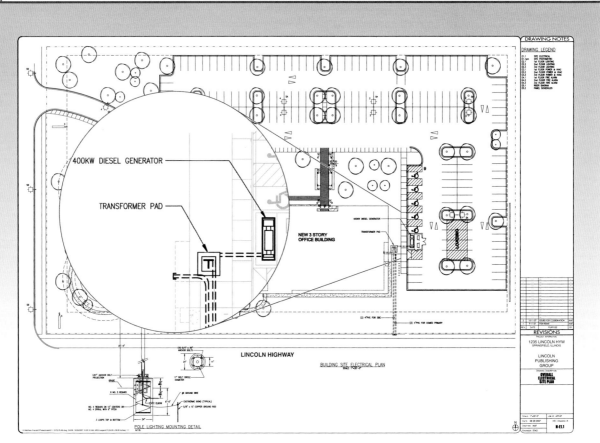

Figure 12-7. Automation of the electrical system may include managing the switching over to secondary power sources, such as diesel engine generators.

The first floor is designed for approximately 9000 sq ft of tenant space, with a 3000 sq ft exercise area in the southeast corner for LPG employees. **See Figure 12-8.** The first floor tenants determine the floor plan required for their specific needs. The building electrical, HVAC, and VDV equipment is designed for flexibility in meeting tenant requirements. The second floor is occupied by LPG's administrative and marketing departments. The third floor will be occupied by the editorial department. The lobby area on the first floor secures access to all building spaces. Employees can access the second and third floor via an elevator or stairs. The reception area for the Lincoln Publishing Group is located on the second floor across from the elevator. The space above the reception area opens to the third floor. The perimeter of the lobby space occupying the second and third floor is enclosed by glass in compliance with fire department code for smoke isolation and containment during a fire.

The second floor contains private offices and workstations associated with the administrative and marketing functions of the company. In addition, the second floor also houses the lunchroom, training rooms, library, business center, and corporate boardroom. The third floor contains private offices and workstations associated with the product development functions of the company. In addition, the third floor also houses a photography studio and multimedia suite.

A voice over Internet protocol (VoIP) system is used to transmit telephone calls over the data network and Internet. The combining of voice and data allows long-distance calling without telephone company charges. The system requires VoIP telephones.

The common areas are heated to a temperature setpoint based on the outside air temperature, sun position, and number of occupants. During office hours on weekdays, electrical demand and consumption are reduced during peak utility demand periods by adjusting temperature setpoints by a few degrees. This small change is usually imperceptible to building occupants but results in significant savings from peak demand utility rates.

||| Floor Plans

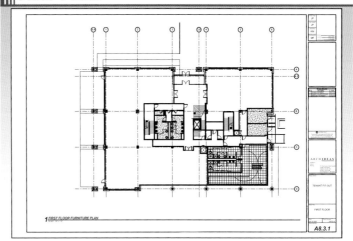

FIRST FLOOR

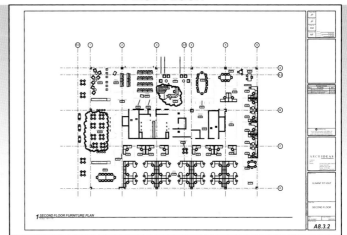

SECOND FLOOR

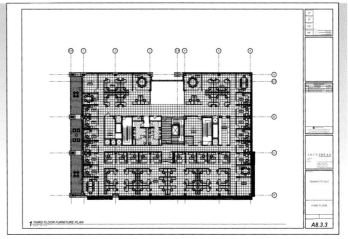

THIRD FLOOR

Figure 12-8. Building interiors made pleasant and comfortable with automated lighting and HVAC systems improve the productivity of the occupants.

The fire alarm system is monitored by an off-premises supervising station, and alarms are also automatically transmitted to the local fire department. This alerts monitoring and emergency personnel immediately and communicates building information prior to their arrival. The fire alarm system takes over the management of other building systems to facilitate safe evacuation and limit the spread of fire and smoke. For example, commands are activated for HVAC shutdown, elevator recall, door unlocking, and smoke door closure.

LEED® Certification

The Leadership in Energy and Environmental Design® (LEED®) Green Building Rating System™ is the nationally accepted standard for the design, construction, and operation of green buildings. LEED® serves as a guide for building owners and operators in realizing the construction and performance of a building. **See Figure 12-9.** The LEED® certification process addresses sustainable site development, water savings, energy efficiency, materials selection, and indoor environmental quality throughout the construction and operation of the building. Two major criteria in the LEED® certification process are the energy performance measures and the renewable energy standard. Both of these criteria are affected by the building automation system.

The building has been designed to meet the criteria for gold certification by the U.S. Green Building Council. This level of certification requires a minimum of 39 points. Each of the categories was analyzed during the designing and construction phase for compliance. **See Figure 12-10.** For example, the building structural steel is recycled, and bricks are supplied by a manufacturer located within a 500-mile radius. In addition to design and construction requirements, the building must also comply with the standards for maintenance and operation, including cleaning procedures, pest control methods, and office waste recycling. **See Appendix.**

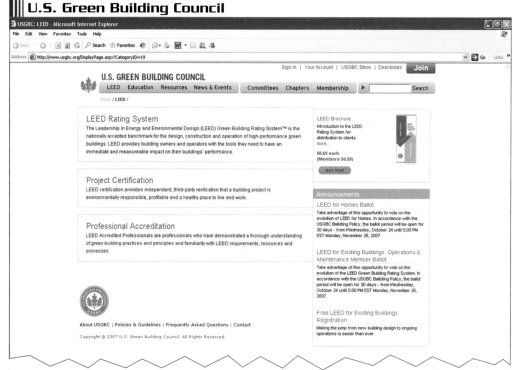

U.S. Green Building Council

Figure 12-9. *The Leadership in Energy and Environmental Design® (LEED®) program establishes criteria for the certification of "green" buildings.*

⫴ LEED® Certification Checklist

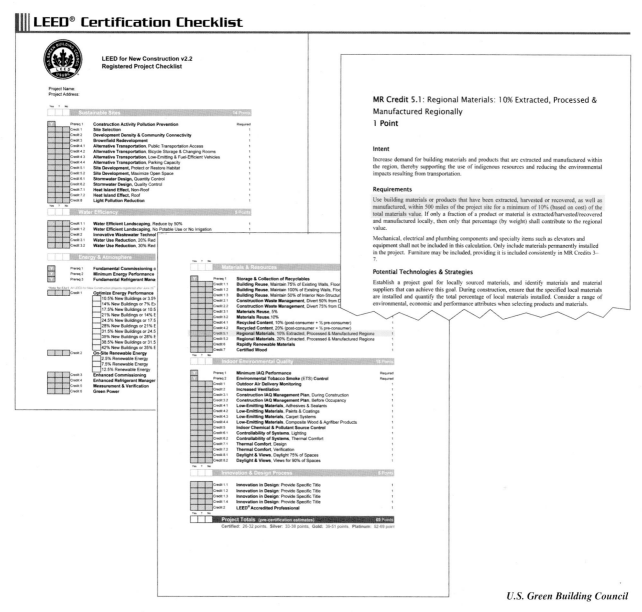

MR Credit 5.1: Regional Materials: 10% Extracted, Processed & Manufactured Regionally
1 Point

Intent

Increase demand for building materials and products that are extracted and manufactured within the region, thereby supporting the use of indigenous resources and reducing the environmental impacts resulting from transportation.

Requirements

Use building materials or products that have been extracted, harvested or recovered, as well as manufactured, within 500 miles of the project site for a minimum of 10% (based on cost) of the total materials value. If only a fraction of a product or material is extracted/harvested/recovered and manufactured locally, then only that percentage (by weight) shall contribute to the regional value.

Mechanical, electrical and plumbing components and specialty items such as elevators and equipment shall not be included in this calculation. Only include materials permanently installed in the project. Furniture may be included, providing it is included consistently in MR Credits 3–7.

Potential Technologies & Strategies

Establish a project goal for locally sourced materials, and identify materials and material suppliers that can achieve this goal. During construction, ensure that the specified local materials are installed and quantify the total percentage of local materials installed. Consider a range of environmental, economic and performance attributes when selecting products and materials.

U.S. Green Building Council

Figure 12-10. The LEED® project checklist defines certification levels that can be achieved through the accumulation of points.

Building Automation System Utilization

Understanding the potential interactions of various building systems is critical to the design and implementation of a building automation system. Selected, potential capabilities of a sophisticated automation system can be highlighted by describing a day in the life of employee Pat Smith of the Lincoln Publishing Group (LPG). Pat is the leader of the new product development staff and has worked for 11 years as Senior Development Editor.

Pat leaves for the office early on Saturday morning in mid-October. It is still quite dark at this time in the morning. The Lincoln Publishing Group headquarters is located just off the interstate in Springfield, Illinois. The building is the first of several buildings being constructed in the new Capitol City Industrial Campus. Construction equipment for other new buildings is located on adjoining properties.

Turning into the drive, Pat's vehicle is sensed by the in-ground loop. The gate arm rises and allows access to the parking facility. **See Figure 12-11.** A CCTV security camera installed in the gate records the entrance of the car and the license plate image. Monitors located in the security office display the video footage captured by these and other CCTV cameras in the facility. The footage is recorded by digital video recorders (DVRs) if needed, but it is otherwise overwritten at scheduled intervals.

A new software release has necessitated a training seminar for key people in the development team. There will be a guest presenter and ten more staff members arriving later at 8:30 AM. The expected staff members have been logged for pre-authorized access to common areas during the seminar. Since the building normally operates in efficiency mode during the weekend, an exception schedule is set to override the Editorial Department HVAC zone operation for Saturday. This was entered into the system via the building management web site.

Approaching the front entrance door, Pat notices that the irrigation system is activated for the morning landscape watering. Stormwater runoff from the building roof is collected in a cistern and supplied to the sprinkler control and distribution unit. A water level sensor in the cistern provides water level readings indicating the amount of available irrigation water. The sprinkler control unit uses a soil moisture sensor so that the landscaping is only watered when necessary.

Pat swipes her access card. An overhead CCTV camera captures her image and the card data is sent through the network to the access control system. Access is granted to the exercise area on the first floor and the second and third floor spaces for management-level employees. In her daily tasks, Pat requires access to most building spaces. Having identified Pat, the building automation system is programmed to turn on the lights in her office and the adjoining work areas. **See Figure 12-12.**

Example Building–Parking Lot Gate

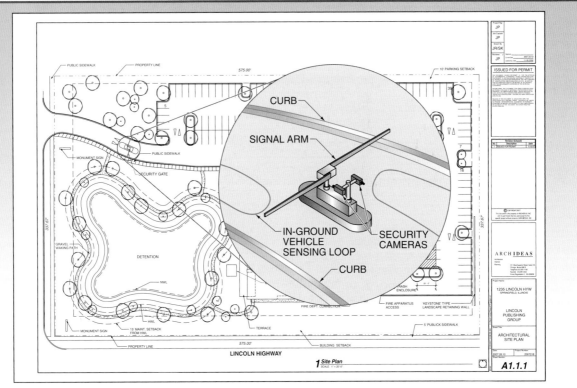

Figure 12-11. Automated security and access control systems provide seamless integration with other building systems.

Selective Lighting Control

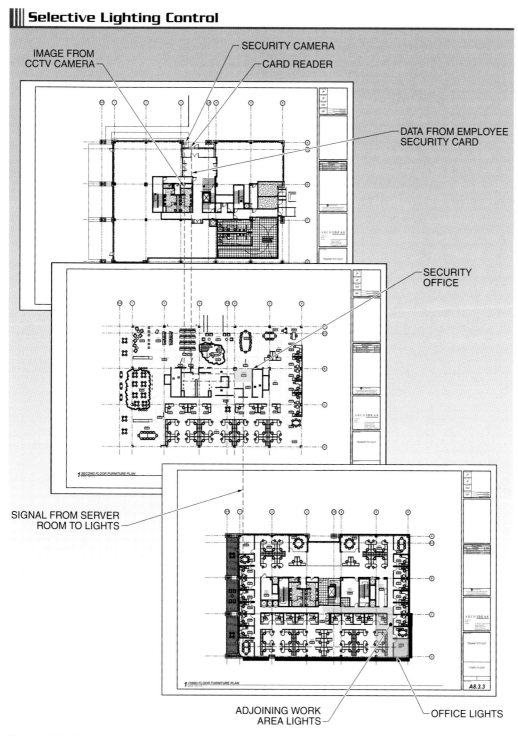

Figure 12-12. Integration of the access control and lighting systems allows certain lighting circuits to be turned on based on the identification of a person entering the area.

The lobby doors to the elevator and second and third level stairway are locked. Typically, the receptionist activates the release of the entrance doors. Pat swipes her card and the doors unlock, allowing access. Anticipating her need for an elevator, the access control system signals the elevator system to send the car to the first floor.

On the third floor, Pat exits the elevator. Fire doors in the glass lobby perimeter are held open by electromagnetic door holders. If there were a fire, the alarm signal from the fire alarm control panel would de-energize the magnetic holders, allowing the doors to close. This is a code requirement to prevent the spread of smoke throughout the building. The elevator is also programmed to be immediately recalled to a designated level in the event of a fire alarm.

The third floor lights are on and the space is comfortable at 72°F and 40% rh, according to the exception schedule. The sun has begun to illuminate the area. Pat walks to her office near the telecommunications room and swipes her access card. The door is unlocked and she enters and hangs up her coat. The office is slightly cooler than the common areas—just how she likes it.

Occupancy sensors throughout the building provide information to the lighting, security, and access control systems.

Security systems can be set up to provide temporary access cards for guests visiting a building. These access cards can be given any type of access rights, though they are typically limited to common areas and special areas assigned to the particular guest.

After responding to e-mail messages, Pat is ready for a second cup of coffee and she walks down the stairway between the second and third floor. Light from the brightening sky has caused the photo control sensors to turn off the parking lot lights. Pat opens the lunchroom door and an occupancy sensor turns on the lighting. After making a large espresso, Pat returns to her office to continue reviewing the new software menus and capabilities.

Back on the third floor, Pat observes that the perimeter lights (nearest the exterior walls) in the common areas are off. Sufficient sunlight entering the building has caused the daylighting sensors to turn those lights off. **See Figure 12-13.** Back in her office, a security system pop-up window shows an image of the front entrance. It is 8:10 AM and Chris Tate, Production Editor, has swiped his access card. Pat catches a glimpse of lights turning on in Chris's office, indicating that the building systems detected his presence. The rest of the seminar attendees will be arriving soon.

A few minutes later, another security window pops up. Jack Morton, Copy Editor, left his access card at home and uses the keypad at the front door to enter the building. Sue Barton, IT Coordinator, arrives with Gary Mandro, the guest presenter and Technical Support Specialist for his software company. The CCTV security camera in the lobby records Gary entering with Sue into the elevator hallway. Sue gives Gary a guest security card to use throughout the day. His card provides limited access to building areas and services.

In preparation for the software upgrade training class, Chris runs copies of the handouts and prepares the training room. A PowerPoint® presentation will provide an overview and the handouts will include a sample application to be completed by the group. The training room has a scene lighting control scheme. The scene settings for the training room correspond to the room functions. In presentation mode, the blinds close and the lighting is dimmed.

The overall lighting is provided by three-lamp fluorescent fixture with bilevel switching. **See Figure 12-14.** The outer two lamps are switched separately from the middle lamp, allowing the user to switch on one, two, or all three lamps. One lamp will be used for the multimedia presentation, two lamps will be used for the question-and-answer session, and all three lamps will be used for the group activity.

> Bilevel switching is a technique that controls general light levels by switching individual lamps or groups of lamps in a fixture separately. This technique requires specially wired multilamp fixtures, so that lamps must be ballasted separately.

Photoelectric Lighting Control

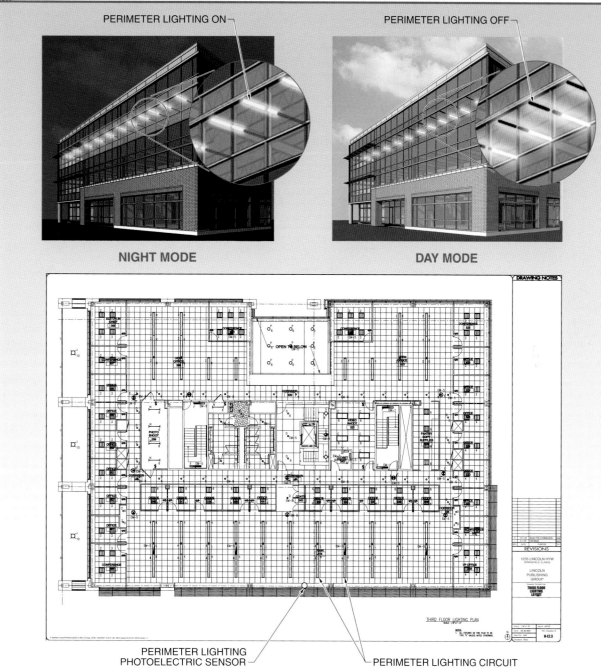

Figure 12-13. Daylighting controls minimize the use of artificial lighting when natural sunlight is available.

Scene Lighting

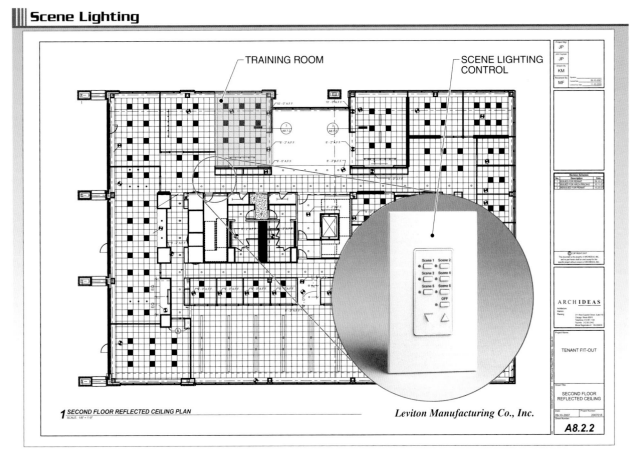

Figure 12-14. Scene lighting controls allow for special lighting schemes optimized for certain tasks.

The presentation is started by Chris at 8:30 AM with all ten attendees present. Chris introduces the presenter, Gary Mandro, and uses the scene controller to reduce the lights to the lowest intensity. The presentation has been underway for an hour when the room lights and the multimedia projector suddenly quit. Pat hears the emergency back-up generator come on. **See Figure 12-15.**

Contractors working at a nearby construction site inadvertently breach the building's lateral electrical service with their construction equipment. The building's transfer switch sensed a loss of power, delayed 10 sec, and then initiated an engine generator start-up sequence. The generator engine takes a short time to come up to speed. When the voltage and frequency stabilize, the automatic transfer switch opens the connection between the building and the utility and closes the connection between the generator output and the building's electrical

system. The uninterruptible power supply (UPS) powers the computer network system, building automation controls, security system, and other critical systems during this interruption, ensuring no data is lost or corrupted.

The presenter, Gary Mandro, sees Chris pointing to his watch, indicating that the power loss may be a good time for a break. A comment is made that "it is good that the coffee maker is on a backed-up circuit." The attendees proceed to the lunchroom on the second floor. Some of the lights along the way are being powered by the back-up generator. On the way, Gary realizes that his cell phone is low and will never make it through his five new voicemail messages. A conference room ahead has a phone on the table. He swipes his guest access card and turns the door latch. It does not open. From down the hall, Chris calls out, "Use the business center—your card will work there."

Back-Up Power

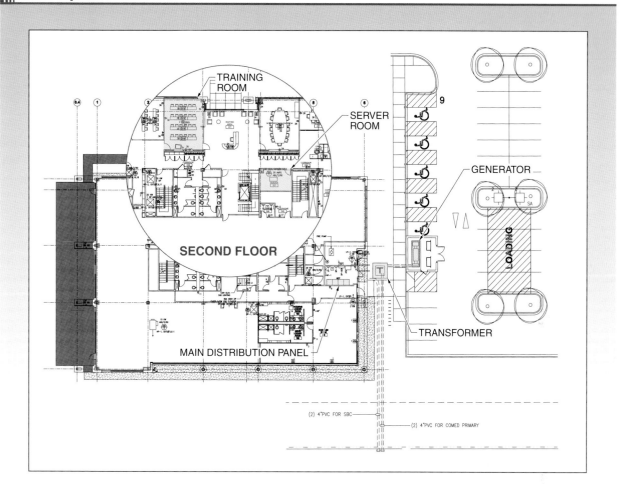

Figure 12-15. *Automatic transfer switches and interruptible power supplies ensure continuous and reliable power to the building in the event of a utility power outage.*

The power is restored in 20 min and the group is reconvened for the rest of the presentation. The next part of the presentation involves how sensitive documents are accessed with the new software using a password. Access to secure files is controlled from the server room. Sue leaves the training room and heads to the server room. A fingerprint scanning biometric device on the wall by the door prevents security breaches such as "card swapping" or other unauthorized access. **See Figure 12-16.** The physical (fingerprint) characteristics unique to Sue Barton are required for authentication and are not easily falsified. Sue accesses the password file, creates a temporary password, and allows limited access to certain secure files for the training room presentation.

A biometric reader analyzes a person's physical or behavioral characteristics for authentication. Fingerprint technology is among the most accurate biometric technologies and analyzes the patterns of the skin on the fingertip. The reader records an image of the fingerprint, which is then analyzed for a unique set of patterns. The patterns are compared to those stored in the access control database for that person. A sufficient match results in authentication.

Fingerprint Access Control

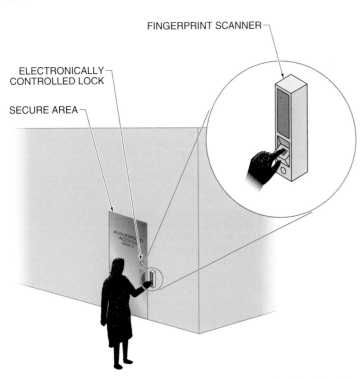

FINGERPRINT SCANNER

ELECTRONICALLY CONTROLLED LOCK

SECURE AREA

AUTHORIZED ACCESS ONLY

Figure 12-16. Biometric access control devices ensure the identity of the user for access to secure areas.

A security CCTV camera near the facility gate records the license plates of vehicles entering and exiting the facility.

The training seminar continues into the afternoon with the participants successfully completing the hands-on activities specified on the handouts. The training seminar adjourns with closing comments by Chris and Pat at 4:30 PM. Pat still has several tasks yet to complete before leaving, so she returns to her office to work.

After everyone else has left the building, she resets the motion sensors located in the common areas away from her office. **See Figure 12-17.** These sensors will now be active input devices for the security system in those security system partitions (instead of just for the lighting system). The lobby area, the entrance door to the workout area, and the hallways on the first and second floor have dual-technology motion sensors to detect intruders. The passive infrared (PIR) sensor senses the heat energy of a person within its field of view in a detection area and the microwave sensor senses movement. The detector only initiates an alarm when both sensors are tripped. Motion sensors would activate a building alarm and a pop-up notice on her computer screen.

With the final tasks completed, Pat disables the motion sensors and shuts down her computer. She turns off her lights, locks the door, and proceeds to the elevator. In the lobby, Pat inputs the night mode security code on the keypad, which activates the exit delay countdown warning. A light on the security system control panel blinks, increasing in frequency to indicate the amount of exit delay time remaining before the system is armed. The programmed exit delay time is 45 sec. This is the amount of time between when a security system is armed and when the control panel begins registering alarm signals from sensing devices.

The exit delay allows Pat time to leave the detection area before her presence causes an alarm. With the alarm system now set for an unoccupied building, the building automation system uses this information to signal the HVAC and lighting systems to return to the weekend energy-saving mode. Pat leaves the building. When leaving the parking lot, the security CCTV camera and DVR records her rear license plate at the gatepost.

BUILDING AUTOMATION SYSTEM PROTOCOLS

Building automation systems structure and share information between devices in a consistent and reliable way. This allows technicians to add devices to an automation system with the expectation that they will operate properly, as long as all the devices on the automation network are compatible with the same protocol. This is because the details of the communication schemes vary between different protocols. A *protocol* is a set of rules and procedures for the exchange of information between two connected devices.

Proprietary Protocols

Protocols have been used by manufacturers for many years as part of proprietary systems. A *proprietary protocol* is a communications and network protocol that is used by only one device manufacturer. For example, early proprietary HVAC control systems were developed by many manufacturers. However, only the specific manufacturer's equipment and related proprietary software can be used with the system. **See Figure 12-18.** The details of proprietary protocols are protected so that other manufacturers cannot market compatible devices.

Open Protocols

In the 1980s, a movement started to create a system that allowed open access to information by devices using a universal communication scheme. This introduces the idea of open protocol. An *open protocol* is a communications and network protocol that is standardized and published for use by any device manufacturer. Systems using open protocols must still use protocol-specific devices, but since the protocol standard is publicly available, the device can be made by any manufacturer. Without being tied to a specific manufacturer, open protocols have also fostered the expansion of automation beyond HVAC systems to include nearly any building system. **See Figure 12-19.**

▌▌Motion Sensors

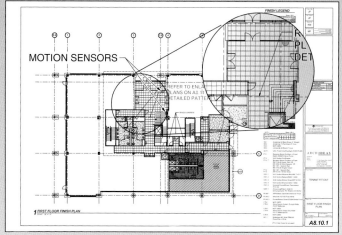

FIRST FLOOR

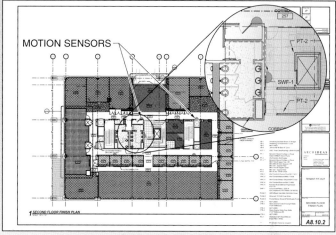

SECOND FLOOR

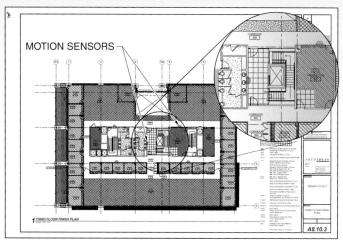

THIRD FLOOR

Figure 12-17. Motion detectors throughout a building provide detection of intruders for the security system.

Proprietary Protocol Systems

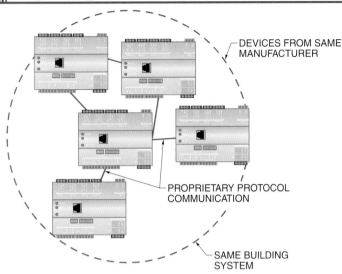

Figure 12-18. Proprietary protocols limit an automation system to control devices from one manufacturer, which also limits the ability to integrate different systems within a building.

A few open protocols evolved from this effort, including the two most commonly used today: LonWorks® and BACnet®. There is also increased interest in wireless open protocols, such as ZigBee®, which are specifically designed for the low-power, low data-rate requirements of wireless technology. Open protocols continue to evolve with technology, but the basic operating concepts are similar. They standardize a common way for devices to organize and share information.

Each of these protocol systems can ultimately be used to achieve similar types of automation, though there are differences in network requirements, device types, communication modes, programming methods, network tools, and other characteristics. Depending on the building and the application, these factors can either be advantages or disadvantages. The protocol chosen for a new building automation system is based on careful analysis of all these considerations.

Open Protocol Systems

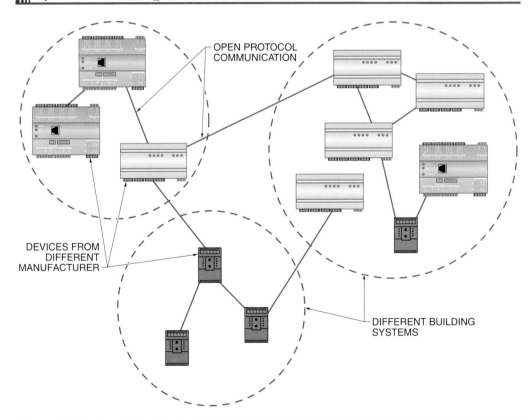

Figure 12-19. The details of open protocols are publicly available for use by any control device manufacturer, making highly flexible and integrated systems possible.

Summary

- Building automation systems use the abundant sharing of system information to provide a highly functional and flexible "living" environment for the building occupants.

- Although the initial cost for a building automation system is typically higher than conventional construction, total life-cycle costs are generally lower due to energy savings, reduced maintenance costs, and increased value to potential tenants.

- An automation device may share its output data for other devices to use, resulting in a significant reduction in redundancy.

- The interoperability of the system components allows for nearly limitless flexibility in system design and component sourcing.

- It is becoming increasingly common for building automation systems to merge with traditional information technology (IT) infrastructures.

- Integration with Internet standards allows the scope of a building automation system to extend beyond the walls of the building.

- By far, the most common implementations of building automation systems are building-wide.

- With a suitable communications network, automation systems can be extended beyond a single building.

- Most worldwide automation applications involve the monitoring of system operations remotely.

- Most buildings do not integrate all possible building systems.

- The Leadership in Energy and Environmental Design® (LEED®) Green Building Rating System™ is the nationally accepted standard for the design, construction, and operation of green buildings.

- Two major criteria in the LEED® certification process are the energy performance measures and the renewable energy standard. Both of these criteria are realized by the building automation system.

- The details of proprietary protocols are protected so that other manufacturers cannot market compatible devices.

- Systems using open protocols must still use protocol-specific devices, but since the protocol standard is publicly available, the device can be made by any manufacturer.

- Without being tied to a specific manufacturer, open protocols have fostered the expansion of automation beyond HVAC systems to include nearly any building system.

Definitions

- *Interoperability* is the ability of diverse systems and devices to work together effectively and reliably.

- A *protocol* is a set of rules and procedures for the exchange of information between two connected devices.

- A *proprietary protocol* is a communications and network protocol that is used by only one device manufacturer.

- An *open protocol* is a communications and network protocol that is standardized and published for use by any device manufacturer.

Review Questions

1. How can automated building systems become interdependent when integrated?

2. What are the primary advantages of interoperability?

3. What is the primary factor in extending an automation system beyond a single building?

4. How can remote building automation system access via the Internet help with monitoring and operating building systems?

5. What general areas does LEED® certification address?

6. What is the difference between a proprietary protocol and an open protocol?

Appendix

Metric to English Equivalents

	Unit	British Equivalent		
LENGTH	kilometer	0.62 mi		
	hectometer	109.36 yd		
	dekameter	32.81″		
	meter	39.37″		
	decimeter	3.94″		
	centimeter	0.39″		
	millimeter	0.039″		
	square kilometer	.3861 sq mi		
AREA	hectacre	2.47 A		
	acre	119.60 sq yd		
	square centimeter	0.155 sq in.		
VOLUME	cubic centimeter	0.061 cu in.		
	cubic decimeter	61.023 cu in.		
	cubic meter	1.307 cu yd		
		cubic	*dry*	*liquid*
CAPACITY	kiloliter	1.31 cu yd		
	hectoliter	3.53 cu ft	2.84 bu	
	dekaliter	0.35 cu ft	1.14 pk	2.64 gal.
	liter	61.02 cu in.	0.908 qt	1.057 qt
	cubic decimeter	61.02 cu in.	0.908 qt	1.057 qt
	deciliter	6.1 cu in.	0.18 pt	0.21 pt
	centiliter	0.61 cu in.		338 fl oz
	milliliter	0.061 cu in.		0.27 fl dr
MASS AND WEIGHT	metric ton	1.102 t		
	kilogram	2.2046 lb		
	hectogram	3.527 oz		
	dekagram	0.353 oz		
	gram	0.353 oz		
	decigram	1.543 gr		
	centigram	0.154 gr		
	milligram	0.015 gr		

English to Metric Equivalents

		Unit	Metric Equivalent
LENGTH		mile	1.609 km
		rod	5.029 m
		yard	0.9144 m
		foot	30.48 cm
		inch	2.54 cm
		square mile	2.590 km^2
AREA		acre	.405 hectacre, 4047 m^2
		square rod	25.293 m^2
		square yard	0.836 m^2
		square foot	0.093 m^2
		square inch	6.452 cm^2
VOLUME		cubic yard	0.765 m^3
		cubic foot	0.028 m^3
		cubic inch	16.387 cm^3
CAPACITY	*U.S. liquid measure*	gallon	3.785 l
		quart	0.946 l
		pint	0.473 l
		gill	118.294 ml
		fluidounce	29.573 ml
		fluidram	3.697 ml
		minim	0.61610 ml
	U.S. dry measure	bushel	35.239 l
		peck	8.810 l
		quart	1.101 l
		pint	0.551 l
	British imperial liquid and dry measure	bushel	0.036 m^3
		peck	0.0091 m^3
		gallon	4.546 l
		quart	1.136 l
		pint	568.26 cm^3
		gill	142.066 cm^3
		fluidounce	28.412 cm^3
		fluidram	3.5516 cm^3
		minim	0.59194 cm^3
MASS AND WEIGHT	*avoirdupois*	short ton	0.907 t
		long ton	1.016 t
		pound	0.454 kg
		ounce	28.350 g
		dram	1.772 g
		grain	0.0648 g
	troy	pound	0.373 kg
		ounce	31.103 g
		pennyweight	1.555 g
		grain	0.648 g
	apothecaries'	pound	0.373 kg
		ounce	31.103 g
		dram	3.888 g
		scruple	1.296 g
		grain	0.648 g

Metric Prefixes

Multiple	Prefixes	Symbols	Meaning
$1,000,000,000,000 = 10^{12}$	tera	T	trillion
$1,000,000,000 = 10^{9}$	giga	G	billion
$1,000,000 = 10^{6}$	mega	M	million
$1000 = 10^{3}$	kilo	k	thousand
$100 = 10^{2}$	hecto	h	hundred
$10 = 10^{1}$	deka	da	ten
$1 = 10^{0}$	——	——	unit
$0.1 = 10^{-1}$	deci	d	tenth
$0.01 = 10^{-2}$	centi	c	hundredth
$0.001 = 10^{-3}$	milli	m	thousandth
$0.000001 = 10^{-6}$	micro	μ	millionth
$0.000000001 = 10^{-9}$	nano	n	billionth
$0.000000000001 = 10^{-12}$	pico	p	trillionth

Metric Conversions

Initial Units	Final Units											
	giga	mega	kilo	hecto	deka	base unit	deci	centi	milli	micro	nano	pico
giga		3R	6R	7R	8R	9R	10R	11R	12R	15R	18R	21R
mega	3L		3R	4R	5R	6R	7R	8R	9R	12R	15R	18R
kilo	6L	3L		1R	2R	3R	4R	5R	6R	9R	12R	15R
hecto	7L	4L	1L		1R	2R	3R	4R	5R	8R	11R	14R
deka	8L	5L	2L	1L		1R	2R	3R	4R	7R	10R	13R
base unit	9L	6L	3L	2L	1L		1R	2R	3R	6R	9R	12R
deci	10L	7L	4L	3L	2L	1L		1R	2R	5R	8R	11R
centi	11L	8L	5L	4L	3L	2L	1L		1R	4R	7R	10R
milli	12L	9L	6L	5L	4L	3L	2L	1L		3R	6R	9R
micro	15L	12L	9L	8L	7L	6L	5L	4L	3L		3R	6R
nano	18L	15L	12L	11L	10L	9L	8L	7L	6L	3L		3R
pico	21L	18L	15L	14L	13L	12L	11L	10L	9L	6L	3L	

Three-Phase Voltage Values

For 208 V × 1.732, use 360
For 230 V × 1.732, use 398
For 240 V × 1.732, use 416
For 440 V × 1.732, use 762
For 460 V × 1.732, use 797
For 480 V × 1.732, use 831
For 2400 V × 1.732, use 4157
For 4160 V × 1.732, use 7205

Ohm's Law

Ohm's law is the relationship between the voltage, current, and resistance in an electrical circuit. Ohm's law states that current in a circuit is proportional to the voltage and inversely proportional to the resistance.

Power Formula

The power formula is the relationship between the voltage, current, and power in an electrical circuit. The power formula states that the power in a circuit is proportional to the voltage and the current.

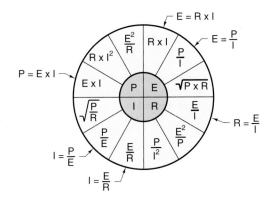

VALUES IN INNER CIRCLE ARE EQUAL TO VALUES IN CORRESPONDING OUTER CIRCLE

OHM'S LAW AND POWER FORMULA

Power Formulas — 1ϕ, 3ϕ

Unit	To Find	British Equivalent	Example		
			Given	Find	Solution
1ϕ	I	$I = \dfrac{VA}{V}$	32,000 VA, 240 V	I	$I = \dfrac{VA}{V}$ $I = \dfrac{32,000 \text{ VA}}{240 \text{ V}}$ $I = \textbf{133 A}$
1ϕ	VA	$VA = I \times V$	100 A, 240V	VA	$VA = I \times V$ $VA = 100 \text{ A} \times 240 \text{ V}$ $VA = \textbf{24,000 VA}$
1ϕ	V	$V = \dfrac{VA}{I}$	42,000 VA, 350 A	V	$V = \dfrac{VA}{I}$ $V = \dfrac{42,000 \text{ VA}}{350 \text{ A}}$ $V = \textbf{120 V}$
3ϕ	I	$I = \dfrac{VA}{V \times \sqrt{3}}$	72,000 VA, 208 V	I	$I = \dfrac{VA}{V \times \sqrt{3}}$ $I = \dfrac{72,000 \text{ VA}}{360 \text{ V}}$ $I = \textbf{200 A}$
3ϕ	VA	$VA = I \times V \times \sqrt{3}$	2 A, 240 V	VA	$VA = I \times V \times \sqrt{3}$ $VA = 2 \times 416$ $VA = \textbf{832 VA}$

Electrical Symbols . . .

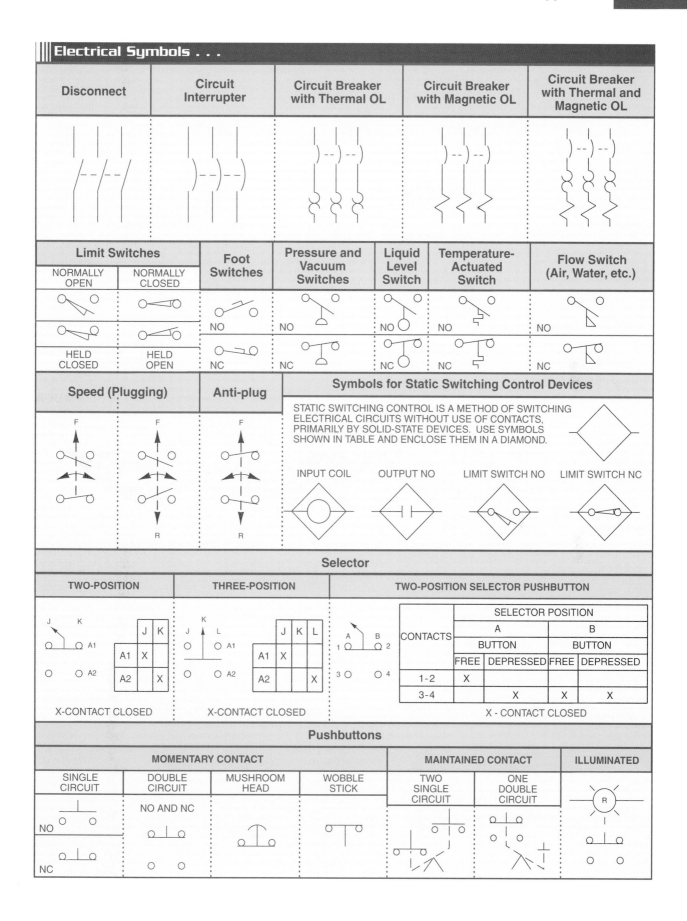

Disconnect	Circuit Interrupter	Circuit Breaker with Thermal OL	Circuit Breaker with Magnetic OL	Circuit Breaker with Thermal and Magnetic OL

Limit Switches		Foot Switches	Pressure and Vacuum Switches	Liquid Level Switch	Temperature-Actuated Switch	Flow Switch (Air, Water, etc.)
NORMALLY OPEN	NORMALLY CLOSED					
HELD CLOSED	HELD OPEN	NO / NC	NO / NC	NO / NC	NO / NC	NO / NC

Speed (Plugging)	Anti-plug	Symbols for Static Switching Control Devices

STATIC SWITCHING CONTROL IS A METHOD OF SWITCHING ELECTRICAL CIRCUITS WITHOUT USE OF CONTACTS, PRIMARILY BY SOLID-STATE DEVICES. USE SYMBOLS SHOWN IN TABLE AND ENCLOSE THEM IN A DIAMOND.

INPUT COIL   OUTPUT NO   LIMIT SWITCH NO   LIMIT SWITCH NC

Selector

TWO-POSITION	THREE-POSITION	TWO-POSITION SELECTOR PUSHBUTTON

TWO-POSITION

	J	K
A1	X	
A2		X

X-CONTACT CLOSED

THREE-POSITION

	J	K	L
A1	X		
A2			X

X-CONTACT CLOSED

CONTACTS	SELECTOR POSITION			
	A		B	
	BUTTON		BUTTON	
	FREE	DEPRESSED	FREE	DEPRESSED
1-2	X			
3-4		X	X	X

X - CONTACT CLOSED

Pushbuttons

MOMENTARY CONTACT				MAINTAINED CONTACT		ILLUMINATED
SINGLE CIRCUIT	DOUBLE CIRCUIT	MUSHROOM HEAD	WOBBLE STICK	TWO SINGLE CIRCUIT	ONE DOUBLE CIRCUIT	
NO	NO AND NC					
NC						

▐▐▌ . . . Electrical Symbols . . .

Contacts								Overload Relays	
INSTANT OPERATING				TIMED CONTACTS - CONTACT ACTION RETARDED AFTER COIL IS:				THERMAL	MAGNETIC
WITH BLOWOUT		WITHOUT BLOWOUT		ENERGIZED		DE-ENERGIZED			
NO	NC	NO	NC	NOTC	NCTO	NOTO	NCTC		

Supplementary Contact Symbols

SPST NO		SPST NC		SPDT		TERMS
SINGLE BREAK	DOUBLE BREAK	SINGLE BREAK	DOUBLE BREAK	SINGLE BREAK	DOUBLE BREAK	SPST SINGLE-POLE, SINGLE-THROW

DPST, 2NO		DPST, 2NC		DPDT		
SINGLE BREAK	DOUBLE BREAK	SINGLE BREAK	DOUBLE BREAK	SINGLE BREAK	DOUBLE BREAK	

TERMS (continued):

SPST — SINGLE-POLE, SINGLE-THROW

SPDT — SINGLE-POLE, DOUBLE-THROW

DPST — DOUBLE-POLE, SINGLE-THROW

DPDT — DOUBLE-POLE, DOUBLE-THROW

NO — NORMALLY OPEN

NC — NORMALLY CLOSED

Meter (Instrument)

INDICATE TYPE BY LETTER	TO INDICATE FUNCTION OF METER OR INSTRUMENT, PLACE SPECIFIED LETTER OR LETTERS WITHIN SYMBOL.

AM or A	AMMETER	VA	VOLTMETER
AH	AMPERE HOUR	VAR	VARMETER
μA	MICROAMMETER	VARH	VARHOUR METER
mA	MILLAMMETER	W	WATTMETER
PF	POWER FACTOR	WH	WATTHOUR METER
V	VOLTMETER		

Pilot Lights

INDICATE COLOR BY LETTER	
NON PUSH-TO-TEST	PUSH-TO-TEST

Inductors

IRON CORE

Coils

DUAL-VOLTAGE MAGNET COILS		BLOWOUT COIL
HIGH-VOLTAGE	LOW-VOLTAGE	

AIR CORE

... Electrical Symbols

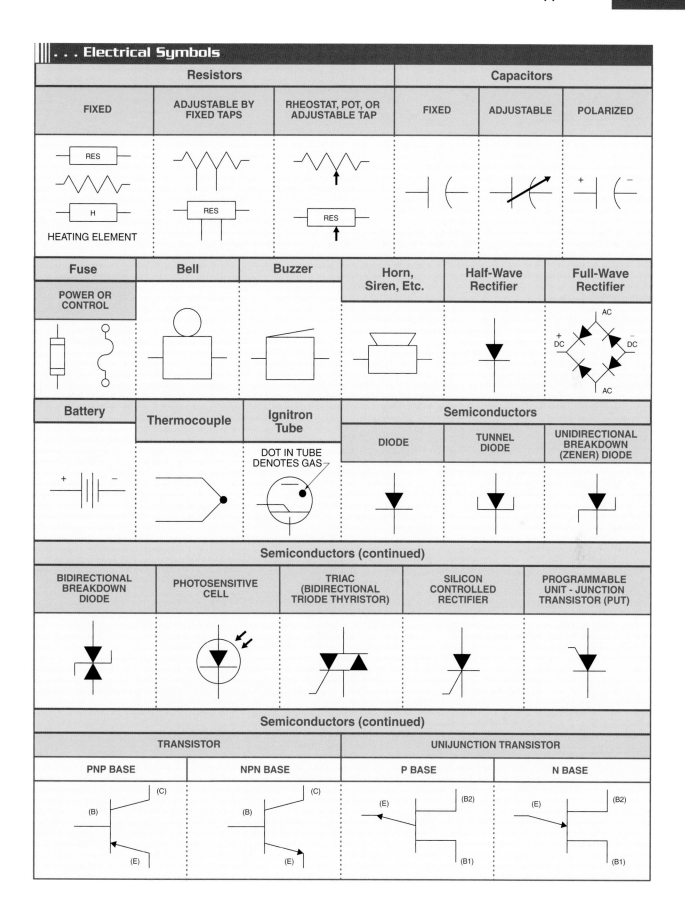

Resistors			Capacitors		
FIXED	ADJUSTABLE BY FIXED TAPS	RHEOSTAT, POT, OR ADJUSTABLE TAP	FIXED	ADJUSTABLE	POLARIZED

HEATING ELEMENT

Fuse	Bell	Buzzer	Horn, Siren, Etc.	Half-Wave Rectifier	Full-Wave Rectifier
POWER OR CONTROL					

Battery	Thermocouple	Ignitron Tube	Semiconductors		
			DIODE	TUNNEL DIODE	UNIDIRECTIONAL BREAKDOWN (ZENER) DIODE

DOT IN TUBE DENOTES GAS

Semiconductors (continued)

BIDIRECTIONAL BREAKDOWN DIODE	PHOTOSENSITIVE CELL	TRIAC (BIDIRECTIONAL TRIODE THYRISTOR)	SILICON CONTROLLED RECTIFIER	PROGRAMMABLE UNIT - JUNCTION TRANSISTOR (PUT)

Semiconductors (continued)

TRANSISTOR		UNIJUNCTION TRANSISTOR	
PNP BASE	NPN BASE	P BASE	N BASE

HVAC Symbols

EQUIPMENT SYMBOLS	DUCTWORK	HEATING PIPING
EXPOSED RADIATOR	DUCT (1ST FIGURE, WIDTH; 2ND FIGURE, DEPTH) — 12 X 20	HIGH-PRESSURE STEAM — HPS —
RECESSED RADIATOR	DIRECTION OF FLOW	MEDIUM-PRESSURE STEAM — MPS —
FLUSH ENCLOSED RADIATOR	FLEXIBLE CONNECTION	LOW-PRESSURE STEAM — LPS —
PROJECTING ENCLOSED RADIATOR	DUCTWORK WITH ACOUSTICAL LINING	HIGH-PRESSURE RETURN — HPR —
UNIT HEATER (PROPELLER) – PLAN	FIRE DAMPER WITH ACCESS DOOR — FD | AD	MEDIUM-PRESSURE RETURN — MPR —
UNIT HEATER (CENTRIFUGAL) – PLAN	MANUAL VOLUME DAMPER — VD	LOW-PRESSURE RETURN — LPR —
UNIT VENTILATOR – PLAN	AUTOMATIC VOLUME DAMPER	BOILER BLOW OFF — BD —
STEAM	EXHAUST, RETURN OR OUTSIDE AIR DUCT – SECTION — 20 X 12	CONDENSATE OR VACUUM PUMP DISCHARGE — VPD —
DUPLEX STRAINER	SUPPLY DUCT – SECTION — 20 X 12	FEEDWATER PUMP DISCHARGE — PPD —
PRESSURE-REDUCING VALVE	CEILING DIFFUSER SUPPLY OUTLET 20" DIA CD 1000 CFM	MAKEUP WATER — MU —
AIR LINE VALVE	CEILING DIFFUSER SUPPLY OUTLET 20 X 12 CD 700 CFM	AIR RELIEF LINE — V —
		FUEL OIL SUCTION — FOS —
STRAINER	LINEAR DIFFUSER 96 X 6-LD 400 CFM	FUEL OIL RETURN — FOR —
THERMOMETER	FLOOR REGISTER 20 X 12 FR 700 CFM	FUEL OIL VENT — FOV —
PRESSURE GAUGE AND COCK		COMPRESSED AIR — A —
RELIEF VALVE	TURNING VANES	HOT WATER HEATING SUPPLY — HW —
AUTOMATIC 3-WAY VALVE	FAN AND MOTOR WITH BELT GUARD	HOT WATER HEATING RETURN — HWR —

AIR CONDITIONING PIPING

		REFRIGERANT LIQUID — RL —
		REFRIGERANT DISCHARGE — RD —
		REFRIGERANT SUCTION — RS —
		CONDENSER WATER SUPPLY — CWS —
		CONDENSER WATER RETURN — CWR —
AUTOMATIC 2-WAY VALVE		CHILLED WATER SUPPLY — CHWS —
		CHILLED WATER RETURN — CHWR —
	LOUVER OPENING 20 X 12-L 700 CFM	MAKEUP WATER — MU —
		HUMIDIFICATION LINE — H —
SOLENOID VALVE		DRAIN — D —

LEED for New Construction v2.2
Registered Project Checklist

Project Name:
Project Address:

Yes	?	No				
			Sustainable Sites			**14** Points
Y			Prereq 1	**Construction Activity Pollution Prevention**		Required
			Credit 1	**Site Selection**		1
			Credit 2	**Development Density & Community Connectivity**		1
			Credit 3	**Brownfield Redevelopment**		1
			Credit 4.1	**Alternative Transportation**, Public Transportation Access		1
			Credit 4.2	**Alternative Transportation**, Bicycle Storage & Changing Rooms		1
			Credit 4.3	**Alternative Transportation**, Low-Emitting & Fuel-Efficient Vehicles		1
			Credit 4.4	**Alternative Transportation**, Parking Capacity		1
			Credit 5.1	**Site Development**, Protect or Restore Habitat		1
			Credit 5.2	**Site Development**, Maximize Open Space		1
			Credit 6.1	**Stormwater Design**, Quantity Control		1
			Credit 6.2	**Stormwater Design**, Quality Control		1
			Credit 7.1	**Heat Island Effect**, Non-Roof		1
			Credit 7.2	**Heat Island Effect**, Roof		1
			Credit 8	**Light Pollution Reduction**		1

Yes	?	No				
			Water Efficiency			**5** Points
			Credit 1.1	**Water Efficient Landscaping**, Reduce by 50%		1
			Credit 1.2	**Water Efficient Landscaping**, No Potable Use or No Irrigation		1
			Credit 2	**Innovative Wastewater Technologies**		1
			Credit 3.1	**Water Use Reduction**, 20% Reduction		1
			Credit 3.2	**Water Use Reduction**, 30% Reduction		1

Yes	?	No				
			Energy & Atmosphere			**17** Points
Y			Prereq 1	**Fundamental Commissioning of the Building Energy Systems**		Required
Y			Prereq 2	**Minimum Energy Performance**		Required
Y			Prereq 3	**Fundamental Refrigerant Management**		Required

*Note for EAc1: All LEED for New Construction projects registered after June 26th, 2007 are required to achieve at least two (2) points under EAc1.

	Credit 1	**Optimize Energy Performance**			1 to 10
			10.5% New Buildings or 3.5% Existing Building Renovations		1
			14% New Buildings or 7% Existing Building Renovations		2
			17.5% New Buildings or 10.5% Existing Building Renovations		3
			21% New Buildings or 14% Existing Building Renovations		4
			24.5% New Buildings or 17.5% Existing Building Renovations		5
			28% New Buildings or 21% Existing Building Renovations		6
			31.5% New Buildings or 24.5% Existing Building Renovations		7
			35% New Buildings or 28% Existing Building Renovations		8
			38.5% New Buildings or 31.5% Existing Building Renovations		9
			42% New Buildings or 35% Existing Building Renovations		10
	Credit 2	**On-Site Renewable Energy**			1 to 3
			2.5% Renewable Energy		1
			7.5% Renewable Energy		2
			12.5% Renewable Energy		3
	Credit 3	**Enhanced Commissioning**			1
	Credit 4	**Enhanced Refrigerant Management**			1
	Credit 5	**Measurement & Verification**			1
	Credit 6	**Green Power**			1

continued…

Yes ? No

				Materials & Resources	**13** Points

Y			Prereq 1	**Storage & Collection of Recyclables**	Required
			Credit 1.1	**Building Reuse**, Maintain 75% of Existing Walls, Floors & Roof	1
			Credit 1.2	**Building Reuse**, Maintain 100% of Existing Walls, Floors & Roof	1
			Credit 1.3	**Building Reuse**, Maintain 50% of Interior Non-Structural Elements	1
			Credit 2.1	**Construction Waste Management**, Divert 50% from Disposal	1
			Credit 2.2	**Construction Waste Management**, Divert 75% from Disposal	1
			Credit 3.1	**Materials Reuse**, 5%	1
			Credit 3.2	**Materials Reuse**, 10%	1
			Credit 4.1	**Recycled Content**, 10% (post-consumer + ½ pre-consumer)	1
			Credit 4.2	**Recycled Content**, 20% (post-consumer + ½ pre-consumer)	1
			Credit 5.1	**Regional Materials**, 10% Extracted, Processed & Manufactured Regional	1
			Credit 5.2	**Regional Materials**, 20% Extracted, Processed & Manufactured Regional	1
			Credit 6	**Rapidly Renewable Materials**	1
			Credit 7	**Certified Wood**	1

Yes ? No

				Indoor Environmental Quality	**15** Points

Y			Prereq 1	**Minimum IAQ Performance**	Required
Y			Prereq 2	**Environmental Tobacco Smoke (ETS) Control**	Required
			Credit 1	**Outdoor Air Delivery Monitoring**	1
			Credit 2	**Increased Ventilation**	1
			Credit 3.1	**Construction IAQ Management Plan**, During Construction	1
			Credit 3.2	**Construction IAQ Management Plan**, Before Occupancy	1
			Credit 4.1	**Low-Emitting Materials**, Adhesives & Sealants	1
			Credit 4.2	**Low-Emitting Materials**, Paints & Coatings	1
			Credit 4.3	**Low-Emitting Materials**, Carpet Systems	1
			Credit 4.4	**Low-Emitting Materials**, Composite Wood & Agrifiber Products	1
			Credit 5	**Indoor Chemical & Pollutant Source Control**	1
			Credit 6.1	**Controllability of Systems**, Lighting	1
			Credit 6.2	**Controllability of Systems**, Thermal Comfort	1
			Credit 7.1	**Thermal Comfort**, Design	1
			Credit 7.2	**Thermal Comfort**, Verification	1
			Credit 8.1	**Daylight & Views**, Daylight 75% of Spaces	1
			Credit 8.2	**Daylight & Views**, Views for 90% of Spaces	1

Yes ? No

				Innovation & Design Process	**5** Points

			Credit 1.1	**Innovation in Design**: Provide Specific Title	1
			Credit 1.2	**Innovation in Design**: Provide Specific Title	1
			Credit 1.3	**Innovation in Design**: Provide Specific Title	1
			Credit 1.4	**Innovation in Design**: Provide Specific Title	1
			Credit 2	**LEED® Accredited Professional**	1

Yes ? No

				Project Totals (pre-certification estimates)	**69** Points

Certified: 26-32 points, **Silver:** 33-38 points, **Gold:** 39-51 points, **Platinum:** 52-69 points

LEED for Existing Buildings v2.0
Registered Building Checklist

Project Name:
Project Address:

Yes	?	No			Sustainable Sites	14 Points
Y			Prereq 1	**Erosion & Sedimentation Control**		Required
Y			Prereq 2	**Age of Building**		Required
			Credit 1.1	**Plan for Green Site & Building Exterior Management -**4 specific actions		1
			Credit 1.2	**Plan for Green Site & Building Exterior Management -**8 specific actions		1
			Credit 2	**High Development Density Building & Area**		1
			Credit 3.1	**Alternative Transportation -** Public Transportation Access		1
			Credit 3.2	**Alternative Transportation -** Bicycle Storage & Changing Rooms		1
			Credit 3.3	**Alternative Transportation -** Alternative Fuel Vehicle:		1
			Credit 3.4	**Alternative Transportation -** Car Pooling & Telecommuting		1
			Credit 4.1	**Reduced Site Disturbance -** Protect or Restore Open Space (50% of site area)		1
			Credit 4.2	**Reduced Site Disturbance -** Protect or Restore Open Space (75% of site area)		1
			Credit 5.1	**Stormwater Management -** 25% Rate and Quantity Reduction		1
			Credit 5.2	**Stormwater Management -** 50% Rate and Quantity Reduction		1
			Credit 6.1	**Heat Island Reduction -** Non-Roof		1
			Credit 6.2	**Heat Island Reduction -** Roof		1
			Credit 7	**Light Pollution Reduction**		1

Yes	?	No			Water Efficiency	5 Points
Y			Prereq 1	**Minimum Water Efficiency**		Required
Y			Prereq 2	**Discharge Water Compliance**		Required
			Credit 1.1	**Water Efficient Landscaping -** Reduce Potable Water Use by 50%		1
			Credit 1.2	**Water Efficient Landscaping -** Reduce Potable Water Use by 95%		1
			Credit 2	**Innovative Wastewater Technologies**		1
			Credit 3.1	**Water Use Reduction -** 10% Reduction		1
			Credit 3.2	**Water Use Reduction -** 20% Reduction		1

Yes	?	No			Energy & Atmosphere	23 Points
Y			Prereq 1	**Existing Building Commissioning**		Required
Y			Prereq 2	**Minimum Energy Performance -** Energy Star 60		Required
Y			Prereq 3	**Ozone Protection**		Required

*Note for EAc1: All LEED for Existing Buildings projects registered after June 26th, 2007 are required to achieve at least two (2) points under EAc1.

Yes	?	No				Points
			Credit 1	**Optimize Energy Performance**		1 to 10
					Energy Star Rating - 63	1
					Energy Star Rating - 67	2
					Energy Star Rating - 71	3
					Energy Star Rating - 75	4
					Energy Star Rating - 79	5
					Energy Star Rating - 83	6
					Energy Star Rating - 87	7
					Energy Star Rating - 91	8
					Energy Star Rating - 95	9
					Energy Star Rating - 99	10
			Credit 2.1	**Renewable Energy -** On-site 3% / Off-site 15%		1
			Credit 2.2	**Renewable Energy -** On-site 6% / Off-site 30%		1
			Credit 2.3	**Renewable Energy -** On-site 9% / Off-site 45%		1
			Credit 2.4	**Renewable Energy -** On-site 12% / Off-site 60%		1
			Credit 3.1	**Building Operation & Maintenance -** Staff Education		1
			Credit 3.2	**Building Operation & Maintenance -** Building Systems Maintenance		1
			Credit 3.3	**Building Operation & Maintenance -** Building Systems Monitoring		1
			Credit 4	**Additional Ozone Protection**		1
			Credit 5.1	**Performance Measurement -** Enhanced Metering (4 specific actions)		1
			Credit 5.2	**Performance Measurement -** Enhanced Metering (8 specific actions)		1

continued…

			Credit 5.3	**Performance Measurement -** Enhanced Metering (12 specific actions)	1
			Credit 5.4	**Performance Measurement** - Emission Reduction Reporting	1
			Credit 6	**Documenting Sustainable Building Cost Impacts**	1

Yes ? No

			Materials & Resources		**16** Points
Y			Prereq 1.1	**Source Reduction & Waste Management** - Waste Stream Audit	Required
Y			Prereq 1.2	**Source Reduction & Waste Management -** Storage & Collection	Required
Y			Prereq 2	**Toxic Material Source Reduction -** Reduced Mercury in Light Bulbs	Required
			Credit 1.1	**Construction, Demolition & Renovation Waste Management -** Divert 50%	1
			Credit 1.2	**Construction, Demolition & Renovation Waste Management -** Divert 75%	1
			Credit 2.1	**Optimize Use of Alternative Materials** - 10% of Total Purchases	1
			Credit 2.2	**Optimize Use of Alternative Materials** - 20% of Total Purchases	1
			Credit 2.3	**Optimize Use of Alternative Materials** - 30% of Total Purchases	1
			Credit 2.4	**Optimize Use of Alternative Materials** - 40% of Total Purchases	1
			Credit 2.5	**Optimize Use of Alternative Materials** - 50% of Total Purchases	1
			Credit 3.1	**Optimize Use of IAQ Compliant Products** - 45% of Annual Purchases	1
			Credit 3.2	**Optimize Use of IAQ Compliant Products** - 90% of Annual Purchases	1
			Credit 4.1	**Sustainable Cleaning Products & Materials** - 30% of Annual Purchases	1
			Credit 4.2	**Sustainable Cleaning Products & Materials** - 60% of Annual Purchases	1
			Credit 4.3	**Sustainable Cleaning Products & Materials** - 90% of Annual Purchases	1
			Credit 5.1	**Occupant Recycling** - Recycle 30% of the Total Waste Stream	1
			Credit 5.2	**Occupant Recycling** - Recycle 40% of the Total Waste Stream	1
			Credit 5.3	**Occupant Recycling** - Recycle 50% of the Total Waste Stream	1
			Credit 6	**Additional Toxic Material Source Reduction** - Reduced Mercury in Light Bulbs	1

Yes ? No

			Indoor Environmental Quality		**22** Points
Y			Prereq 1	**Outside Air Introduction & Exhaust Systems**	Required
Y			Prereq 2	**Environmental Tobacco Smoke (ETS) Control**	Required
Y			Prereq 3	**Asbestos Removal or Encapsulation**	Required
Y			Prereq 4	**PCB Removal**	Required
			Credit 1	**Outside Air Delivery Monitoring**	1
			Credit 2	**Increased Ventilation**	1
			Credit 3	**Construction IAQ Management Plan**	1
			Credit 4.1	**Documenting Productivity Impacts** - Absenteeism & Healthcare Cost Impacts	1
			Credit 4.2	**Documenting Productivity Impacts** - Other Productivity Impacts	1
			Credit 5.1	**Indoor Chemical & Pollutant Source Control** - Reduce Particulates in Air System	1
			Credit 5.2	**Indoor Chemical & Pollutant Source Control** - Isolation of High Volume Copy/Print/Fa	1
			Credit 6.1	**Controllability of Systems** - Lighting	1
			Credit 6.2	**Controllability of Systems** - Temperature & Ventilation	1
			Credit 7.1	**Thermal Comfort** - Compliance	1
			Credit 7.2	**Thermal Comfort** - Permanent Monitoring System	1
			Credit 8.1	**Daylight & Views** - Daylight for 50% of Spaces	1
			Credit 8.2	**Daylight & Views** - Daylight for 75% of Spaces	1
			Credit 8.3	**Daylight & Views** - Views for 45% of Spaces	1
			Credit 8.4	**Daylight & Views** - Views for 90% of Spaces	1
			Credit 9	**Contemporary IAQ Practice**	1
			Credit 10.1	**Green Cleaning** - Entryway Systems	1
			Credit 10.2	**Green Cleaning** - Isolation of Janitorial Closets	1
			Credit 10.3	**Green Cleaning** - Low Environmental Impact Cleaning Policy	1
			Credit 10.4	**Green Cleaning** - Low Environmental Impact Pest Management Policy	1
			Credit 10.5	**Green Cleaning** - Low Environmental Impact Pest Management Policy	1
			Credit 10.6	**Green Cleaning** - Low Environmental Impact Cleaning Equipment Policy	1

Yes ? No

			Innovation & Design Process		**5** Points
			Credit 1.1	**Innovation in Upgrades, Operation & Maintenance**	1
			Credit 1.2	**Innovation in Upgrades, Operation & Maintenance**	1
			Credit 1.3	**Innovation in Upgrades, Operation & Maintenance**	1
			Credit 1.4	**Innovation in Upgrades, Operation & Maintenance**	1
			Credit 2	**LEED™ Accredited Professional**	1

Yes ? No

			Project Totals (pre-certification estimates)	**85** Points

Certified: 32-39 points, **Silver:** 40-7 points, **Gold:** 48-63 points, **Platinum:** 64-85

A

absolute humidity: The amount of water vapor in a particular volume of air.

access card: A plastic card encoded with some type of identification number that, when read by an access control device, allows access.

access control: The ability to permit or deny the use of something by someone.

access control system: A system used to deny those without proper credentials access to a specific building, area, or room.

access group: A group of users with the same access rights.

access level: The level of security for a group or single card holder.

access rights: The collection of access control information about a user related to the user's access to certain areas, what times the user is allowed into those areas, and which doors are accessible.

acoustic glass break detector: A sensing device that detects the noise (sound frequencies) of breaking glass.

actuator: A device that accepts a control signal and causes a mechanical motion.

addressable circuit: A circuit of addressable devices that communicate using digital signals.

addressable device: A device with a unique identifying number that can exchange messages with other addressable devices.

air changes per hour (ACH): The measure of the number of times the entire volume of air within a building space is circulated through the HVAC system in one hour.

air conditioner: A self-contained cooling unit for forced-air HVAC systems.

airflow station: A sensor that measures the velocity of the air in a duct system.

air-handling unit (AHU): A forced-air HVAC system device consisting of some combination of fans, ductwork, filters, dampers, heating coils, cooling coils, humidifiers, dehumidifiers, sensors, and controls to condition and distribute supply air.

air velocity: The speed at which air moves from one point to another.

alarm: 1. A signal that indicates a fire or other hazardous conditions. **2.** A security system signal that indicates an intrusion or other alarm condition.

algorithm: A sequence of instructions for producing the optimal result to a problem.

analog signal: A signal that has a continuous range of possible values between two points.

annunciator: A fire alarm notification appliance that displays fire protection system modes, status, and alarms.

anti-passback: An access control function that does not allow a credential to be used twice within a certain period or until the credential has been used to exit the area.

appliance: A plumbing fixture that performs a special function and is controlled and/or energized by motors, heating elements, or pressure- or temperature-sensing elements.

armed (away) mode: A security system mode that makes all security-sensing devices active.

armed (stay) mode: A security system mode that makes all perimeter-sensing devices active while signals from the sensing devices (that were defined as being stay/away zone types) are ignored.

audible/visual exit delay countdown warning: A security system control panel feature that audibly and visually indicates the amount of exit delay time remaining before a system is armed.

authentication: The process of identifying a person and verifying their credentials.

authority having jurisdiction (AHJ): The organization, office, or individual responsible for approving the equipment and materials used for building automation installation.

automatic phone dialer: A notification device that dials telephone numbers and communicates a trouble or alarm condition to the recipients.

autotransformer: A transformer with only one winding, which is tapped by both the primary and secondary sides at various points to produce varying voltages.

B

ballast: A device with a circuit that controls the flow of current to gas discharge lamps while providing sufficient starting voltage.

barrier: A manmade obstacle strategically placed to prevent or limit access.

baud rate: A measure of the number of symbols transmitted per unit of time, typically a second.

bi-level switching: A technique to control general light levels by switching individual lamps or groups of lamps in a multilamp fixture separately.

biometric device: An interface device that evaluates specific human physical characteristics for authentication.

biometric reader: An access control reader that analyzes a person's physical or behavioral characteristics and compares them to a database for authentication.

bit rate: A measure of the number of binary bits transmitted per unit of time, typically a second.

boiler: A closed metal container that heats water to produce steam or hot water.

boost pump: A pump in a water supply system used to increase the pressure of the water while it is flowing to fixtures.

branch: A water distribution pipe that routes a water supply horizontally to fixtures or other pipes at the same approximate level.

branch circuit: The circuit in a power distribution system between the final overcurrent protective device and the associated end-use points, such as receptacles or loads.

building automation: The control of the energy- and resource-using devices in a building for optimization of building system operations.

building automation system: A system that uses a distributed system of microprocessor-based controllers to automate any combination of building systems.

building drain: The lowest part of the drainage system; it receives the discharge from all drainage pipes in the building and conveys it to the building sewer.

building main: A water distribution pipe that is the principal pipe artery supplying water to the entire building.

building sewer: The part of the drainage system that connects the building drain to the sanitary sewer.

C

camera: A device that converts an image into a time-varying electrical signal.

call: A request for an elevator to stop at a certain floor to either pick up or drop off passengers.

call button: A user-operated pushbutton input device that sends a call signal to an elevator controller.

carbon dioxide sensor: A sensor that detects the concentration of carbon dioxide (CO_2) in air.

card holder: The information identifying an access control system user and the user's access rights.

card reader: An interface device that evaluates the magnetic or electrical characteristics of a special handheld keycard for authentication.

carrier frequency: The frequency of the ON/OFF voltage pulses that simulate the fundamental frequency.

cathode: A tungsten coil coated with a material that releases electrons when heated.

centralized generation: An electrical distribution system in which electricity is distributed through a utility grid from a central generating station to millions of customers.

centralized lighting control: A lighting system that controls lights around a building via one main control panel.

central station security system: A security system that is monitored by an off-site station.

centrifugal pump: A pump with a rotating impeller that uses centrifugal force to move water.

chiller: A refrigeration system that cools water.

circuit breaker: An overcurrent protective device with a mechanism that automatically opens a switch in the circuit when an overcurrent condition occurs.

circulation: The continuous movement of air through a building and its HVAC system.

closed-circuit television (CCTV) system: A system of video capture and display devices where the video signal distribution is limited to within a facility.

closed-loop circuit: A security system circuit with devices that are all normally closed.

closed-loop control system: A control system in which the result of an output is fed back into a controller as an input.

color depth: The number of possible colors for each display pixel.

comfort: The condition of a person not being able to sense a difference between themselves and the surrounding air.

composite video signal: An analog video signal encoded with all of the component parts of a video signal.

condensation: The formation of liquid (condensate) as moisture or other vapor cools below its dewpoint.

condenser: A heat exchanger that removes heat from high-pressure refrigerant vapor.

conditioned air: Indoor air that has been given desirable qualities by the HVAC system.

constant-air-volume air-handling unit: An air-handling unit that provides a steady supply of air and varies the heating, cooling, or other conditioning functions as necessary to maintain the desired setpoints within a building zone.

construction elevator: A temporary elevator for the transportation of materials and workers during the building construction.

consulting-specifying engineer: A building automation professional that designs the building automation system from the owner's list of desired features.

contactor: A heavy-duty relay for switching circuits with high-power loads.

contract document: A set of documents produced by the consulting-specifying engineer for use by a contractor to bid a project.

contract specification: A document describing the desired performance of the purchased components and means and methods of installation.

control device: A building automation device for monitoring or changing system variables, making control decisions, or interfacing with other types of systems.

controller: A device that makes decisions to change some aspect of a system based on sensor information and internal programming.

control logic: The portion of controller software that produces the necessary outputs based on the inputs.

control loop: The continuous repetition of the control logic decisions.

control panel: A panel that contains electrical connections and control interfaces for a control system.

control point: A variable in a control system.

control signal: A changing characteristic used to communicate building automation information between control devices.

conveyance: A means by which people and materials are transported, such as an elevator.

conveying system: A system for the transporting of people and/or materials between points in a building or structure.

cooling coil: A heat exchanger that removes heat from the air surrounding or flowing through it.

cooling tower: A device that uses evaporation and airflow to cool water.

credential: A method of authenticating an access request.

cross zoning: A security system control panel feature that requires two predetermined zones to trip before an alarm signal is transmitted.

D

damper: A set of adjustable metal blades used to control the amount of airflow between two spaces.

daylighting: A lighting control scheme that measures the total amount of illumination in a space from all sources and adjusts the artificial lighting to maintain a minimum level.

deadband: The range between two setpoints in which no control action takes place.

dead zone: An area within the field-of-view of an occupancy sensor that is not covered by one of the detection zones on the lens.

dehumidifier: A device that removes moisture from air by causing the moisture to condense.

delay zone: The zone type that allows a certain period of time to elapse before a triggered sensor initiates an alarm.

demand limiting: The automated shedding of loads.

derivative control algorithm: A control algorithm in which the output is determined by the instantaneous rate of change of a variable.

dewpoint: The air temperature below which moisture begins to condense.

differential pressure: The difference between two pressures.

differential pressure switch: A switch that activates at a differential pressure either above or below a certain value.

digital signal: A signal that has only two possible states.

dimming: The intentional reduction of electrical power to a lamp in order to reduce its light output.

direct digital control (DDC) system: A control system in which electrical signals are used to measure and control system parameters.

disarmed mode: A security system mode in which the control panel ignores all inputs from sensing devices.

display resolution: The maximum number of pixels that can be displayed on a monitor screen.

distributed generation: An electrical distribution system in which many smaller power-generating systems create electrical power near the point of consumption.

distributed generator interconnection relay: A specialized relay that monitors both a primary power source and a secondary power source for the purpose of paralleling the systems.

distributed lighting control: A lighting system that controls lights directly from local control devices.

door closer: A device that pulls a door shut after an access has been gained.

door position switch: A magnetic contact that indicates an abnormal condition of a door or barrier.

door/window contact: A sensing device that indicates whether a door or window is fully closed.

down peak mode: An elevator system operation mode that stations all available cars on a certain building floor in anticipation of many downward traveling passengers.

dry-bulb temperature: The temperature of air measured by a thermometer freely exposed to the air but shielded from radiation and moisture.

dual-tone multifrequency (DTMF) signaling: A method used to transmit numbers with a pair of tones.

dumbwaiter: An elevator for transporting very small loads, primarily consisting of domestic items.

E

economizing: A cooling strategy that adds cool outside air to the supply air.

electrical demand: The amount of electrical power drawn by loads at a specific moment.

electrical service: The electrical power supply to a building or structure.

electrical system: A combination of electrical devices and components, connected by conductors, that distributes and controls the flow of electricity from its source to a point of use.

electricity: The energy resulting from the flow of electrons through a conductor.

electricity consumption: The total amount of electricity used during a billing period.

electric lock: A lock that uses a solenoid to actuate the bolt.

electric motor: A device that converts electrical energy into rotating mechanical energy.

electric strike: An electrically operated lock that unlocks by moving the door strike.

electronic keypad: An authentication device that uses a manually entered numeric or alphanumeric code as a credential.

elevator: A conveying system for transporting people and/or materials vertically between floors in a building.

elevator car buffer: A piston and cylinder arrangement designed to cushion the fall of an elevator car or counterweight.

end-of-line (EOL) resistor: A resistor installed at the far end of an initiating-device circuit for the purpose of monitoring circuit integrity.

end switch: A switch that indicates the fully actuated damper positions.

environmental sensor: A sensing device that detects abnormal conditions in the environment.

equipment grounding conductor (EGC): A conductor that provides a low-impedance path from electrical equipment and enclosures to the grounding system.

evaporation: The process of a liquid changing to a vapor by absorbing heat.

evaporator: A heat exchanger that adds heat to low-pressure refrigerant liquid.

exhaust air: Air that is ejected from a forced-air HVAC system.

exit delay: The programmed amount of time between when a security system is commanded to be armed and when the control panel begins registering alarm signals from sensing devices.

express elevator: An elevator specifically intended to service a single floor or a specific group of floors.

F

facsimile machine: A device used for transmitting image information over telephone lines.

fail-safe lock: An electric door lock that is normally unlocked when unpowered.

fail-secure lock: An electric door lock that is normally locked when unpowered.

false alarm: An activation of an alarm without any evidence of an actual or attempted intrusion.

fan: A mechanical device with spinning blades that move air.

feeder: The circuit conductors between a building's electrical supply source, such as a switchboard, and the final branch-circuit overcurrent device.

filament: A conductor with a high resistance that causes it to glow white-hot from electrical current.

filtration: The process of removing particulate matter from air.

fire alarm control panel (FACP): An electrical panel that is connected to fire protection system circuits and interfaces with other building systems.

fire alarm system: A system that detects hazardous conditions associated with fires (such as smoke or heat) and notifies building occupants.

fire alarm verification: A security system control panel feature that requires two trips of a fire alarm-initiating device within a specified amount of time before it transmits a fire alarm signal.

fire protection system: A building system for protecting the safety of building occupants during a fire.

fire safety control function: An integration of a fire protection system with other building systems that is intended to make the building safer for evacuating occupants and/or control the spread of fire hazards.

fire suppression agent: A substance that can extinguish a fire.

fire suppression system: A system that releases fire suppression agents to control or extinguish a fire.

fitting: A device used to connect two lengths of pipe.

fixture: A receptacle or device that is connected to the water distribution system, demands a supply of potable water, and discharges the waste directly or indirectly into the sanitary drainage system.

fixture branch: A water supply pipe that extends between a water distribution pipe and fixture supply pipe.

fixture drain: A drainage pipe that extends from the trap of a fixture to the junction of the next drainage pipe.

fixture supply pipe: A water supply pipe connecting the fixture to the fixture branch.

fixture trim: The set of water supply and drainage fittings installed on a fixture or appliance to control the water flowing into a fixture and the wastewater flowing from the fixture to the sanitary drainage system.

floor mat: A sensing device that detects the steps of a person walking on it.

flow meter: A device used to measure the flow rate and/or total flow of water flowing through a pipe.

flow pressure: The water pressure in the water supply pipe near a fixture while it is wide open and flowing.

flow rate: The volume of water passing a point at a particular moment.

flow switch: A switch with a vane that moves from the force exerted by the water or air flowing within a duct or pipe.

fluorescent lamp: An electric lamp that produces light from the ionization of mercury vapor.

foot-candle (fc): The illuminance from 1 lumen per square foot (lm/ft^2) of surface.

forced-air HVAC system: A system that distributes conditioned air throughout a building in order to maintain the desired conditions.

freight elevator: An elevator primarily intended to transport equipment or materials.

frequency: The number of AC waveforms per interval of time.

Fresnel lens: A lens with special grooved patterns that provide strong focusing ability in a compact shape.

full-way valve: A valve designed to operate in only the fully open or fully closed positions.

fundamental frequency: The voltage frequency simulated by the changing pulse widths of the carrier frequency.

furnace: A self-contained heating unit for forced-air HVAC systems.

fuse: An overcurrent protective device with a fusible link that melts and opens the circuit when an overcurrent condition occurs.

G

gas discharge lamp: An electric lamp that produces light by establishing an arc through ionized gas.

grid: The utility's network of conductors, substations, and equipment that distributes electricity from a central generation point to the consumers.

grounded conductor: A current-carrying conductor that has been intentionally grounded.

grounding The intentional connection of all exposed noncurrent-carrying metal parts to the earth.

grounding electrode conductor (GEC): A conductor that connects the grounding system to the buried grounding electrode.

H

halogen lamp: An incandescent lamp filled with a halogen gas (iodine or bromine).

head: The difference in water pressure between points at different elevations.

heat detector: An initiating device that is activated by the high temperatures of a fire.

heat exchanger: A device that transfers heat from one fluid to another fluid without allowing the fluids to mix.

heating coil: A heat exchanger that adds heat to the air surrounding or flowing through it.

heat pump: A mechanical compression refrigeration system that moves heat from one area to another area.

high-intensity discharge (HID) lamp: An electric lamp that produces light by striking an electrical arc across tungsten electrodes housed inside an arc tube.

high-pressure sodium lamp: An HID lamp that produces light when current flows through sodium-vapor under high pressure and high temperature.

hot water loop: A closed circuit of distribution pipes, including the water heater, through which hot water is continuously circulated.

human control: A form of access control that requires human involvement.

human-machine interface (HMI): An interface terminal that allows an individual to access and respond to building automation system information.

humidifier: A device that adds moisture to the air by causing water to evaporate into the air.

humidistat: A switch that activates at humidity levels either above or below a certain setpoint.

humidity: The amount of moisture present in the air.

HVAC system: A building system that controls a building's indoor climate.

hydraulic elevator: An elevator that lifts the elevator car from below using a hydraulic ram.

hydraulic ram: A piston that is driven into or out of a hollow cylinder by fluid pressure.

hydronic HVAC system: A system that distributes water or steam throughout a building as the heat-transfer medium for heating and cooling systems.

hydropneumatic tank: A water tank incorporating an air-filled bladder that raises the pressure of water as it is pumped into the tank.

hygrometer: A device that measures the amount of moisture in the air.

hygroscopic element: A material that changes its physical or electrical characteristics as the humidity changes.

I

illuminance: The quantity of light per unit of surface area.

impeller: The bladed, spinning hub of a fan or pump that forces fluid to its perimeter.

incandescent lamp: An electric lamp that produces light by the flow of current through a tungsten filament inside a gas-filled, sealed glass bulb.

independent service mode: An elevator system operation mode that releases the elevator from control by the central elevator controller.

inductor: A coil of wire that creates a magnetic field, which resists changes in the current flowing through the coil.

initial lumens: The rated intensity of light produced by a lamp when it is new.

initiating device: A fire protection system component that signals a change-of-state condition.

initiating-device circuit (IDC): An electrical circuit consisting of fire alarm initiating devices, any of which can activate an alarm signal by closing its contacts.

inspect mode: An elevator system operating mode in which the elevator ignores all control inputs.

instant zone: A zone type that indicates an immediate alarm condition when a sensing device in the zone is activated.

integral control algorithm: A control algorithm in which the output is determined by the sum of the offset over time.

integration: A function that calculates the amount of offset over time as the area underneath a time-variable curve.

interface device: A device that facilitates communication between a person and the security system for sharing information or controlling the system.

interior zone: A zone type used to differentiate interior-sensing devices.

interoperability: The ability of diverse systems and devices to work together effectively and reliably.

K

keypad: An interface device used to display the status and control the operation of the security system.

key switch: An interface device that arms or disarms a security system with a special key.

key system: A multiline telephone system in which each telephone has direct access to each outside (PSTN) line.

L

lamp: An electrical output device that converts electrical energy into visible light and other forms of energy.

LED driver: A circuit that provides a constant DC voltage source and protection from line voltage transients, such as surges and sags.

lens: A clear optical element for focusing a pattern of light, or image, onto a smaller area.

level switch: A switch that activates at liquid levels either above or below a certain setpoint.

lift station: A submersible pump in a sump pit that pushes wastewater to a higher elevation.

light: The portion of the electromagnetic spectrum that the human eye can perceive.

light-emitting diode (LED): A semiconductor device that emits a specific color of light when DC voltage is applied in one direction.

light fixture: An electrical appliance that holds one or more lamps securely and includes the electrical components necessary to connect the lamp(s) to the appropriate power supply.

lighting contactor: A solenoid-operated, high-power relay used with lighting circuits.

lighting system: A building system that provides artificial light for indoor areas.

light-level sensor: A control device with a photocell that measures the intensity of light it is exposed to.

load shedding: The deactivation or modulation of non-critical loads in order to decrease electrical power demand.

local security system: A self-contained security system that does not require full-time surveillance or off-premises wiring.

low-pressure sodium lamp: An HID lamp that produces light by an electrical discharge through low-vapor-pressure sodium.

lumen (lm): The measure of the intensity of light radiating from a light source.

luminous efficacy: The ratio of a lamp's light output (lumens) to the electrical power input (watts).

lux: The illuminance from 1 lumen per square meter (lm/m^2) of surface.

M

machine room: A space directly above an elevator shaft to house the motor, sheave, and elevator controls.

machine room-less elevator: A traction elevator that system using special materials and improved electric motors that require little space, eliminating the need for the machine room.

magnetic lock: A lock that uses magnetic force to hold a door closed.

magnetic motor starter: A specialized contactor for switching electrical power to a motor; includes overload protection.

main bonding jumper (MBJ): A connection at the service equipment that connects the equipment grounding conductor, the grounding electrode conductor, and the grounded conductor (neutral conductor).

mantrap: A barrier configuration that holds a person in a small area until the door or gate just passed through is secured.

manual fire alarm pull station: An initiating device that is manually operated by a person to cause a fire alarm signal.

mean lumens: The rated average intensity of light produced by a lamp after it has operated for approximately 40% of its rated life.

mechanical control: A form of access control that uses a mechanical device to maintain a closed area.

mechanical keypad: A mechanical locking device that can be released by pressing a sequence of numbers on the keypad.

mercury-vapor lamp: An HID lamp that produces light by an electrical discharge through mercury vapor.

metal-halide lamp: An HID lamp that produces light by an electrical discharge through mercury vapor and metal halide.

micron: A unit of measure equal to one-millionth of a meter, or 0.000039″.

microwave sensor: A sensor that activates when it senses changes in the reflected microwave energy caused by people moving within its field-of-view.

mixed air: The blend of return air and outside air that is combined inside the air-handling unit and goes on to be conditioned.

modem: A device that modulates (changes) an analog carrier signal into digital information for transmission over a medium.

momentary unlock: An access control function that keeps a door unlocked for a certain amount of time after a valid authentication.

monitor: A device that converts video signals into monochrome or color motion images.

motion sensor: A sensing device that detects movement within the coverage area of the detector.

N

nonsupervised circuit: A security system circuit of field devices that cannot monitor the integrity of the wiring.

notification appliance: A fire alarm system component that provides an audible and/or visible indication of alarm signals.

notification appliance circuit (NAC): An electrical circuit of fire alarm notification appliances that are activated by the fire alarm control panel (FACP) under certain signal conditions.

notification device: A device that alerts building occupants of an alarm condition.

null zone: An unused zone.

O

occupancy sensor: A sensor that detects whether an area is occupied by people.

offset: The difference between the value of a control point and its corresponding setpoint.

open-loop circuit: A security system circuit with devices that are all normally open.

open-loop control system: A control system in which decisions are made based only on the current state of the system and a model of how it should work.

open protocol: A communications and network protocol that is standardized and published for use by any device manufacturer.

outlet: An end-use point in the power distribution system.

outside air: Fresh air from outside the building that is incorporated into the forced-air HVAC system.

overcurrent protective device: A device that prevents conductors or devices from reaching excessively high temperatures from high currents by opening the circuit.

P

panelboard: A wall-mounted power distribution cabinet containing overcurrent protective devices for lighting, appliance, or power distribution branch circuits.

panic/duress button: A sensing device that causes an alarm signal when manually activated.

pan-tilt-zoom (PTZ) camera: A camera that can be remotely controlled to change its field-of-view.

partition: A portion of a security system that operates under a different program than the rest of the system.

passenger elevator: An elevator primarily intended to transport people.

passive infrared (PIR) sensor: A sensor that activates when it senses the heat energy of people within its field-of-view.

phase control: The frequent switching of AC voltage to limit the power in a circuit.

phase one mode: An elevator system operating mode in which the elevator car automatically moves to a designated floor during a fire.

phase two mode: An elevator system operating mode that enables rescue personnel to manually operate an elevator that is in phase one mode.

photocell: A small semiconductor component that changes its electrical characteristics, such as output current or resistance, in proportion to the light level.

photo control: An automatic lighting control that uses a light-level sensor to turn lamps ON around dusk and turn lamps OFF at dawn.

photoresistor: A photocell that changes resistance in proportion to its exposure to light.

phototransistor: A photocell that controls a current in proportion to the light level.

piggybacking: The action of a person following directly behind another person to avoid presenting a credential for authentication.

plumbing system: A system of pipes, fittings, and fixtures within a building that conveys a water supply and removes wastewater and waterborne waste.

pneumatic control system: A control system in which compressed air is the medium for sharing control information and powering actuators.

point-of-entry protection: A type of security system protection that monitors the specific points that might allow entry or exit from a secure area.

point-of-entry: A potential opening in the envelope or perimeter of a building or area through which a person may enter.

pole: A set of contacts that belong to a single circuit.

port: An opening in a valve that allows a connection to a pipe.

positive-displacement pump: A pump that creates flow by trapping a certain amount of water and then forcing that water through a discharge outlet.

potable water: Water that is free from impurities that could cause disease or other harmful health conditions.

power quality: The measure of how closely the power in an electrical system matches the nominal (ideal) characteristics.

pressure gauge: A pressure-sensing device that indicates the pressure of a fluid on a numeric scale. Pressure switches are used in some applications where exact pressure is not needed.

pressure loss due to friction: The loss of water pressure resulting from the resistance between water and the interior surface of a pipe or fitting.

pressure loss due to head: The loss of 0.434 psi of water pressure for every foot of height.

pressure sensor: A sensor that measures the pressure exerted by a fluid, such as air or water.

pressure switch: A switch that activates at pressures either above or below a certain value.

private branch exchange (PBX): A telephone exchange that serves a small, private network.

programmable abort: A security system control panel feature that delays the transmission of an alarm signal for a short period to allow a manual cancellation.

project closeout information: A set of documents produced by the controls contractor for the owner's use while operating the building.

proportional control: A control algorithm in which the output is in direct response to the amount of offset in the system.

proprietary protocol: A communications and network protocol that is used by only one device manufacturer.

protocol: A set of rules and procedures for the exchange of information between two connected devices.

pseudopoint: A control point that exists only in software.

pulse meter: A meter that outputs a pulse for every predetermined amount of flow in the circuit or pipe.

pump: A device that moves water through a piping system.

pyroelectric sensor: A sensor that generates a voltage in proportion to a change in temperature.

R

rainwater leader: A vertical drainage pipe that conveys rainwater from a drain to the building storm drain or to another point of disposal.

reader: A device that reads access cards or biometric patterns.

receptacle: An outlet for the temporary connection of corded electrical equipment.

refrigerant: A fluid that is used for transferring heat.

register: A cover for the opening of ductwork into a building space.

relative humidity: The ratio of the amount of water vapor in the air to the maximum moisture capacity of the air at a certain temperature.

relay: An electrical switch that is actuated by a separate electrical circuit.

remote-controlled circuit breaker: A circuit breaker that switches a circuit ON or OFF when it receives a control signal.

resistance temperature detector (RTD): A temperature-sensing element made from a material with an electrical resistance that changes with temperature.

resolution: The number of pixels displayed on a monitor screen.

return air: The air from within the building space that is drawn back into the forced-air HVAC system to be exhausted or reconditioned.

request-to-exit device: A device that automatically authorizes an exit from a secure area without needing credentials or without causing an alarm condition.

riser: A water distribution pipe that routes a water supply vertically one full story or more.

S

sanitary drainage system: A plumbing system that conveys wastewater and waterborne waste from the plumbing fixtures and appliances to a sanitary sewer.

sanitary sewer: A sewer that carries sewage but does not convey rainwater, surface water, groundwater, or similar nonpolluting wastes.

scene: A group of settings for all the lamps in a space that correspond to a certain use for the space.

scene lighting: A lighting control scheme that switches and/or dims groups of lamps together to a mixture of predetermined levels.

scheduling: The automatic control of devices according to the date and time.

security system: A building system that protects against intruders, theft, and vandalism.

sensing device: A security system device that sends an input signal to the control panel when triggered by some security condition.

sensor: A device that measures the value of a variable and transmits a signal that conveys this information.

service elevator: An elevator reserved exclusively for the use of building service personnel.

setpoint: The desired value to be maintained by a system.

sewage: Any liquid waste containing animal or vegetable matter in suspension or solution and/or chemicals in solution.

sewer gas: The mixture of vapors, odors, and gases found in sewers.

shock sensor: A sensing device that detects the physical vibration of breaking glass.

shop drawing: A document produced by the controls contractor with the details necessary for installation.

signaling line circuit (SLC): An electrical circuit of fire alarm initiating devices that communicate using digital signals.

siren driver: A circuit that produces the frequency, volume, and pattern of a desired sound.

smoke detector: An initiating device that is activated by the presence of smoke particles.

solenoid: A device that converts electrical energy into a linear mechanical force.

solenoid valve: A full-way valve that opens when supplied with an electric current and closes when the current stops.

speaker: A device that produces sound from electrical audio signals.

specific-area protection: A type of security system protection that monitors only one defined location.

spot protection: A type of security system protection that monitors one particular object.

stack: A vertical drainage pipe that extends one or more floors.

stale air: Air with high concentrations of carbon dioxide and/or other vapor pollutants.

stay/away zone: A zone type that is programmed to ignore signals from its sensing devices when the system is armed, but when there is no exit from the premises.

storm sewer: A sewer used for conveying groundwater, rainwater, surface water, or similar nonpolluting wastes.

stormwater drainage system: A plumbing system that conveys precipitation collecting on a surface to a storm sewer or other place of disposal.

stress sensor: A sensing device that measures a weight load over a specific area.

strobe: A visual notification device that flashes a bright light at regular intervals.

supervised circuit: A security system circuit of field devices that can monitor the integrity of the wiring. This type of circuit requires end-of-line resistors.

supervisor first: An access control function that allows no one into a secure area until a supervisor has made a valid read into that area.

supervisory computer: A computer that stores access control database information and makes intelligent decisions for the system.

supervisory security system: A security system that is monitored continuously by on-site security personnel.

supervisory signal: A signal that indicates an abnormal condition in a fire suppression system.

supply air: Newly conditioned mixed air that is distributed to the building space.

swinger shutdown: A security system control panel feature that disables a zone after repeated triggering during an armed period.

switch: A device that isolates an electrical circuit from a power source.

switchboard: The last point on the power distribution system for the power company and the beginning of the power distribution system for the property owner's electrician.

switching: The complete interruption or resumption of electrical power to a device.

T

telephone: A device used to transmit and receive sound (most commonly speech) between two or more people.

temperature: the measurement of the intensity of the heat of a substance.

temperature sensor: A device that measures temperature.

temperature stratification: An undesirable variation of air temperature between the top and bottom of a space.

terminal unit: The end point in an HVAC distribution system where the conditioned medium (air, water, or steam) is added to or directly influences the environment of the conditioned building space.

thermistor: A temperature-sensing element made from a semiconductor material that changes resistance in response to changing temperatures.

thermocouple: A temperature-sensing element consisting of two dissimilar metal wires joined at the sensing end.

thermostat: A switch that activates at temperatures either above or below a certain setpoint.

thermostatic mixing valve (TMV): A valve that mixes hot and cold water in proportion to achieve a desired temperature.

thermowell: A watertight and thermally conductive casing for immersion temperature sensors that mounts the sensing element inside the pipe, vessel, or fixture containing the water to be measured.

three-way valve: A valve with three ports that can control water flow between them.

throttling valve: A valve designed to control water flow rate by partially opening or closing.

throw: A position that a switch can adopt.

time clock: A switch that automatically changes state (switches) at certain times.

time zone: A specific range of time for an access control function.

total flow: The volume of water that passes a point during a specific time interval.

traction elevator: An elevator system that raises and lowers the elevator car with cables operated by an electric motor.

transducer: A device that converts one form of energy into another form of energy.

transfer switch: A switch that allows an electrical system to be switched between two power sources.

transformer: An electric device that uses electromagnetism to change AC voltage or electrically isolate two circuits.

transient voltage: A temporary, undesirable voltage in an electrical circuit, ranging from a few volts to several thousand volts and lasting from a few microseconds up to a few milliseconds.

traveling cable: A cable that provides power, control, communication, and other wiring to an elevator car.

trouble signal: A signal that indicates a wiring or power problem that could disable part or all of the system.

tuning: The adjustment of control parameters to the optimal values for the desired control response.

24-hour fire zone: A zone type that is always active and initiates a fire alarm.

24-hour supervisory zone: A zone type that is always active and initiates a trouble signal.

24-hour zone: A zone type that is always active (never disarmed).

U

ultrasonic sensor: A sensor that activates when it senses changes in the reflected sound waves caused by people moving within its detection area.

uninterruptible power supply (UPS): An electrical device that provides stable and reliable power, even during fluctuations or failures of the primary power source.

up peak mode: An elevator system operation mode that stations all available elevator cars at the building entry floor in anticipation of many upward traveling passengers.

utility: A company that generates and/or distributes electricity to consumers in a certain region or state.

V

valve supervisory switch: An initiating device that indicates when a valve is not fully open.

valve: A fitting that regulates the flow of water within a piping system.

variable: Some changing characteristic in a system.

variable-air-volume air-handling unit: An air-handling unit that provides air at a constant air temperature but varies the amount of supplied air in order to maintain the desired setpoints within the building zone.

variable-air-volume (VAV) terminal box: A device located at the building zone that provides heating and airflow as needed in order to maintain the desired setpoints within the building zone.

variable-frequency drive: A motor controller that is used to change the speed of AC motors by changing the frequency of the supply voltage.

ventilation: The process of introducing fresh outdoor air into a building.

vent piping system: A plumbing system that provides for the circulation of air in a sanitary drainage system.

videocassette recorder (VCR): A video storage device that records information on magnetic tapes.

voice-data-video (VDV) system: A building system used for the transmission of information.

Voice over Internet Protocol (VoIP): The transmission of audio telephone calls over data networks.

voice verification: A security system control panel feature that allows two-way communication between the central station and the facility where the security system is installed.

voltage: The difference in electrical potential between two points in an electrical circuit.

vortex damper: A pie-shaped damper at the inlet of a centrifugal fan that reduces the ability of the fan to grip and move air.

W

water flow switch: An initiating device that detects the flow of water in an automatic sprinkler fire suppression system.

water heater: A plumbing appliance used to heat water for a plumbing system's hot water supply.

water supply fixture unit (wsfu): An estimate of a plumbing fixture's water demand based on its operation.

water supply system: A plumbing system that supplies and distributes potable water to points of use within a building.

wet-bulb temperature: The temperature measured by a thermometer that has its bulb kept in contact with moisture.

wireless key fob: An interface device that arms or disarms a security system from a handheld wireless transmitter.

Z

zone: 1. An area within a building that shares the same HVAC requirements. **2.** A defined area protected by a group of security system sensing devices.

zone type: A definition of the type of alarm a zone initiates and the corresponding reaction of the control panel.

Index

Numbers in *italic* indicate an illustration.

USING THE *BUILDING AUTOMATION: CONTROL DEVICES AND APPLICATIONS* CD-ROM

Before removing the CD-ROM from the protective sleeve, please note that the book cannot be returned for refund or credit if the CD-ROM sleeve seal is broken.

System Requirements

To use this Windows®-compatible CD-ROM, your computer must meet the following minimum system requirements:
* Microsoft® Windows Vista™, Windows XP®, Windows 2000®, or Windows NT® operating system
* Intel® Pentium® III (or equivalent) processor
* 256 MB of available RAM
* 90 MB of available hard-disk space
* 800 × 600 monitor resolution
* CD-ROM drive
* Sound output capability and speakers
* Microsoft® Internet Explorer 5.5, Firefox 1.0, or Netscape® 7.1 web browser and Internet connection required for Internet links

Opening Files

Insert the CD-ROM into the computer CD-ROM drive. Within a few seconds, the home screen will be displayed allowing access to all features of the CD-ROM. Information about the usage of the CD-ROM can be accessed by clicking on USING THIS CD-ROM. The Quick Quizzes®, Illustrated Glossary, Flash Cards, Media Clips, and ATPeResources.com can be accessed by clicking on the appropriate button on the home screen. Clicking on the American Tech web site button (www.go2atp.com) accesses information on related educational products. Unauthorized reproduction of the material on this CD-ROM is strictly prohibited.

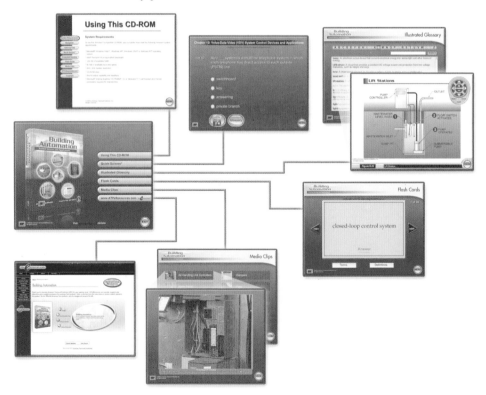